IFMBE Proceedings

Volume 52

The International Federation for Medical and Biological Engineering, IFMBE, is a federation of national and transnational organizations representing internationally the interests of medical and biological engineering and sciences. The IFMBE is a non-profit organization fostering the creation, dissemination and application of medical and biological engineering knowledge and the management of technology for improved health and quality of life. Its activities include participation in the formulation of public policy and the dissemination of information through publications and forums. Within the field of medical, clinical, and biological engineering, IFMBE's aims are to encourage research and the application of knowledge, and to disseminate information and promote collaboration. The objectives of the IFMBE are scientific, technological, literary, and educational.

The IFMBE is a WHO accredited NGO covering the full range of biomedical and clinical engineering, healthcare, healthcare technology and management. It is representing through its 60 member societies some 120.000 professionals involved in the various issues of improved health and health care delivery.

More information about this series at http://www.springer.com/series/7403

James Goh · Chwee Teck Lim
Editors

7th WACBE World Congress on Bioengineering 2015

6th to 8th July, 2015, Singapore

Springer

Editors
James Goh
Faculty of Engineering
National University of Singapore
Singapore

Chwee Teck Lim
National University of Singapore
Singapore

ISSN 1680-0737 ISSN 1433-9277 (electronic)
IFMBE Proceedings
ISBN 978-3-319-19451-6 ISBN978-3-319-19452-3 (eBook)
DOI 10.1007/978-3-319-19452-3

Library of Congress Control Number: 2015941131

Springer Cham Heidelberg New York Dordrecht London

The IFMBE Proceedings is an Official Publication of the International Federation for Medical and Biological Engineering (IFMBE)

Printed on acid-free paper

Springer International Publishing AG Switzerland is part of Springer Science+Business Media (www.springer.com)

Preface

On behalf of the Organizing Committee of the WACBE World Congress on Bioengineering 2015 (WACBE 2015), we would like to warmly welcome you to this meeting. This congress is part of a series that began in 2002 and we had six meetings thus far. This year's meeting is jointly organized by the Department of Biomedical Engineering of the National University of Singapore (NUS) and the Biomedical Engineering Society (Singapore) (BES). This congress is also endorsed by the International Federation for Medical and Biological Engineering (IFMBE).

We are glad to report that we have more than 200 participants from 17 countries. In this congress, we have received close to 300 abstracts with 50 being keynote/invited presentations. The rest of the contributed abstracts were peer-reviewed, with 124 accepted for oral and 99 accepted for poster presentations. Out of these accepted abstracts, we received 55 final papers submitted for the proceedings, with 2 being invited papers, 25 oral presentation papers and 28 Poster papers.

We are very honoured to have Prof Kam Leong as the inaugural Savio Woo Distinguished Lecturer, as well as other very prominent speakers as our Plenary and Keynote Speakers. Each of these speakers is an authority in their field of research and we are grateful that they are able to participate in this congress.

We do know that the success of a congress lies in the participation of the delegates and the quality of papers presented. Nevertheless, we also know that this will not be possible without the help and effort put in by the volunteers, reviewers, as well as the members of the Organizing Committee, Scientific Committee and the International Advisory Committee. Their dedicated contributions to this meeting are very much acknowledged and appreciated. We would also like to sincerely thank our sponsors, supporters and exhibitors for contributing to the success of this congress.

Finally, we would like to thank the staff of INMEET CMS Pte Ltd who has ensured the smooth running of the congress. Finally, to all our delegates, I hope this WACBE 2015 meeting will not only be one where excellent scientific ideas are exchanged and shared, but also friendships are renewed and new friends made. Do enjoy the congress as well as the sights and sounds of Singapore!

Best wishes

Prof James Goh
Chairman

Prof Chwee Teck LIM
Scientific Program Chair

WACBE 2015 Organizing Committee

Committees

Organising Committee

Chairman

Goh James	National University of Singapore, Singapore

Vice Chair

Toh Siew Lok	National University of Singapore, Singapore

Secretary

Lim Sierin	Nanyang Technological University, Singapore

Treasurer

Buist Martin	National University of Singapore, Singapore

International Advisory Committee Chair

Chew Yong Tian	National University of Singapore, Singapore

Scientific Programme Chair

Lim Chwee Teck	National University of Singapore, Singapore

Scientific Programme Co-Chair

Zhang Yong	National University of Singapore, Singapore

International Advisory Committee

Members

An, Kai-Nan	Mayo Clinic, USA
Cheng, Cheng-Kung	National Yang Ming University, Taiwan
Cheng, Dong	Penn State University, USA
Chien, Shu	University of California, San Diego, USA
Fan, Yubo	Beihang University, China
Guo, Edward	Columbia University, USA
Hsing, I-Ming	The Hong Kong University of Science and Technology, Hong Kong
Ji, Baohua	Beijing Institute of Technology, China
King, Albert	Wayne State University, USA
Lee, Abraham	University of California, Irvine, USA
Lee, Peter Vee Sin	The University of Melboume, Australia
Leong, Kam W.	Columbia University, USA
Li, Guoan	Massachusetts General Hospital / Harvard Medical School, USA
Li, Zong-Ming	Cleveland Clinic, USA

Lin, Feng-Huei	National Taiwan University, Taiwan
Lin, Kang-Ping	Chung Yuan Christian University, Taiwan
Lu, Helen	Columbia University, USA
Lu, William	The University of Hong Kong, Hong Kong
Mak, Arthur FT	The Chinese University of Hong Kong, Hong Kong
Ning, Gangmin	Zhejiang University, China
Ouyang, Hongwei	Zhejiang University, China
Phee, Louis	Nanyang Technological University, Singapore
Su, Fong-Chin	National Cheng Kung University, Taiwan
Sung, Hsing-Wen	National Tsing-Hua University, Taiwan
Teoh, Swee Hin	Nanyang Technological University, Singapore
Wang, Jawlin	National Taiwan University, Taiwan
Wu, Benjamin	University of California, Los Angeles, USA
Yang, Changhuei	California Institute of Technology, USA
Zhang Fan	Fudan University, China
Zhang Ming	The Hong Kong Polytechnic University, Hong Kong

Scientific Program Committee

Chair

Lim Chwee Teck	National University of Singapore, Singapore

Co-Chair

Zhang Yong	National University of Singapore, Singapore

Members

Li Jun	National University of Singapore, Singapore
Leo Hwa Liang	National University of Singapore, Singapore
Chen Chia Hung	National University of Singapore, Singapore
Chen Nanguang	National University of Singapore, Singapore
Alberto Corrias	National University of Singapore, Singapore
Huang Zhiwei	National University of Singapore, Singapore
Kah Chen Yong James	National University of Singapore, Singapore
Kanchanawong Pakorn Tony	National University of Singapore, Singapore
Tan Kian Lee	National University of Singapore, Singapore
Yu Haoyong	National University of Singapore, Singapore
Ren Hongliang	National University of Singapore, Singapore
Mrinal Kanti Musib	National University of Singapore, Singapore
Toh Siew Lok	National University of Singapore, Singapore
Yen Shih Cheng	National University of Singapore, Singapore
Yim Evelyn	National University of Singapore, Singapore

Themes and Topics

Bioimaging and Biosignals

- Bioelectronics
- Biosignal Processing
- Image Guided Intervention
- Medical Imaging
- Optical Imaging
- Biophotonics

Biomaterials and Tissue Engineering

- Biological Materials
- Biopolymers
- Bioceramics
- Biometals
- Nanobiomaterials
- Composite Materials
- Biomaterial-tissue Interaction
- Implant Materials
- Regenerative Medicine

Biomechanics and Computational Bioengineering

- Biomimetics
- Mechanobiology
- Molecular Biomechanics
- Cell Mechanics
- Tissue Mechanics
- Cardiovascular Mechanics
- Organ Mechanics
- Musculoskeletal Mechanics
- Orthopaedics
- Sports Mechanics
- Insect and Animal Biomechanics
- Computational Mechanics

Biomedical and Micro/Nano-Devices

- Biosensors
- BioMEMS
- Microarrays
- Organ-on-Chip
- Lab-on-a-Chip
- Micro/Nanofluidics
- Micro/Nanofabrication
- Micro-Total Analysis Systems
- Diagnostic Devices
- Implantable Devices
- Surgical Tools
- Interventional Devices
- Biomedical Device Innovation

Nanomedicine

- Nanomaterial Design and Synthesis
- Nanotoxicology
- Diagnosis and Imaging
- Controlled/Targeted Drug Delivery
- Controlled Gene Delivery
- Therapeutic Applications
- Theranostics

Biomedical Robotics and Surgical Technology

- Bio- and Medical Robotics
- Bioinstrumentation and Control
- Soft Robotics
- Computer Aided Surgery

Healthcare Information Sensing and Analytics

- Wireless Sensors
- E-Health System

Neural and Rehabilitation Engineering

- Clinical and Translational Neuroengineering
- Cognitive Engineering
- Neurotechnology
- Neurotherapeutics
- Assisted Technologies
- Rehabilitation

Special Topics

- Bioethics
- Biomedical Engineering Education
- Manuscript Writing
- Medical Device Regulation

Speakers

Savio Woo Distinguished Lecture Speaker

Kam W. Leong
Samuel Y. Sheng Professor
Columbia University, USA
"Bioengineering of Direct Cellular Reprogramming"
Monday, 6 July 2015 | 09:45 - 10:45 hrs
Auditorium

Plenary Speakers

Hsueh-Chia Chang
Bayer Professor of Engineering
Director, Center for Microfluidics and Medical Diagnostics
Department of Chemical and Biomolecular Engineering
University of Notre Dame, USA
"A BioChip Platform for Point-of-Care Diagnostic Devices"
Monday, 6 July 2015 | 11:15 - 12:00 hrs
Auditorium

Michael Sheetz
Director, Mechanobiology Institute
Distinguished Professor of the Department of Biological Sciences,
National University of Singapore &
Professor of the Department of Biological Sciences,
Columbia University, USA
"Rigidity Sensing Through Pinching Sarcomeres"
Tuesday, 7 July 2015 | 09:30 - 10:15 hrs
Auditorium

Guang-Zhong Yang
Director and Co-founder of the Hamlyn Centre for Robotic Surgery
Deputy Chairman of the Institute of Global Health Innovation,
Imperial College, UK
"Robotic Surgery - Current State-of-the-Art and Emerging Trends"
Wednesday, 8 July 2015 | 09:30 - 10:15 hrs
Auditorium

Keynote Speakers

Peter So
Professor of Mechanical Engineering and Biological Engineering,
Massachusetts Institute of Technology, USA
"Advances in Quantitative Phase Microscopy for Cell Mechanics Measurements"
Monday, 6 July 2015 | 13:30 - 14:00 hrs
Auditorium

Nitish Thakor
Director
SINAPSE, National University of Singapore, Singapore
"Frontiers of Neurotechnology"
Monday, 6 July 2015 | 16:15 - 16:45 hrs
Auditorium

Dong Sun
Chair Professor of Biomedical Engineering,
City University of Hong Kong, Hong Kong
"Single Cell Fusion with Optical Tweezers Manipulation System"
Tuesday, 7 July 2015 | 10:45 - 11:15 hrs
Auditorium

Yubo Fan
Professor of Biological and Medical Engineering,
Beihang University, China
"Computational Methodology in Biomechanics"
Tuesday, 7 July 2015 | 14:30 - 15:00 hrs
Auditorium

Kirk Shung
Dean's Professor in Biomedical Engineering,
University of Southern California, USA
"High Frequency Ultrasound and Its Biomedical Applications"
Wednesday, 8 July 2015 | 10:45 - 11:15 hrs
Auditorium

Guy Genin
Professor,
Washington University in St. Louis, USA
"Reattaching Tendon to Bone"
Wednesday, 8 July 2015 | 14:00 - 14:30 hrs
Auditorium

7th WACBE World Congress on Bioengineering
6 - 8 July 2015 | Singapore

世界華人生物醫學工程協會
World Association for Chinese Biomedical Engineers

Jointly Organised by

BES SINGAPORE

NUS National University of Singapore | Department of Biomedical Engineering Faculty of Engineering

Endorsed by

IFMBE

Sponsor

ThermoFisher SCIENTIFIC

Supporters

金基氏李 LEE FOUNDATION

Table of Contents

Optofluidics: Guiding the Nanoparticles and Biomolecules with Light

L.K. Chin[1], Y.Z. Shi[1], H.T. Zhao[1], Patricia Y. Liu[1], P.H. Yap[2], K. Wang[3], G. Wang[1], B. Liedberg[4], W. Ser[1], and A.Q. Liu[1]

[1] School of Electrical and Electronic Engineering, Nanyang Technological University, Singapore 639798
[2] Lee Kong Chian School of Medicine, Nanyang Technological University, Singapore 639798
[3] Institute of Biological Chemistry, Academia Sinica, Taipei 115, Taiwan
[4] School of Materials Science and Engineering, Nanyang Technological University, Singapore 639798

Abstract— **Optofluidics has recently gained huge research attentions because of its synergic manipulations of light and liquids. Several optofluidic components and devices have been demonstrated in the last 10 years, leading to its applications in guiding nanoparticles and biomolecules in the microchannel.**

Keywords— **Optofluidics, nanoparticle manipulation, micro/nanofluidics and nanophotonics.**

I. Introduction

Microfluidics represent the science and technology that process or manipulate small amount of fluids (10^{-9} to 10^{-18} liters) with dimensions of tens-to-hundreds of micrometers in microfluidic chip. Optofluidics aims to manipulate light and liquids at microscale, exploiting their interaction to create highly versatile devices that is significant scientifically and interests in biological and biomedical research areas. The novelties of the optofluidics are twofold [1]. First, liquids are used to carry substances to be analyzed in highly sensitive optical microdevices. Second, liquids is exploited to control optical microdevices, making them tunable, reconfigurable and adaptive. It is a new breakthrough research area that provides new opportunities for a wide range of traditional photonic devices, allowing tuning and reconfiguration at the micrometer scale using microfluidic manipulation.

II. Nanoparticle and Biomolecule Guiding

Many novel innovations have been demonstrated, such as liquid-liquid waveguide [2], liquid lens [3-4], liquid gratings [5-7] and liquid prism [8] etc. Light propagation can be tuned by varying the refractive index contrast or the curvature of the interface. These components are operated in a fast flow rate or two-phase flow conditions whereby diffusion is insignificant. Several devices have been demonstrated such as microfluidic waveguide laser [9], evanescent wave sensor [10] and cell refractometers [11-13].

Recently, diffusion between liquids in a microchannel is exploited for light manipulation. Diffusion within the liquid flow streams in the microchannel is studied to design different photonic components. When two or more miscible liquids are flowing in a microchannel, diffusion between the flow streams occur and a bidirectional gradient index profile (concentration and refractive index) is formed. We have demonstrated how lightwave can be bent in a bidirectional gradient index profile in the liquid waveguide, which supports novel wave-focusing and interference phenomena [14]. In addition, using the same approach with different bidirectional gradient index profile in the microchannel, an optofluidic Y-branch splitter with large-angle bending and tuning is demonstrated [15].

The manipulation and sorting of a small size of particle/molecule with dimensions of tens to hundreds of nanometers in a microfluidic chip is one of the most significant research approaches. In this talk, light is shaped efficiently to generate distinctive interference patterns in the optofluidic chip, which can be used to sort and assemble biological samples. For example, the optical field can be switched from the Bessel profile for particle sorting to the discrete interference patterns for particle assembly.

III. Conclusions

In conclusion, optofluidics is a perfect candidate for nanoparticle and biomolecule guiding because it is capable of not only manipulate liquid mixing and diffusion, but also shaping light propagation and optical force in the microchannel. It has high potential applications for nanomedicine, virus disease diagnosis and biomolecule studies.

Acknowledgment

The authors would like to acknowledge the financial support from Environmental and Water Industry (EWI) Development Council of Singapore (Grant No. 1102-IRIS-05) and Ministry of Education (Grant No. Tier-1 RG89/13).

J. Goh and C.T. Lim (eds.), *7th World Congress on Bioengineering 2015*,
IFMBE Proceedings 52, DOI: 10.1007/978-3-319-19452-3_1

References

1. A. Q. Liu. Preface to special topic: Optofluidics. Biomicrofluidics 4:042901
2. D. B. Wolfe, R. S. Conroy, P. Garstecki et al. (2004) Dynamic control of liquid-core/liquid cladding optical waveguides. Proceed. Natl. Acad. Sci. 101:12434-12438
3. Y. C. Seow, A. Q. Liu, L. K. Chin et al. (2008) Different curvature of tunable liquid microlens via the control of laminar flow rate. Appl. Phys. Lett. 93:084101
4. L. K. Chin, A. Q. Liu, C. S. Lim et al. (2010) An optofluidic volume refractometer using Fabry-Pérot resonator with tunable liquid microlenses. Biomicrofluidics 4:024107
5. L. K. Chin, A. Q. Liu, Y. C. Soh et al. (2010) A reconfigurable optofluidic Michelson interferometer using tunable droplet grating. Lab Chip 10:1072-1078
6. L. K. Chin, A. Q. Liu, C. S. Lim et al. (2008) An on-chip liquid tunable grating using multiphase droplet microfluidics. Appl. Phys. Lett. 93:164107
7. J. Q. Yu, Y. Yang, A. Q. Liu et al. (2010) Microfluidic droplet grating for reconfigurable optical diffraction. Optics Lett. 35:1890-1892
8. S. Xiong, A. Q. Liu, L. K. Chin et al. (2011) An optofluidic prism tuned by two laminar flows. Lab Chip 11:1864-1869
9. Y. Yang, A. Q. Liu, L. K. Chin et al. (2012) Optofluidic waveguide as a transformation optics device for lightwave bending and manipulation. Nature Communications 3:651.
10. X. C. Li, J. Wu, A. Q. Liu et al. (2008) A liquid waveguide based evanescent wave sensor integrated onto a microfluidic chip. App. Phys. Lett. 93:193901.
11. W. Z. Song, X. M. Zhang, A. Q. Liu et al. (2006) Refractive index measurement of single living cells using on-chip Fabry-Pérot cavity. Appl. Phys. Lett. 89:203901.
12. L. K. Chin, A. Q. Liu, X. M. Zhang et al. (2007) Differential single living cell refractometry using grating resonant cavity with optical trap. Appl. Phys. Lett. 91:243901.
13. P. Y. Liu, L. K. Chin, W. Ser et al. (2014) An optofluidic imaging system to measure the biophysical signature of single waterborne bacteria. Lab Chip 14:4237-4243.
14. Y. Yang, A. Q. Liu, L. Lei et al. (2011) A tunable 3D optofluidic waveguide dye laser via two centrifugal Dean flow streams. Lab Chip 11:3182-3187
15. Y. Yang, L. K. Chin, J. M. Tsai et al. (2012) Transformation optofluidics for large-angle light bending and tuning. Lab Chip 12:3785-3790.

Author: Ai-Qun Liu
Institute: Nanyang Technological University
Street: 60 Nanyang Avenue
City: Singapore
Country: Singapore
Email: eaqliu@ntu.edu.sg

Our Trial on Biomedical Engineering Research and Education System Related to Medical Regulatory Science

Mitsuo Umezu

Center for Advanced Biomedical Sciences (TWIns), Waseda University, Tokyo, Japan

Abstract— **I would like to introduce our new biomedical engineering research and education system related to medical regulatory science. This system has been operated in TWIns, which is an abbreviation for Tokyo Women's Medical University and Waseda University Joint Institution for Advanced Biomedical Sciences. TWIns opened in 2008 as the first collaborative research institute between a school of medicine and a school of science and engineering in Japan. The present number of residents at TWIns is 600, including 300 graduate students.**

I personally had a critical experience on "First in Human (FIH)" twice throughout my 40-year developmental life on ventricular assist devices. Therefore, I strongly believed an importance to establish a practical environment to cultivate human resources on medical regulatory science. As for a biomechanical engineering research, we have established dry-laboratories to provide practical tools to assess safety and effectiveness of medical treatments and to set up reliable test platforms for assessing the performances of keep-coming new medical devices. In addition, in 2010 we opened a joint graduate school that offers the first government-approved PhD degree program in Medical Regulatory Science. We hope this unique environment will contribute to the future success of those who experience our new concept, called another EBM: Engineering-Based Medicine.

Keywords— **Medical regulatory science, Biomedical engineering, First in Human, Dry laboratory, Engineering-Based Medicine**

I . Introduction

TWIns was opened in 2008 as the first collaborative research institute between a school of medicine(Tokyo Women's Medical University) and a school of science and engineering (Waseda University) in Japan. TWIns is an abbreviation for Tokyo Women's Medical University and Waseda University Joint Institution for Advanced Biomedical Sciences. [1]Waseda University had following two synergistic objectives to construct TWIns; 1) As research laboratories for life science, biomedical engineering and biology had been broadly dispersed in different campuses, we tried to consolidate facilities in one place. 2) As there is no medical faculty in Waseda, it was a great opportunity to be in official partnership with Tokyo Women's Medical University to create new fields of innovative medical treatments. It was not a hard job to create a synergetic philosophy, because we had a 50-year history on biomedical engineering collaboration between both universities. TWIns new building is constructed next to Tokyo Women's Medical University Hospital with a total floor area of 20,000 square meter.

Present number of residents at TWIns is 600; 450 from Waseda University and 150 from Tokyo Women's Medical University, including 300 graduate students.

I originally belong to a faculty member of the department of "Modern Mechanical Engineering" for Undergraduate program and "Integrative Bioscience and Biomedical Engineering" for Master program. There are three mechanical-engineering based laboratories at TWIns, and total number of academic staffs and graduate students with a background of biomedical engineering and bio-robotics are 100.

Fig. 1 Front view and Map (upper), and side view (bottom) of TWIns building

J. Goh and C.T. Lim (eds.), *7th World Congress on Bioengineering 2015*, IFMBE Proceedings 52, DOI: 10.1007/978-3-319-19452-3_2

II. SET UP AN IDEAL ENVIRONMRNT OF "ANOTHER EBM; ENGINEERING BASED MEDICINE"

A. Background to establish dry laboratories

I personally had a critical experience on "First in Human (FIH)" twice throughout my 40-year developmental life on ventricular assist system (VAS). The first FIH was experienced with our original device in 1981 at the National Cardiovascular Center, Osaka, Japan. [2]The technology of our pulsatile device was transferred to Toyobo Company and commercialized later (now distributed by Nipro Corporation). It was approved as a medical product of short term use (less than one month) by the Japanese government in 1991. Toyobo pumps have already been used over 900 cases in Japan to restore patients with poor blood circulatory condition. Due to a limited number of cardiac transplantation donors in Japan, an average waiting duration to receive a donor heart was over two years. So, it was strongly required to maintain a high quality of life during a usage of LVAS. Toyobo VAS is placed at the outside of the chest wall using two large diameter conduits, which may be a major part to induce infection problem for a long-term use. To solve the problem, several trials to install a pump inside of the body were conducted in 1980's. Finally, the development of non-pulsatile VAS was considered to be the most realistic method for long-term use. EVAHEART is one of compact centrifugal pumps which enable to implant into a small thoracic cavity for better quality of life. Sun Medical Research Corporation, Tokyo Women's Medical University, and Waseda University worked together to create a mechanical circulatory support device for patients with end-stage heart failure [3]. In 2005, the FIH with EVAHEART, that was my 2nd FIH experience, was conducted at Tokyo Women's Medical University Hospital. After eighteen clinical trials that included pivotal and pilot studies, the Japanese Ministry of Health, Labor and Welfare finally approved the first commercial implantable ventricular assist device. Those experiences led us to conclude that bioengineering laboratories should be established to design, construct, test, and perfect prototype medical devices including robots. Complex devices manufactured with precision components could be brought to market with a high degree of confidence in their reliability and durability. Unexpected problems with commercial products could be solved by computer modeling and simulation technologies. Academic team members, especially students, would gain hands-on experience and real-world responsibilities to prepare them for successful careers.

Therefore, I strongly believed an importance to establish a practical environment to cultivate human resources on medical regulatory science. As for a bio-mechanical engineering research, we have established dry-laboratories to provide practical tools to assess safety and effectiveness of medical treatments and to set up reliable test platforms for assessing the performances of keep-coming devices.

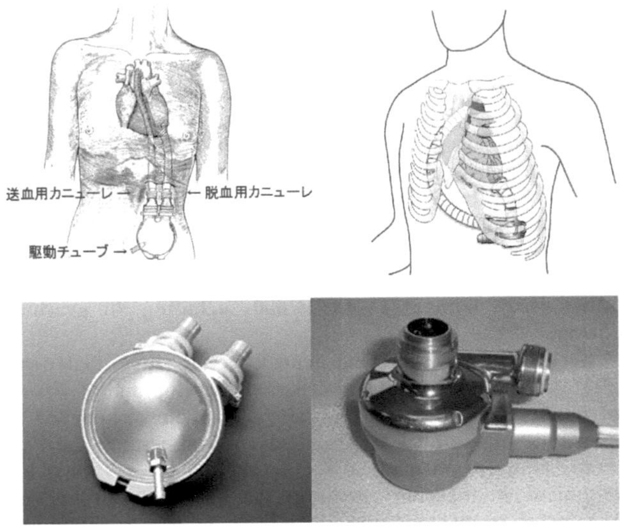

Fig. 2 Schematic drawings of para-corporeal, pneumatically-driven pulsatile VAS (upper left) and implantable, motor-driven non-pulsatile VAS (upper right)
Bottom two photographs are indicated commercial VAS pumps; Toyobo (Nipro) pulsatile VAS (left) and Non-pulsatile EVAHEART (right)

B. Establishment of a practical facilities

The dry-laboratories based on a philosophy of "Another EBM"are dedicated to three purposes [4] (Fig.3-6):

1) Testing medical devices such as artificial heart, prosthetic heart valves and vascular stents for safety and effectiveness. For this purpose, we also assemble mock circulatory systems to test the basic performance, durability and biocompatibility of devices. Fig.3 shows anti- thrombogenicity test circuits.

2) Training surgeons to implant or to control new medical devices to achieve the best performance of the devices. Fig.4 shows a training machine to improve a surgical skill on coronary bypass surgery and mitral valve repair.(Fig.4,5)

3) Analyzing to select the most effective patient with new medical devices using flow visualization method and/or computer fluid dynamics (CFD). We devise instruments that enable clinicians and surgeons to visualize blood flow and compute hemodynamics to help them determine optimal patient-specific treatments. We are studying cerebral aneurysms (Fig.6) to validate engineering, pathological, and clinical data.

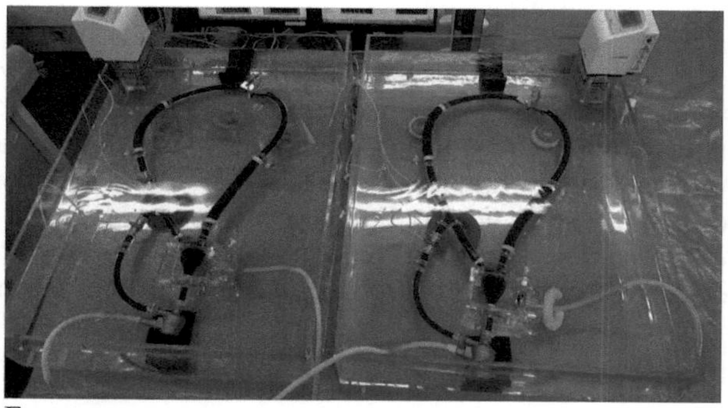

Fig.3 Experimental set up to evaluate the anti- thrombogenicity of medical devices under the same hemodynamic condition. This study is conducted to clarify the effect of surface roughness on thrombogenicity, using two identical blood circuits with different inlet cannula of EVAHEART.

Fig. 4 Beating heart simulator for coronary bypass surgery (left) OPCAB Boot camp(right)

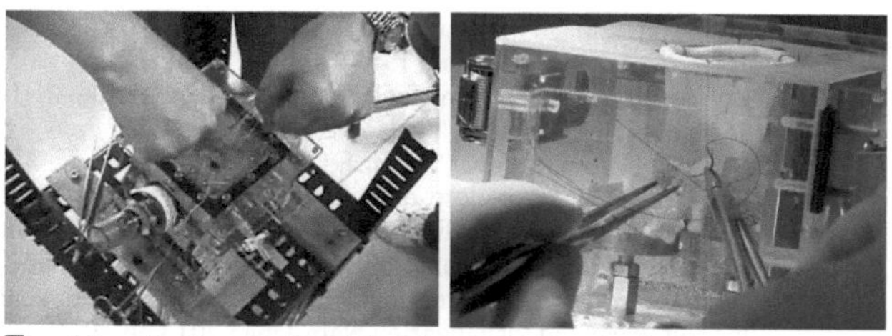

Fig. 5 Practical surgical training simulator for mitral valve repair and/or replacement

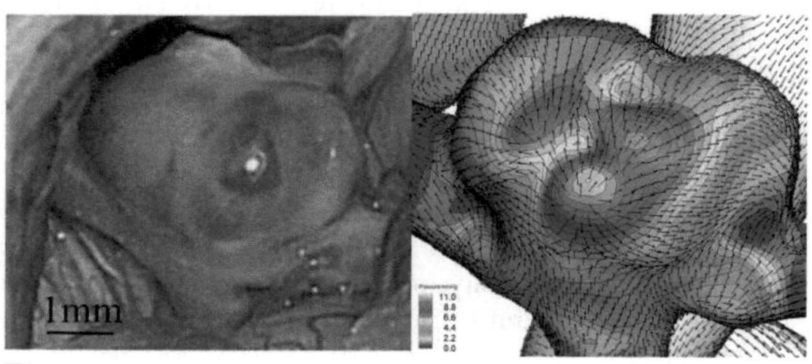

Fig. 6 Cerebral aneurysms of a patient (left) and CFD result to indicate thickness distribution of the aneurysms (right)

III. New graduate school program on medical regulaory science (MRS)

A. Background and Concept to propose a new PhD program

Medical regulatory science (MRS) within the healthcare industries pertains to governmental regulation of pharmaceuticals, medical devices, in vitro diagnostics, biologics, and nutritional products. Professional responsibilities in these areas include monitoring research and development, clinical trials, premarket approval, manufacturing, labeling, advertising, and post-market surveillance. MRS aims to improve decision-making in in vitro tests, animal experiments, and clinical trials of drugs, diagnostics, devices, and therapy protocols based on previous experience.

Post-approval proper use is another issue MRS seeks to clarify.

B. Approach

To strengthen the foundation of this new subject, we propose to consolidate MRS with three other academic disciplines: evaluation, prediction, and determination. [4] Gaining approval of a product can be a lengthy, tedious, and costly process. Regulatory agencies and their procedures vary widely between different countries, and efforts are underway to establish global standards. Regulatory affairs departments within healthcare companies are evolving and expanding to ensure the delivery of safe products to market in a reasonable time. So, professionals in regulatory affairs are increasingly in demand. Financial compensation for their employment also is increasing.

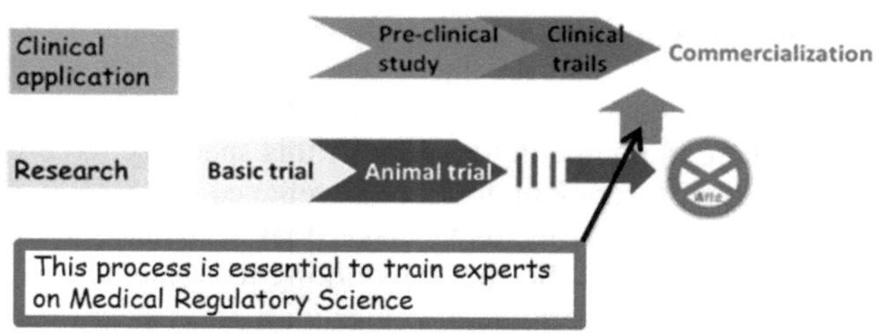

Fig. 7 Relationship between Basic medical science and clinical application towards commercialization

C. Cooperative Major in Advanced Biomedical Sciences

In 2010, Tokyo Women's Medical University and Waseda University opened a joint graduate school department that offers the first government approved PhD degree program for students of MRS. The program is called Cooperative Major in Advanced Biomedical Sciences. The number of students proposed for this doctoral degree course is limited to ten, five from each university. Candidates must earn at least thirty credits and pass a thesis review. Fourteen credits must be earned from the following subjects:

1) Biostatistics
2) Clinical Studies
3) Bioethics
4) Introduction to GLP/GCP/GMP
5) Medical Regulatory Science
6) Seminar in the Bioscience and Biomedical Laboratories
7) Practical training in the Advanced Medical Institute

Thesis titles of the first PhD recipients, who completed their work at the end of 2013 are as follows:
a) Study on Global Harmonization Initiatives for Medical Device Regulation
b) Study on International Radiation of High-intensity Therapeutic Ultrasound Devices
c) Study on Post-market Risk Assessment of Implantable Arrhythmia Devices in Japan
d) Study on Visualization and Evaluation of Medical Device Development using System Dynamics[5]
e) Study on Optimization of Therapeutic Parameters for Photodynamic Therapy
f) Study on Hemodynamic Evaluation Method of the Vascular Grafts by Computational Fluid Dynamics[6]
g) Study on the Current Status and Challenges of Pre-market Clinical Trials for Medical Devices in Japan
h) Study on Health Technology Assessment in Japan and Cost-Utility Analysis of Molecular Targeting Agents
j) Study on Regulation and Clinical Evaluation for Human Cells and Tissue Products

Up to now, total number to pass the entrance examination of this PhD program is 66 for these six years, and 18 completed a PhD course, then successfully received PhD. (Fig.8) Students with a degree in MRS will have a decided advantage in competing for jobs in this specialty. We want to reduce or supplant traditional reliance on animal tests with in vitro experiments to make final decisions with government approval. So, we are working to create new guidelines by holding periodic meetings with government regulators, academic researchers, clinicians, and MRS section members in the healthcare industries.

Fig. 8 Memorial photograph of the first PhD students, received on 26 March 2013. Their background, occupation and ages are different, but all became Medical Regulatory Science experts.

IV. Final remarks

I expect our synergistic strategy and trial at TWIns will help us create the next generation of new medical treatments for future applications in medicine: this approach is named "Another EBM: Engineering Based Medicine". We hope the unique experience of Engineering-Based Medicine at TWIns will help them contribute to the future success of the medical instrument and pharmaceutical industries in Japan.

Acknowledgment

This work was supported by Health Labor Sciences Research Grant (A11539000) of MHLW and MEXT KAKENHI (12018543). Some part of our project was funded by Research Institute of Science & Engineering (13L01) and Research Institute on Medical Regulatory Science, Waseda University.

Conflict of Interest

The author declare that he has no conflict of interest.

References

1. TWIns at http://www.twins.sci.waseda.ac.jp/
2. Noda H, Takano H, et al.(1988)Clinical consideration of life-saving effect of left ventricular assist device. Artificial Heart 2(Akutsu T ed).Springer Verlag.p205-214
3. Umezu M, Yamazaki K, Yamazaki S, Iwasaki K, Kitano T, Tokuno T(2013) Japanese-made implantable artificial heart (EVAHEART) true medical-engineering collaboration towards successful clinical outcome, FULUCOME 2013 Program and Abstract book,12th International Conference on Fluid control measurement and visualization,Nara,Japan,2013,pp 19-28
4. Umezu M, Iwasaki K, Yagi T, Park Y, Sakaguchi K, Shiurba R, Kasanuki H(2013) Engineering-based medicine: A new way to collaborate, IREMD, The 2013 Hirosaki University International Symposium,Hirosaki,Japan,2013,pp 13-18
5. Kato T, Umezu M, Iwasaki K, Kasanuki H, Takahashi Y(2013) Preliminary study on the development of a system dynamics model:the case of EVAHEART, Journal of Artificial organs,vol.16, no.2, 2013,pp 242-247
6. Konoura C, Shimizu T, Nakamura M, Iwasaki K, Qian Y, Okuda S, Yoshitake A, Shimizu H, Yozu R, Umezu M(2013)Numerical analysis pf blood flow distribution in 4-and 3-branch vascular grafts, Journal of Artificial organs,vol,16 no.2, 2013, pp 157-163

Author: Mitsuo Umezu
Institute: Center for Advanced Biomedical Sciences Waseda University (TWIns)
Street: 2-2 Wakamatsu -cho
City: Shinjuku –ku, Tokyo
Country: Japan
Email: umezu@waseda.jp

Induced Pluripotent Stem Cell Differentiation under Constant Shear Stress

N.K. Mohd Zin[1], K. Sakaguchi[1], Y. Haraguchi[2], T. Yagi[1], K. Matsuura[2], T. Shimizu[2], and M. Umezu[1]

[1] Graduate School of Advance Science and Engineering, Waseda University, Tokyo Japan
[2] Institute of Advance Biomedical Engineering and Science, Tokyo Women Medical University, Japan

***Abstract*—Mechanotransduction in in-vitro studies has yet to be done extensively using human induced pluripotent (hiPS) cells. These external mechanical factors are important controlling factor for cell differentiation at a cellular level particularly in cardiomyocytes cells. Cells in a living body are being exposed to many kind of stresses; shear stress in particular modulates cellular function in a living body. Similarly, we are trying to understand the link between the shear stress in the bioreactor and the differentiation effects it has on iPS cells. In this study, we investigate the difference of strictly uniform laminar shear stress with a non-uniform laminar flow and its effects on the cell survival rate and differentiation. Additionally, the elimination of biochemical factors were done to satisfy the need of demand for clinically usable cells for the near future use. By eliminating cytokine induced differentiation. In addition, information pertaining to relationship between fluid shear stress, cellular deformations, cell differentiation and cell survival rate could provide a more optimised condition for cultivation specific type of cells within a shorter time period.**

Keywords*—iPSC; Shear stress; Cell viability;

I. INTRODUCTION

Interdisciplinary collaboration amongst mechanical engineers and medical scientist are inevitable. In this collaborative field, complete understanding of principles of each distinctive field are in crucial demand. This is especially true in the case of mechanically viable tissues from biocompatible compound. Mechanically viable compound such as scaffolds in which is constructed with a specific message, encoded chemically or mechanically created in vitro for the use and or aid the growth in vivo. On their own, scaffolds has show quite a potential it could bring to the regenerative medicine world. Potential in that it shows latent promise in clinical needs and demands [1,2]. Even with the potential shown by these constructs, time constraint is one of the worries that are still to be over came by engineers and scientist alike. The length of time needed to attain complete healing is much too slow for patient need. The time needed for migration of cells into the scaffold and the conditioning to engineer the peak micro environment is much too long.

Growth of tissue and tissue remodelling, cell embedded scaffold has to posses multipotent progenitor. Multipotent progenitor possesses the ability to differentiate cells to different linages. These different linages are determined by the various different cues such as mechanical stimuli, chemical agitation and hypoxia. The aim of most study now is to study the basics, the specific differentiation stimulus and the time span taken to be differed. By knowing these detailed stimulus, the studies could help with preventing unwanted tumour and knowing the cell-matrices reaction due to bio-mechanotransduction [3]. On an upper note, the available studies now showed that multipoint progenitor cells has an advantage over wound healing [4]. To up hold the demands that are getting higher and higher each day by lessening the steps in the animal based studies, three-dimentional (3D) studies has the potential to flourish in this field. Not only to grow as a method, but as a staple for the near future cultivation of multipotent and pluripotent cells.

Pluripotent cells are cells that has the embryonic expression, which to say is at it early stages which is yet to be differentiated to a linage. Induced pluripotent stem cells (iPSC) however are generated by the forcing of expression of embryonic transcription on a differentiated cell or an adult cell. IPSC has several attribute that made them more suited for cellular differentiation pathological and functional studies. These cell are rather resilient compared to an embryonic cell in which they could be cultured for several passages without losing their normal karyotype, are readily transfected and effortlessly differentiated. IPSC including human iPSC (hiPSC) are able to be differentiated to any line of adult cell in fairly short amount of time. The ease of care and ease of experimentation using hiPSC would hold a bright future for regenerative engineering.Barring that reprogramming of cells does require a little time, the possibility to program adult cells makes patient-specific regenerative treatment more likely in the near future.

Post natal somatic hiPSC also exhibit immense potential in cellular therapy especially pathological studies, practical and diseased cellular therapy. In fact, integration free reprogramming technology as those which uses plasmids, promises an autologous and higher grade hiPSC line, and if were manufactured under proper care and procedure the possibility to be used as therapeutic application is going to be very prominent [2,5,6].

Cultivating the cells are one thing, but maintaining the cells under proper procedure and care would greatly effect the differentiation of the cells and the condition as well as the viability of them. According to Mehta, the optimum dosage for a seventy kilograms (70 kg) adult patient are 4.2×10^8 to 5.6×10^8 $CD34^+$ cells for hematopoietic stem cell (HSC) [7]. For a clinically applicable usage of hiPSC and their progenies are estimated to be around 1 to 2 billion cells which are a huge multiplication compared to laboratory experiments [8]. In a constrained laboratory budget, maintaining and reproducing

J. Goh and C.T. Lim (eds.), *7th World Congress on Bioengineering 2015*,
IFMBE Proceedings 52, DOI: 10.1007/978-3-319-19452-3_3

hiPSC economically remains a huge challenge. Yet this is one of the challenge that will be brought fourth in order to realise the hiPSC for clinical use.

The older standard of procedure, hiPSC were induced and expand on a feeder cells and was cultured in a culture medium containing sera or serum replacement. The significant for these serum or sera is to mimic the chemical compound or nutrition in a human body [9,10]. This was also done in order to abide the systematically tighter rules over the year for clinical studies. In order not only to be safe economically but also clinically, a rather lesser additive medium were created by a few groups that are utilising hiPSC and mouse iPSC. There are viable improvements that could be seen compared with the earlier days when iPSC was just becoming a staple in the news. Feeder free and serum free medium gave yet another promising result [11,12]. There is however another hurdle that we, scientist have to over come which the way to control the differentiation and separation of cells to cope to the demand for clinically viable cells.

In the topic of cell viability, another method that is currently being used is a suspension culture or 3-D culture. This kind of culture provides a solution to both the economic problems and also expansion problems. Although the turbulent flow and stress it has on the cells are currently still being researched by our group. For the suspension culture, Rho-associated- coiled-coil kinase (ROCK) inhibitor Y27632 was said to encourage the survival of human Embryonic Stem Cells (hESC) on the firstly of seeding [13]. A rather detailed standard of procedure is also available for single cell inoculation and various types of suspension culture for hiPSC [14,15]. There are also reports that praises commercially available suspension medium such as StemPro and mTeSR that is more complex and due to it's contents, it is also more expensive [16,17,18]. It is great that these media is available in the commercial market yet, due to its unknown contents and the price, these kind of media is less likely to be used as an initial experiments especially for students and new researchers.

A significantly refined hiPSC culture medium was recorded by Chen et all recently. This hiPSC media, E8 which contains seven clearly described components and are completely xeno-free. In addition to that this additive supplements the standard DMEM/F-12 medium [19]. E8 also does not need any additive such as bovine serum albumin, Fraction V or even human albumin to support the growth of iPSC [20]. This media is being used in the feeder-free environment. Based on this, the significance of simplified E8 medium could support a robust and economic suspension culture system in a stirred bioreactor for large-scale expansion and cryopreservation of hiPSCs was tested.

Another problem that is faced by researchers are that static culturing of progenitor cells results in porous scaffolds. Maturing these cells in a differentiation media or by using chemical agitation, it might help with the development of the methodology for osteogenic tissue construct for in vivo or clinical usages. With static culturing and chemical agitation, the imitation is not complete without mechanical stimuli. It is a well known fact that biomechanics stimuli affect the differentiation especially the osteogenic differentiation. The biomechanical stimuli is necessary for bone remodelling especially because as in in-vivo remodelling, the simulation is needed to elicit correct cellular differentiation and function. This argument is being supported by the loss of bone mass when mechanical load is lessen or absent such as space flight or after significant periods of bed rest. It back the argument of needing biomechanical simulation in bone homeostasis [21-24]. Also according to McCoy and O'Brien, bone is predominantly subjected to two different forms of biomechanical stimulations that controls turnover of cells especially strain levels that was predicted for humans in vivo to be around <2000 $\mu\varepsilon$ for physical deformations and 0.8 - 3 Pa from fluid shear stress that are generated solely by fluid moving in the interstitial space through cavity caused by tension and compression while being under load bearing pressure.

With this in mind, our team thought up an idea to not only to find a way to incorporate mechanical agitation but also take in consideration the distribution of shear stress. With this, we came up with a device so that has less turbulence flow and easily calculated shear stress at almost all point. The device is made by utilising a 3D printer to printout the mould for the spinning inner cylinder inside the spinning flask. Utilising mechanical theory of Coutte flow and using only the basic media for differentiation, the growth rate, viability and lineage differentiation are being studied.

II. MATERIALS AND METHODS

A. Maintenance of hiPSCs in adhesion

Human iPSC (253G1) were initially maintained on mouse embryonic fibroblast (MEF) feeders in standard Primate ES Cell Medium (ReproCELL, Japan). Primate ES Cell Medium were changed every 24-36 hours for 4 days or to 80 percent (80%) confluency. For the use of proliferating or passaging, cells were harvested by using cells were harvested by using 0.5 mL prewarmed at 37°C ReproCELL Dissociation Medium (ReproCELL, Japan) for 2 minutes. Cells were procured by pipetting the dish gently and centrifuged at 1000 RPM for 3 minutes.

However, for the use of suspension culture, human induced pluripotent stem cells were again harvested by using ReproCELL Dissociation Medium (ReproCELL, Japan) with as stated in the paragraph above. They were then scraped gently off the dishes and was not centrifuged, instead were let to settle in the centrifuging tube for 15 minutes under room temperature and were plated on MatriGel surface. ReproFF2 media were used to supplement the feeder free cells. In both the media, 5ng/µL basic Fibroblast Growth Factor (bFGF) were added.

B. Suspension culture of hiPSCs

Suspension media made of a mix of DMEM, 20% Foetal Bovine Serum (FBS), 1% Non-essential amino acid (NEAA) supplements and 0.18% b-mercaptoethanol were pre-warmed at 37°C in a water bath before 18mL transferred to the spinning flask and being conditioned in an incubator with 37°C, 5%CO_2 concentration. Two (2) pieces of the 96 well, hanging drop processed Embryonic bodies (EB) were used for each spinning flask. Cells were spun for four (4) days with the media being changed every two (2) days.

C. *Hanging drop*

Cells plated on MatriGel were harvested as mentioned in the paragraph above. Feeder free cell adhesion were done at least twice (2 passages) before being counted to 5x10^4 cells/mL. 20 mL of phosphate buffered saline (PBS) were added to 100 mm culture dish. 20µL of harvested cells were then pipetted on to the lid of the dish. Cells were incubated at 37°C with 5% CO_2 concentration for two (2) days.

On day three (3) the 20µL drop were then moved to 96-well plate with 180µL pre-warmed suspension culture media. The EBs were then left for two (2) more days to develop before transfering them to the 3D suspension culture.

D. *Bioreactor*

Fig. 1. Newly Proposed Model

Fig. 2.Conventional Method

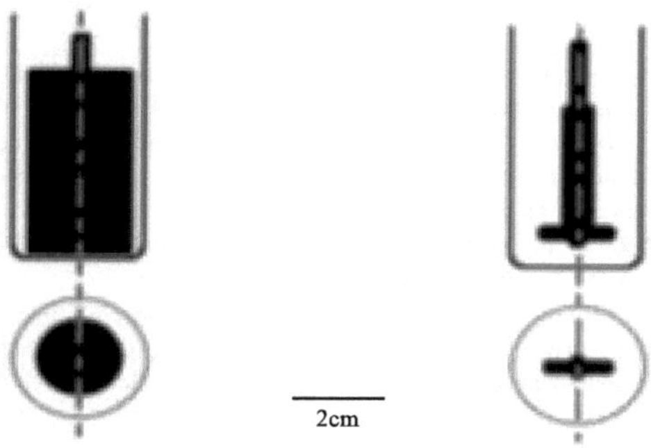

Fig. 3. Illustration of difference between Newly Proposed Method and Conventional Method.

The bioreactor used is from Wheaton Inc., USA which is represented in Fig. 1. By adding a small cylinder, incorporating the theory that are being similarly used by a viscometer, in theory, a controlled shear stress bioreactor could be made that is being shown in Fig. 2. Fig. 3 is the illustration to further clarify Fig. 1 and Fig. 2.

Cells in both the bioreactors were conditioned at 37°Cwith 5%CO_2 concentration. Both the new and conventional bioreactors were ran at the same time with five (5) different speeds; 40, 45, 50, 55 and 60 RPM.

These reactors ran for four (4) days after hanging drop part of the procedure. Media used for this step was explained earlier in the suspension culture of iPSC in materials and method section of the paper.

III. Results

Analysis for aggregates

Cell aggregates were counted by utilising a haemocytometer and microscope. Based on the observation made and calculation, below is the result for the haemocytometer.

Table 1: Result for cell count

Speed	Cell Count (x10^4 cells)	
	Conventional ± error	*New ± error*
40RPM	942.5 ± 53.8	2397.5 ± 35.2
45RPM	1685 ± 40.2	3610 ± 20.4
50RPM	1470 ± 33.7	3022.5 ± 25.6
55RPM	1632.5 ± 25.8	3345 ± 20.8
60RPM	1392.5 ± 98.5	2200 ± 50.6

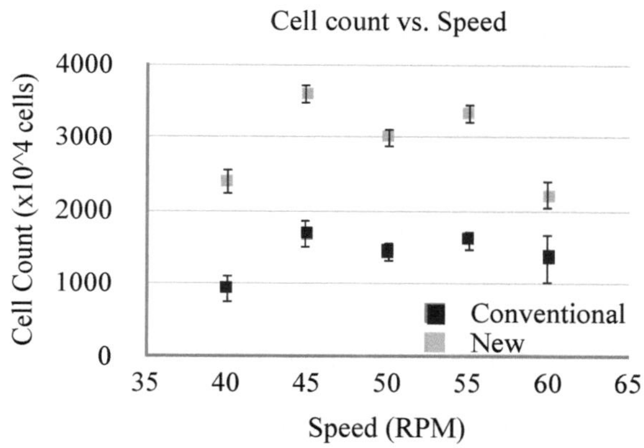

Fig. 4. Graphical representation of cell count

According to the result procured, cell count for the newly proposed method is at it's highest when the speed is at 45RPM and the lowest at 60RPM. As for the conventional method, the highest count is also obtained at 45RPM however for the lowest count is acquired at 40RPM.

IV. Discussion

A. *Analysis for aggregates*

Based on the result of aggregates analysis, the number of cells in 45RPM of the new smooth version yielded as the highest and at 40RPM of the conventional is the lowest. From the graph in Fig. 4. It could be hypothesised that the new bioreactor might have a more distributed shear stress that therefore it has higher proliferation of cells. The conventional version might have more back flow, resulting in non-uniform stress making proliferation harder and results in more cell death.

With this result, the interest to evaluate on the metabolism rate would further explain the difference between the cell counts and cell viability as well as the glucose consumption of the cells.

B. RNA quantification

With this in mind, the question of cell viability, RNA concentration and differentiation of cell type also became a point of interest. It is expected that for cell differentiation result, the higher speed would differentiate into an osteogenic cells and the lower speed into an angiogenic cells. There are also the possibility that the cells are not all being differentiated and are likely to be only proliferated.

V. CONCLUSION

Based on the hypothesis although the result is not as conclusive yet, but it is believed that the newly proposed method would have a more distributed shear stress compared to the conventional method. It is also expected for the cells to be differentiatied into osteogenic cells when the reactor is spun at higher speed and at lower speed, it should be differentiated into angiogenic cells.

With the results that we have obtained, it is supposed that the hypothesis is flawed but still viable to a certain degree. Although some adjustment should be continuous and more through experiment are being conducted, for the time being, our team are still left with a promising start.

ACKNOWLEDGEMENTS

This research have been funded by grants from the Ministry of Education, Culture, Sports, Science & Technology (MEXT) through the Funding Program of Advanced Interdisciplinary Center for the Establishment of Regenerative Medicine: Cell Sheet Tissue Engineering Center (CSTEC)

CONFLICT OF INTEREST

The authors declare that they have no conflict of interest.

REFERENCES

1. Chen, V.C., et al., 2012. Scalable GMP compliant suspension culture system for human ES cells. Stem Cell Res. 8 (3), 388–402.
2. Dowey, S.N., et al., 2012. Generation of integration-free human induced pluripotent stem cells from postnatal blood mononuclear cells by plasmid vector expression. Nat. Protoc. 7 (11), 2013–2021.
3. Yoshida Y., Yamanaka S. Recent stem cell advances: induced pluripotent stem cells for disease modelling and stem cell-based regeneration. Circulation. 2010;122:80–87.
4. Carvajal-Vergara X., Sevilla A., D'Souza S.L., Ang Y.S., Schaniel C., Lee D.F. Patient-specific induced pluripotent stem-cell-derived models of LEOPARD syndrome. Nature. 2010;465:808–812.
5. Chou, B.K., et al., 2011. Efficient human iPS cell derivation by a non-integrating plasmid from blood cells with unique epigenetic and gene expression signatures. Cell Res 21 (3), 518–529.
6. Ye, Z., et al., 2009. Human-induced pluripotent stem cells from blood cells of healthy donors and patients with acquired blood disorders. Blood 114 (27), 5473–5480.
7. Mehta, J., et al., 2009. Optimizing the CD34 + cell dose for reduced intensity allogeneic hematopoietic stem cell transplantation. Leuk. Lymphoma 50 (9), 1434–1441.
8. Kehoe, D.E., et al., 2010. Scalable stirred-suspension bioreactor culture of human pluripotent stem cells. Tissue Eng. Part A 16 (2), 405–421.
9. Okita, K., Ichisaka, T., Yamanaka, S., 2007. Generation of germline-competent induced pluripotent stem cells. Nature 448 (7151), 313-U1.
10. Yu, J.Y., et al., 2007. Induced pluripotent stem cell lines derived from human somatic cells. Science 318 (5858), 1917–1920.
11. Li, Y., et al., 2005. Expansion of human embryonic stem cells in defined serum-free medium devoid of animal-derived products. Biotechnol. Bioeng. 91 (6), 688–698
12. Ludwig, T.E., et al., 2006. Derivation of human embryonic stem cells in defined conditions. Nat. Biotechnol. 24 (2), 185–187.
13. Watanabe, K., et al., 2007. A ROCK inhibitor permits survival of dissociated human embryonic stem cells. Nat. Biotechnol. 25 (6), 681–686.
14. Amit, M., et al., 2011. Dynamic suspension culture for scalable expansion of undifferentiated human pluripotent stem cells. Nat. Protoc. 6 (5), 572–579.
15. Zweigerdt, R., et al., 2011. Scalable expansion of human pluripotent stem cells in suspension culture. Nat. Protoc. 6 (5), 689–700.
16. Olmer, R., et al., 2012. Suspension culture of human pluripotent stem cells in controlled, stirred bioreactors. Tissue Eng. Part C Methods 18 (10), 772–784
17. Singh, H., et al., 2010. Up-scaling single cell-inoculated suspension culture of human embryonic stem cells. Stem Cell Res. 4 (3), 165–179.
18. Steiner, D., et al., 2010. Derivation, propagation and controlled differentiation of human embryonic stem cells in suspension. Nat. Biotechnol. 28 (4), 361–364.
19. Chen, G.K., et al., 2011. Chemically defined conditions for human iPSC derivation and culture. Nat. Methods 8 (5), 424-U76.
20. Wang, Y., Chou B. K., et al., 2013 Scalable expansion of human induced pluripotent stem cells in the defined xeno-free E8 medium under adherent and suspension culture conditions. Stem Cell Research (2013) 11, 1103–1116
21. Burr, D.B., Robling, A.G., and Turner, C.H. Effects of bio- mechanical stress on bones in animals. Bone 30, 781, 2002. 18. Turner, C., Forwood, M., and Otter, M. Mechanotrans- duction in bone: do bone cells act as sensors of fluid flow? FASEB J 8, 875, 1994.
22. Bikle, D.D., and Halloran, B.P. The response of bone tounloading. J Bone Miner Metab 17, 233, 1999.
23. Morey, E., and Baylink, D. Inhibition of bone formationduring space flight. Science 201, 1138, 1978.
24. Zerwekh,J.E.,Ruml,L.A.,Gottschalk,F.,andPak,C.Y.C.The effects of twelve weeks of bed rest on bone histology, bio- chemical markers of bone turnover, and calcium homeostasisin eleven normal subjects. J Bone Miner Res 13, 1594, 1998.
25. McCoy RJ, O'Brien, FJ. Influence of shear stress in perfusion bioreactor cultures for the development of three-dimensional bone tissue constructs: a review. Tissue Engineering Part B Review. 2010;16(6): 577-585.

Design of Unobtrusive Wearable Mental Stress Monitoring Device Using Physiological Sensor

T. Salafi and J.C.Y. Kah

Department of Biomedical Engineering, National University of Singapore, Singapore

Abstract— **Stress is recognized to be the most prevalent life risk factor all over the world. Chronic stress can elicit various mental disorders and increase the risk of cardiovascular disease. However, stress measurement from psychiatrist is instantaneous, qualitative and quarrelsome. Furthermore, there is a need to measure the stress level ubiquitously to prevent the chronic stress from happening. This study proposed an inconspicuous stress monitoring system using physiological sensors with accurate algorithm to measure the stress level of the user. We fabricated a stress monitoring patch prototype with mobile app user interface and conducted experiment with 50 participants to classify the stress and relax data distribution from three physiological signals namely heart rate, skin temperature, and galvanic skin response. We applied support vector machine algorithm and K-means clustering to classify the obtained training data and index the stress level of the user which resulted in the overall accuracy of 91.26%. The system also has feature to alert the users if their stress level is more than 80% and provide animated deep breathing guide bar to assist the users in alleviating their stress.**

Keywords— **Mental Stress Monitoring, Physiological Sensor, Support Vector Machine, Mobile App**

I. Introduction

WHO in 1986 defined health as the condition which is not merely the absence of disease but a positive state of physical, mental and social well-being [3]. One of the mental disorders that contribute to the most pressing lifestyle factor worldwide is stress. Stress is an outcome of the body's fight or flight response that can be beneficial in a dangerous situation but prolonged stress could reduce performance and physical health. Anxiety and depression are the most prominent outcome of the prolonged stress which will increase the risk of cardiovascular diseases. Stress disorders are highly treatable, yet only about one-third of those suffering receive treatment. When a stress patient is treated by psychiatrist, there are some difficulties in monitoring the progress of the treatment. Firstly, the patient subjectivity level in measuring their own mental health condition. Secondly, the usage of questionnaires and interviews for measuring the stress level of patients is qualitative and prone to error. In addition, these patients need therapy to be able to adapt with their stressor and relax themselves in the presence of the stressor.

A number of physiological markers are widely utilized to measure the stress level of the patients such as heart rate, skin conductance and temperature, muscle activity, and respiration [4]. With the advances in wearable sensor, it is possible to assess the stress level of the patient quantitatively with these physiological markers. With the advances of mobile computing, these data can be sent to the psychiatrist for remote monitoring of the therapy progress. Mobile application also could provide biofeedback by guiding the patient to do a deep breathing exercise to enhance the therapy. In this paper, an unobtrusive stress monitoring patch is proposed to monitor the stress level and provides feedback to the patient to lower their stress level.

II. Background

A. Stress and Health

Stress will trigger the autonomic nervous system (ANS) to affect the body physiology by entering the fight or flight response. This ANS stimulation will increase heart rate, perspiration, respiration rate, blood pressure, muscle activity, dilate pupils and decrease skin temperature. When the stress is acute, it can boost performance level of a person. However, when the stress is chronic, it can lead to serious consequences such as elevated blood pressure that can make a blood clot in arterial wall which will lead to coronary heart disease, elevated ambulatory blood pressure, and an increased risk of myocardial infarction [5]. Over time, stress can damage the body as it results in suppression of the immune system, inhibition of the inflammatory response, increased blood pressure, damaging muscle tissue, infertility, and diabetes. Suppression of the immune system leads to increase in the severity of common cold and susceptibility to infectious disease.

B. System Consideration for Wearable Stress Monitoring Device

Wearable sensors must be minimally cumbersome and inconspicuous to avoid apprehension associated with wearing medical devices in public. Therefore it is imperative to choose unobtrusive small sensors and avoid any dangling

J. Goh and C.T. Lim (eds.), *7th World Congress on Bioengineering 2015*,
IFMBE Proceedings 52, DOI: 10.1007/978-3-319-19452-3_4

wires for the sensor connections. Several devices such as AutoSense [2] and SHIMMER [7] system have wires from the wrist to the sensor on the finger which hinder the movement of the user's hand. Besides being minimally obtrusive, the display and user interface must be friendly and capable of alerting the user when they are stress. Furthermore, it is important that the user interface can guide the user to relieve and reduce their stress level. More importantly, the accuracy of the stress classification based on the physiological sensor must be at least 90% high to enhance the reliability of the system. The result of the classification should also provide the user stress level in the scale of 0-100% so they can quantitatively approximate their stress level. Upon the determination of stress index, a biofeedback mechanism must be available to guide the user to lessen their stress index.

III. Description of the System

A design of unobtrusive mental stress monitoring device is fabricated. The system comprised of a sticky patch which composed of physiological sensor, microcontroller and Bluetooth transmitter. We developed a smartphone application for the display system and user interface. The sticky patch is attached on the user palm to avoid any unnecessary wires that hinder the user hand movement. Three physiological sensors are utilized for the system, namely heart rate sensor, skin temperature sensor and galvanic skin response. Figure 1 depicts the overall architecture of the system.

A. Heart Rate Monitor Sensor

Current available technologies for monitoring the heart rate include electrocardiogram (ECG) and photoplethysmography (PPG). The PPG sensor is very sensitive and therefore motion artifacts can decrease the accuracy of the heart rate reading. In addition, the PPG sensor is best located on the finger or earlobe where the skin thickness is low, but these two places can obstruct the user activity due to the wire connections. Therefore, a chest strap with heart rate receiver was utilized to ensure the device does not hinder any user movement.

B. Skin Temperature Sensor

In a state of increased exertion, excitement and stress, human muscle fibers will contract, causing a stenosis of vasculature. This leads to a reduction of skin temperature, since blood circulation through the tissue is reduced. Studies about the average skin temperature alteration showed that hand palm temperature appeared to be significantly decreased in response to stress [8]. Therefore, we incorporated a thermistor into the patch to sense the user skin temperature change for stress measurement.

C. Galvanic Skin Response Sensor (GSR)

Galvanic skin response was utilized to measure the perspiration or electro dermal activity (EDA) of the skin. This GSR sensor was position on the palm or finger because these places have the highest eccrine gland density. However, the electrode can be easily detached from the skin due to more sweat produced on these areas. Therefore, a sticky patch design was proposed to ensure the electrode is in place and avoid any detachment. Additionally, The GSR electrode was located on the palm to avoid any obstruction as the wires and patch will interfere with the user hand movement if the GSR is placed around the finger.

D. Smartphone User Interface

Smartphone application was developed for user interface to display the stress level so as to make use of smartphone's current technology. In addition, the application will be able to provide a more interactive experience compared to through the use of embedded LED or LCD screen. A game can be added to decrease the stress level of the patient. In the proposed device, collected data from patch sensors were sent via bluetooth to the smartphone and an alert system was created to inform the user when their stress level is higher than 80%. After alerting the user, the app will display a simple deep breathing page which automatically guides the user to alleviate their stress with an animated bar.

IV. Data processing and algorithm

A. Support Vector Machine (SVM)

In the system, support vector machine algorithm is employed to classify the user data. It is a powerful learning algorithm that is able to sort data into two or more classes based on prior data [1]. SVM generates the optimal hyper plane, which is the line that best splits the unstressed and stressed data with a boundary. This is done by finding all potential boundaries for the data and using the one with the largest margin of difference as that would indicate the most accurate separation between the two sets of data points. Since the design measures three parameters, the hyper plane equation would have three variables. In order to classify data with SVM, experiment data and the state are preprocessed into the form of sensor value and state (xi : [HR; ST;GSR]; yi : State). The SVM for the binary linear classification problem require the following optimization model including the error-tolerant margin.

$$Minimize\,\frac{1}{2}w^T w + C\sum \xi_i \quad (1)$$

$$subject\ to\ y_i(w^T x_i + b) \geq 1 - \xi_i and\ \xi_i > 0 \quad (2)$$

Where w is weight vector and b is a bias. C is the error penalty and ξ_i are slack variables, measuring the degree of miscalculation of the sample x_i. The maximum margin is obtained by minimizing the first term of objective function, while the minimum total error of all training example is assured by minimizing the second term. For the efficient purposes, we trained the data with sequential minimal Optimization using kernel function [6].

B. Stress Index Generator

The decision boundary obtained by the SVM is only able to classify whether the user is stressed. To create a stress index from 0-100%, the instantaneous physiological signal was employed to the Eq. 4 followed by Eq. 5 or 6. The distance of the instantaneous values to the hyper plane is first calculated and the distance of the point was compared to the centroid of each state cluster. The centroid of each cluster is obtained from K-means iteration algorithm by applying the training data to the following equation

$$argmin\ S \sum_{i=1}^{k} \sum_{x_j \epsilon S_i} \left\| x_j - \mu_i \right\|^2 \quad (3)$$

Where x_j are sets observations and μ_i are the cluster center i. The distance of the user values to the hyper plane is calculated from this equation

$$D_k = \frac{|Ahr_k + Bst_k + Cgsr_k + D|}{\sqrt{A^2 + B^2 + C^2}} \quad (4)$$

The obtained distance D_k is subsequently applied to the following equation to generate the stress index.

For the point classified as unstressed,

$$SI_k = \left|1 - \frac{D_k}{\overline{D}_u}\right| * 0.5 \quad (5)$$

For the point classified as stressed,

$$SI_k = \left|1 + \frac{D_k}{\overline{D}_s}\right| * 0.5 \quad (6)$$

Where SI_k is the stress index at instant k and $\overline{D}_u$ and $\overline{D}_s$ is the distance of the centroid of unstressed and stressed training data to the hyper plane respectively. These equations above calculate the proportion of distance between the mean or centroid values and the measured values. This proportion gives the stress index based on where the measured values lie between the mean value and the decision boundary from a distance point of view. For instance, if the measured value's distance is exactly half of that from the mean values to the decision boundary, the stress index would be 25% or 75% if classified under unstressed or stressed respectively. The overall stress index generator flow diagram is shown in figure 1 below.

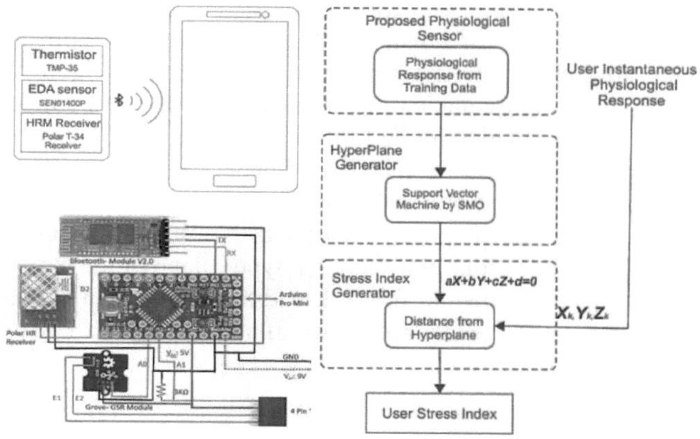

Fig. 1 overall hardware design and flow diagram of stress index generator

C. Experiment

To obtain training data for the proposed system, total of 50 participants consists of 23 Male and 27 female were randomly selected from National University of Singapore's Student. There were two stages of data collection in order to obtain relaxed and stressed state data of participants. First, for relax data collection, the participants were required to listen to soothing music while doing the deep breathing technique to ensure they are in relax state. After which, the participants will undergo a stroop color test for two minutes and subsequently rigorous arithmetic test with limited time to elicit the stress state. The heart rate, skin temperature and galvanic skin response data are subsequently recorded and processed using the proposed algorithm.

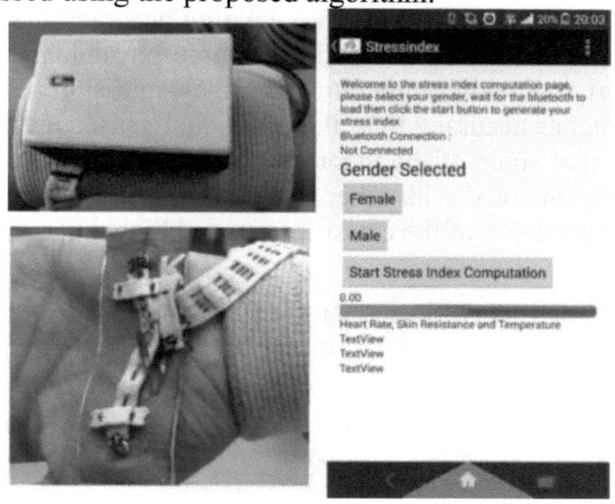

Fig. 2 the prototype comprises of the holster unit to house the electrical component, the sticky patch with sensors and the mobile app

V. Result and Discussion

A prototype of the wearable system was fabricated and utilized to collect the experiment data. The prototype consists of sticky patch with wires connected to the holster unit. The final design of holster unit, the sticky patch and the android app for the user interface is depicted in figure 2. The mobile app contains the stress index page and simple deep breathing guide in the form of animated bar to help the user relieve their stress. The experiments data were collected during the participants' relaxed and stressed state and trained with the proposed method using WEKA SMO library [9]. The scatter plot with the hyper plane of the experiment data are shown in figure 3, and the accuracy of classification for each gender are shown in table 1.

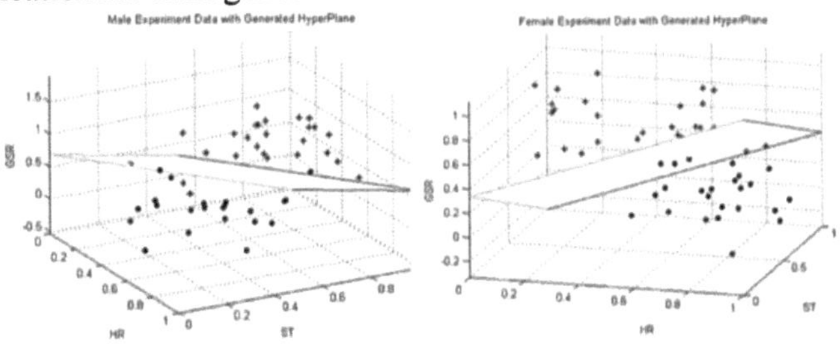

Fig. 3 Plot of male (left) and female (right) experiment data with the generated SVM hyperplane.

As the electronics component of prototype is quiet huge, it is still not feasible to incorporate the entire component inside the sticky patch, instead a holster is created to store several components of the electronics such as microcontroller and Bluetooth module. This holster unit is meant to be worn on the wristband of the user. With this arrangement, the overall system is still inconspicuous and does not hinder any user movement as it does not use finger as the sensor data collection point. The classification accuracy graphs in table 1 shows that combination of three physiological data of heart rate, skin temperature and galvanic skin response have higher accuracy than the individual sensor used. This means that some of the incorrectly classified parameter can be complemented with other two parameters which resulted in the correction of the classification and higher accuracy.

Table 1 the classifications accuracy based on gender and sensors

Subject	HR	ST	GSR	HR,ST,GSR
Male	78.25%	84.35%	79.68%	89.81%
Female	75.22%	80.52%	81.96%	92.70%
Overall	76.74%	82.44%	80.82%	91.26%

VI. Conclusion

In this study, a wearable unobtrusive stress monitoring system using three physiological sensors which are heart rate, skin conductance and temperature is designed. The support vector machine algorithm was applied to the experiment's data which results in an accuracy of 91.26%. The chosen sensors, connections, and user interface system ensure that the device does not interfere with the user daily activities. Furthermore, the index generated from the algorithm assists the users to understand their stress level quantitatively and the animated deep breathing guide bar can help the user to reduce their stress level.

References

1. Boser, B. E., Guyon, I. M., and Vapnik, V. N. A training algorithm for optimal margin classifiers. In Proceedings of the Fifth Annual Workshop on Computational Learning Theory, COLT '92, ACM (New York, NY, USA, 1992), 144–152.
2. Ertin, E., Stohs, N., Kumar, S., Raij, A., al'Absi, M., and Shah, S. Autosense: Unobtrusively wearable sensor suite for inferring the onset, causality, and consequences of stress in the field. In Proceedings of the 9th ACM Conference on Embedded Networked Sensor Systems, SenSys '11, ACM (New York, NY, USA, 2011), 274–287.
3. Jr, J. E. W. Standards for validating health measures: Definition and content. Journal of Chronic Diseases 40, 6 (1987), 473 – 480. The Portugal Conference: Measuring Quality of Life and Functional Status in Clinical and Epidemiologic Research.
4. Lin, H.-P., Lin, H.-Y., Lin, W.-L., and Huang, A. C.-W. Effects of stress, depression, and their interaction on heart rate, skin conductance, finger temperature, and respiratory rate: sympathetic-parasympathetic hypothesis of stress and depression. Journal of Clinical Psychology 67, 10 (2011), 1080–1091.
5. Nabi, H., Kivimki, M., Batty, G. D., Shipley, M. J., Britton, A., Brunner, E. J., Vahtera, J., Lemogne, C., Elbaz, A., and Singh-Manoux, A. Increased risk of coronary heart disease among individuals reportingadverse impact of stress on their health: the whitehall ii prospective cohort study. European Heart Journal (2013).
6. Song, X.-f., Chen, W.-m., Chen, Y.-P., and Jiang, B. Candidate working set strategy based smo algorithm in support vector machine. Information Processing and Management 45, 5 (2009), 584–592. cited By (since 1996)9.
7. Sun, F.-T., Kuo, C., Cheng, H.-T., Buthpitiya, S., Collins, P., and Griss, M. Activity-aware mental stress detection using physiological sensors. In Mobile Computing, Applications, and Services, M. Gris and G. Yang, Eds., vol. 76 of Lecture Notes of the Institute for Computer Sciences, Social Informatics and Telecommunications Engineering. Springer Berlin Heidelberg, 2012, 211–230.
8. Vinkers, C. H., Penning, R., Hellhammer, J., Verster, J. C., Klaessens, J. H. G. M., Olivier, B., and Kalkman, C. J. The effect of stress on core and peripheral body temperature in humans. Stress 16, 5 (2013), 520–530. PMID: 23790072.
9. Witten, I. H., Frank, E., Trigg, L., Hall, M., Holmes, G., and Cunningham, S. J. Weka: Practical machine learning tools and techniques with java implementations, 1999.

Assessment Method for Bacterial Virulence Based on Ultraviolet and Visible (UV-Vis) Spectroscopy Analysis

Cholid Badri[1], Budiman Bella[2], Anwar Soefi Ibrahim[1], and Sutarmo Setiadji[1]

[1] Graduate Program, University of Indonesia, Jakarta, Indonesia
[2] Faculty of Medicine, University of Indonesia, Jakarta, Indonesia

Abstract-- **Purpose : To develop assessment method of bacterial virulence based on UV-Vis spectroscopy analysis. The method was based on the measurement of absorbance of DNA extracted from the bacterial cultures during the growth phases.**

Methods: Non-pathogenic E. coli samples were taken from bacterial cultures in the exponential phase as well as the stationary phase of the bacteria. DNA extraction from the bacteria in growth phases were measured their normalized optical density (OD) at UV-Vis spectroscopy in the range of 240 - 320 nm. The same procedures were performed with pathogenic E. coli. Therefore, four groups containing 15 samples in each group were measured of their OD.

Results : In non-pathogenic E. coli, the optical densities at the exponential phase were significantly lower as compared with the stationary phase, while in pathogenic E. coli the optical densities at the exponential phase were not significantly different than the stationary phase. In non-pathogenic E. coli groups, the optical densities at the exponential phase were significantly lower than in the corresponding phase in pathogenic E. coli, while the optical densities at the stationary phase of non-pathogenic E. coli were significantly lower than in the correspondence phase of pathogenic E. coli.

Conclusion: DNA quantities of pathogenic E. coli were significantly higher than in non-pathogenic E. coli in the both exponential as well as stationary phases while DNA quantities of pathogenic - E. coli in the exponential phases were not different than in the stationary phase. Therefore the pattern of DNA quantities in growth phases could be considered as virulence indicator for E. coli. This method could be adopted as an alternative to other methods used for determination of bacterial virulence.

Keywords-- **UV-Vis Spectroscopy, optical density, DNA quantitative analysis, Escherichia coli, bacterial virulence.**

I. Introduction

Escherichia coli (E. coli) bacteria is composed of various strains that generally do not cause disease, but virulent strains can cause a variety of diseases including gastroenteritis, urinary tract infections, and neonatal meningitis for it is often necessary to distinguish between virulent or non-virulent strains for prevention or treatment of related diseases [1,2].

It has long been recognized that the growth and multiplication of bacteria can be divided into phases. It demonstrated the marked changes in cell volume that take place in the course of cultivation. A clear distinction between growth, i.e. increase in weight of protoplasm and multiplication [3]. The growth curve for a bacterial population is usually divided into four phases: lag phase; exponential or log phase; stationary phase and death phase. During the lag phase, bacteria are acclimating to their environment by synthesizing enzymes required to achieve growth under the culture conditions. In the lag phase: nutrients are transported into the cell; enzymes are being synthesized; the cells recover from damage caused by heat or temperature changes; preparation for cell division takes place (increase in size, replication of DNA etc.).

The exponential or log phase is the phase where all growing cells in the population are actively growing and dividing. During exponential growth there is a time of balanced growth, when there are comparable increases in all measurable parameters of the population (biomass, protein, DNA, RNA, intracellular water, viable count, etc.) [4,5].

In the stationary phase the number of growing and dying cells balance each other out. In the death phase the number of cells that are dying is greater than the number of cells that are growing. Quantitative measurement of non-pathogenic E. coli microorganism suspension has been carried out by calculating all the parameters of the bacterial population i.e biomass, proteins, DNA, RNA, intracellular water and others. It has been obtained the result of differences in the absorbance of the phases of cell growth which generates patterns of the overall cell quantity of each phase. It is known that the quantity of DNA is a dominant parameter in the calculation of the cell quantity in the phases of bacterial growth.

UV-Vis spectroscopy, which measures the amount of light that is attenuated by a solution or a suspension of particles, is a quantitative, reliable, and rapid analytical tool that can be immediately applied as a biosensor for the detection, identification, and enumeration of cells UV-Vis spectroscopy of cells has been used to estimate the number of cells in a solution and their chemical composition, including nucleic acid and protein concentrations [6,7] Based on the current uses of UV-Vis spectroscopy, it is

J. Goh and C.T. Lim (eds.), *7th World Congress on Bioengineering 2015*, IFMBE Proceedings 52, DOI: 10.1007/978-3-319-19452-3_5

feasible that a detection scheme could be developed to distinguish nucleic acid and protein composition of cells in growth phases of the bacteria.

It is estimated that there are differences in the amount of DNA of pathogenic and non-pathogenic E. coli in cell growth phase, especially in the exponential phase.

In this study only the DNA quantity will be measured so that the results were not influenced by other factors. In addition the use of spectrophotometry in microbiology for identification of pathogenic E. coli can also be applied in monitoring the food industry so that it can be a method used by the institutions concerned.

The aim of this study is demonstrated that the spectral changes observed during the growth of microorganism can be used to obtain quantitative information of DNA on their growth behavior. Further motivation for this study stems from the fact that multi wave length UV-Vis spectroscopy can be easily implemented as a real-time continuous monitoring system for cell cultures [8,9]

II. MATERIALS & METHODS

Pathogenic as well as non pathogenic E. coli cells were obtained from the Laboratory of Department of Microbiology, University of Indonesia, Jakarta. For each species, a bacterial suspension was prepared by culturing 0.1 mL of bacterial colony inoculated in 15 mL of nutrient broth for 24 hour or more. The nutrient broth was prepared by dissolving 5 g of peptone and 3 g of meat extract in 1000 mL of sterilized deionized water and then sterilizing the mixture in an autoclave.

From culture that has been grown 2 ml was taken and centrifuged at 8000 rpm for 10 minutes. The supernatant was discarded, the precipitate was added with 200 uL Phosphate Buffer Saline (PBS). Then 20 uL Proteinase K was added and homogenized with vortex mixer for few seconds. 200 uL Binding Buffer was added and homogenized with vortex mixer for few seconds. The solution was incubated at 60 ° C for 10 minutes then 100 uL of Isopropanol was added and homogenized with vortex mixer for few seconds. The liquid was transferred into the Binding Column with filter and then centrifuged at 8000 rpm for 1 minute. The liquid was discarded, and the filter was removed to a new sample cup. 500 uL Washing Buffer 1 (W1) was added and centrifuged at 8000 rpm for 1 minute. The liquid was discarded and 500 uL Washing Buffer 2(W2) was added (without replacing new sample cup) and was centrifugated at 8000 rpm 1 minute. The liquid was discarded and the precipitate was centrifugated at 12000 rpm for 1 minute. The binding column was moved to a new sample cup 200 uL of Elution Buffer (EL) / TE 50 μL of E. coli DNA was added into 450 μL of TE Buffer in 1,5 ml sample column and homogenized. The liquid was transfered to a micro cell for measurement their normalized optical density (OD) with UV-Vis Spectrophotometer (Cary-5000, Agilent Technologies) at 240-320 nm wavelength. The same procedures were performed with pathogenic as well as non-pathogenic E. coli at the log and stationary phases. Therefore, four groups containing 15 samples in each group were measured of their OD.

Non-pathogenic E. coli samples were taken from bacterial cultures in the exponential phase as well as the stationary phase of the bacterial growth. The samples were taken out from the cultures at 8 and 24 hour cultivation representing log and lag phases and were done on pathogenic as well as non pathogenic E. coli cultures.

III. RESULTS & DISCUSSION

Four experimental groups, each group consisting of 15 samples, taken from pathogenic E. coli cultures at 8 and 25 hour incubation respectively, as well as from non pathogenic E. coli were measured . Group I and II were samples taken at 8 and 25 hour cultivation respectively from non- pathogenic E. coli while group III and IV samples were taken at 8 and 25 hour cultivation from pathogenic E. coli. They were observed for the variability in the amount of the DNA contents on their spectral features and to assess how the parameters obtained from the spectra vary as function of growth phase. At every examination a graph was obtained showing the levels of DNA concentration absorbance at wavelengths between 240-320 nm. As a correction to the purity of the examined DNA it was taken into account also the concentration of DNA at a wavelength of 280 nm.

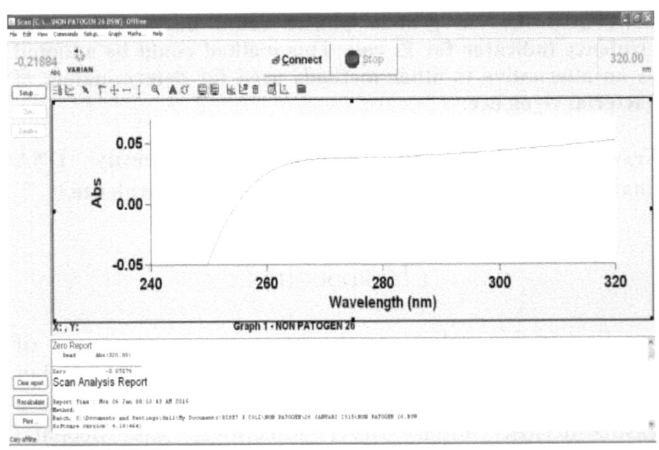

Figure 1. Graphs absorbance of DNA quantity on a wavelength of 240 nm to 320 nm at 8 hours incubation of non-pathogenic E coli.

A typical presentation of measurement of one sample of non-pathogenic E. coli in the UV-VIS Spectrophotometer with visible chart of DNA concentration absorbance at a wavelength of 240 -320 nm was demonstrated in figure 1. On this graph the Y axis representing of optical density while the X axis representing the wave length used.

On this figure the Y axis represents DNA absorbance (optical density) in the range of – 0,05 to 0,05 and the X axis represents the wavelength from 240 – 320 nm.

Numerical data of DNA absorbance of non pathogenic E coli at 8 hour incubation can be seen in Table 1. In this table the Optical Densities were seen at 260, 280 and 320 nm wavelengths with corresponding DNA concentration and purity.

Table 1 The absorbance value (optical density) of DNA concentration of non-pathogenic E. coli at 8 hour incubation

Non-pathogen	A/260	A/ 280	A/ 320	Spectrophotometer	
				Concentration	Purity
1	0.019	0.032	0.043	9.5	2.18
2	0.008	0.024	0.040	4.0	2.00
3	0.011	0.026	0.042	5.5	1.94
4	0.011	0.026	0.041	5.5	2.00
5	0.010	0.026	0.041	5.0	2.07
6	0.008	0.024	0.040	4.0	2.00
7	0.009	0.027	0.041	4.5	2.29
8	0.015	0.029	0.042	7.5	2.08
9	0.007	0.024	0.041	3.5	2.00
10	0.010	0.026	0.042	5.0	2.00
11	0.013	0.028	0.043	6.5	2.00
12	0.010	0.026	0.041	5.0	2.07
13	0.013	0.028	0.043	6.5	2.00
14	0.013	0.029	0.044	6.5	2.07
15	0.012	0.028	0.043	6.0	2.07

To evaluate the purity of DNA, computation was performed for the ratio between the absorbance at 260 nm divided by the absorbance at 280 nm. Good quality DNA is A260 / A280 = 1.7 to 2.0 . The ratio of less than 1.7 could not be used for precise calculation because it shows sufficient DNA contamination.

Corresponding figure of one sample of pathogenic E. coli in UV-Vis Spectrophotometer of DNA concentration absorbance at a wavelength of 240 – 320 nm at 8 hour after incubation of E. coli was presented on Figure 2.

Similar DNA absorbance measurement of the other two groups werealso performed but not presented in this paper.

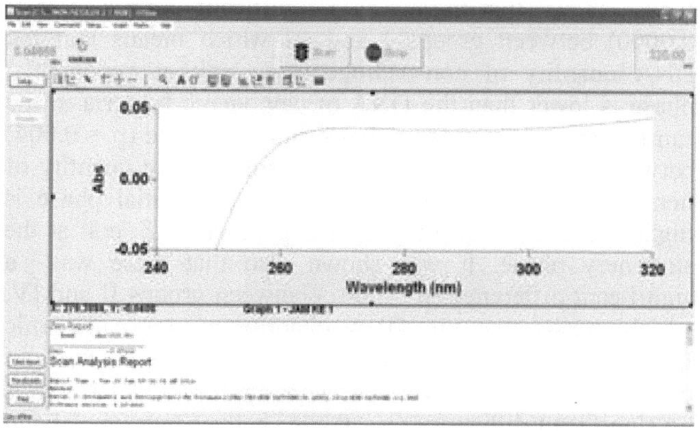

Figure 2. Graphs absorbance of DNA quantity on a wavelength of 240 nm to 320 nm at 8 hours incubation of pathogenic E coli.

Numerical data of DNA absorbance of pathogenic E. coli at 8 hour incubation were shown on table 2.

Table 2. The absorbance value (optical density) of DNA concentration of pathogenic E. coli at 8 hour incubation

Pathogen	A/ 260	A/ 280	320	Spectrophotometer	
				Concentratic	Purity
1	0.034	0.041	0.049	17.0	1.88
2	0.039	0.044	0.049	19.5	2.00
3	0.032	0.040	0.048	16.0	2.00
4	0.033	0.040	0.047	16.5	2.00
5	0.031	0.04	0.049	15.5	2.00
6	0.018	0.030	0.043	9.0	1.92
7	0.019	0.031	0.042	9.5	2.09
8	0.024	0.034	0.042	12.0	2.25
9	0.025	0.034	0.043	12.5	2.00
10	0.023	0.033	0.044	11.5	1.91
11	0.026	0.035	0.046	13.0	1.82
12	0.030	0.040	0.049	15.0	2.11
13	0.018	0.03	0.042	9.0	2.00
14	0.017	0.029	0.041	8.5	2.00
15	0.017	0.030	0.042	8.5	2.08

In the non-pathogenic E. coli at 8 hour incubation the Mean DNA concentration was 5.633 with Standard Deviation 1.5407. In the non-pathogenic bacteria E. coli at 25 hour incubation, Mean DNA concentration was 7.700 with standard deviation 2.0336. In the pathogenic bacteria E. coli at 8 hour the Mean DNA concentration was 12.867

with standard deviation 3.5630, while in the pathogenic bacteria E. coli at 25 hour incubation the Mean DNA concentration was 11.067 with standard deviation 2.9992 It was shown that there was a significant difference (p = 0.0000) between groups I and III which means that the DNA quantity of non-pathogenic E. coli at exponential phase is lower than the DNA of pathogenic bacteria in the same phase. There was a significant difference (p = 0.004) between groups I and II, which means that the quantity of non-pathogenic E. coli DNA at the exponential phase is higher than DNA quantity of non-pathogenic E. coli at the stationery phase. It was shown also that there was a significant difference (p = 0001) between groups II and IV, which means that the DNA quantity of non-pathogenic DNA at stationery phase was lower than DNA of pathogenic bacteria in the corresponding phase. There was no significant difference (p = 0.146) between groups III and IV, which means that the DNA quantity of pathogenic E. coli at the exponential phase was not significantly different with the DNA of pathogenic E. coli at the stationery phase. It was demonstrated that there was a significant difference (p. 0000) between groups I and III which means that the DNA quantity of non-pathogenic DNA at exponential phase is lower than the DNA of pathogenic bacteria in the same phase. 5

. There was a significant difference (p = 0.004) between groups I and II, which means that the quantity of non-pathogenic bacterial DNA in the exponential phase is higher than the non-pathogenic bacterial DNA in the stationery phase. DNA quantities of pathogenic E. coli were significantly higher than in non-pathogenic E. coli in the both exponential as well as stationary phases while DNA quantities of pathogenic - E. coli in the exponential phases were not different than in the stationary phase. Therefore the pattern of DNA quantities in growth phases could be considered as virulence indicator for E. coli. This method could be adopted as an alternative to other methods used for determining bacterial virulence. .

. It was seen that DNA quantitative measurement the data from the spectrophotometric analysis showed that the pathogenic and non-pathogenic E. Coli have different patterns of DNA quantities during their growth phase particularly in exponential and stationary phases. Our data are preliminary, and further studies should be performed to better define the factors that can affect DNA quantities during bacterial growth [10].

V. Conclusions

DNA quantities of pathogenic E. coli were significantly higher than in non-pathogenic E. coli in the both exponential as well as stationary phases while DNA quantities of pathogenic - E. coli at the exponential phases were not different than at the stationary phase. Therefore the pattern of DNA quantities in growth phases could be considered as virulence indicator for E. coli. In conclusion, quantitative analysis appears to be a valid tool for assessment of bacterial virulence and for better microbiological characterization of E. coli bacteria for control and prevent health risks. The present study demonstrates how UV-vis spectroscopy measurements can yield information on bacterial growth with a degree of detail and precision. This method could be adopted as an alternative to other methods used for determination of bacterial virulence.

VI. Conflicts of Interest

The authors declare that they have no conflict of interest.

VII. References

1. Eckburg PB, Bik EM, Bernstein CN, Purdom E, Dethlefsen L, Sargent M,Gill SR, Nelson KE, Relman DA, (2005), Diversity of the human intestinal microbial flora. Science 308 (5728): 1635–8, Copenhagen.
2. Feng P, Weagant S, Grant, M . ((2002), Enumeration of Escherichia coli and the Coliform Bacteria. Bacteriological Analytical Manual (8th ed.). FDA/Center for Food Safety & Applied Nutrition, Boston.
3. Wade, H. E. (1952). Observations on the Growth Phases of Escherichia coli, American Type ' B 'J. gen. Microbiol. 7, 18-23
4. Chul Woo Park, Ki Young Yoon, Jeong Hoon Byeon, Kyoungsik Kim, Jungho Hwang (2012) Development of Rapid Assessment Method to Determine Bacterial Viability Based on Ultraviolet and Visible (UV-Vis) Spectroscopy Analysis Including Application to Bioaerosols.
5. Catalina E, Alupoaei Luis H, García-Rubio (2003). Growth Behavior of Microorganisms Using UV-Vis spectroscopy: Escherichia coli Manuscript 03-336.R1 in Biotechnology & Bioengineering
6. Alupoaei, C. E., J. A. Olivares, and García-Rubio, L. H (2002) Quantitative Analysis of Prokaryotic Cells: Vegetative Cells and Spores, Report No: LA-UR026586, Los Alamos National Laboratory, Los Alamos, MN.
7. Alupoaei, C. E., García-Rubio, L. H., Heler, L., Davis, C. A., Cannons, P., Amuso, and Cattani, J. (2003) Novel Method for the Discrimination of Microorganisms using Spectroscopy, Paper No 1998, American Society for Microbiology, Washington, DC.
8. Callahan, M. C., J. B. Rose and García-Rubio, L. H (2003) The Use of Multiwavelength Transmission Spectroscopy for the Characterization of Cryptosporidium parvum Oocysts: - Quantitative Interpretation, in Environmental Science & Technology
9. Feng P, Weagant S, Grant, M . 2002, Enumeration of Escherichia coli and the Coliform Bacteria. Bacteriological Analytical Manual (8th ed.). FDA/Center for Food Safety & Applied Nutrition, Boston.
10. Fotadar U, Zaveloff P, Terracio L. 2005, Growth of Escherichia coli at elevated temperatures. J. Basic Microbiol. New York University, New York.

Human Heart Oscillatory Behavior during Atrial Fibrillation Based on Second Order System

N.A. Abdul-Kadir[1], N. Mat Safri[1], M.A. Othman[1], and A.M. Embong[2]

[1] Department of Electronics and Computer Engineering, Faculty of Electrical Engineering, Universiti Teknologi Malaysia, 81310 UTM Johor Baru, Johor, Malaysia

[2] Department of Humanities, Universiti Teknologi PETRONAS, Bandar Seri Iskandar, 31750 Tronoh, Perak, Malaysia

Abstract— **In this paper, the viability of a second order system to characterize the oscillatory behavior of human heart of atrial fibrillation patient was monitored and analysed. Sampling were patients who experienced atrial fibrillation. This study used the MIT-BIH Atrial Fibrillation dataset (MIT-afdb) from the Physiobank ECG database. ECG recordings of normal sinus rhythm (N) and atrial fibrillation (AF) which occurred sequentially, were analyzed for both ECG's Lead I and Lead II. From here, the oscillatory behavior of human heart was characterized in accordance to the extracted parameters for each rhythm, i.e. natural frequency, damping coefficient and forcing input from the second order system. Results show that there were significant differences in mean value of natural frequency (ω), ratios of natural frequency to damping coefficient (ω/ζ), derivative of the natural frequency with respect to time (dω/dt) and derivative of the forcing input with respect to time (dμ/dt), between N and AF from Lead I (P < 0.01). Each parameter provides more than 95 % accuracy using artificial neural networks.**

Keywords— **second order system, oscillatory behavior, atrial fibrillation, natural frequency, forcing input**

I. Introduction

During atrial fibrillation, the atria can no longer pump blood to the ventricles, which may cause a heart attack, high blood pressure, coronary heart disease (CHD), or heart valve disease [1]. It has been reported that during this atrial fibrillation, heart rhythms of up to 600 beats per minute can occur, with ventricular rates above 100 beats per minute [1], [2]. This heart rate can cause blood to stagnate, and may enlarge or move freely in the vein or artery. Besides, it can also cause ischemic stroke. Stroke is reported as a leading cause of death in most countries [3], [4].

The concept of a second-order system has evolved from gaming [5], pattern recognition [6], and detection of ventricular arrhythmias [7]. A few algorithms exist to detect, characterize, and classify atrial fibrillation using ECG signals, but they do not incorporate second-order systems, i.e. concept of oscillatory behavior. Some of the previous researches related to atrial fibrillation are based on P-wave occurrences [8], RR intervals [9], [10] or both methods [11], QRS detection [12], and statistical analysis [13].

In 2013, P-wave occurrences were found in 34 of 68 stroke patients, 17 of whom developed atrial fibrillation. Non-atrial fibrillation contraction occurred with a frequency of 88.2 %, with atrial fibrillation occurring at a rate of 37.3 % in each group [8]. An algorithm for atrial fibrillation detection based on RR-interval time series and has achieved sensitivity and specificity rates of 94.1 % and 95.1 %, respectively [10]. The MIT-BIH Atrial Fibrillation Database and the MIT-BIH Arrhythmia Database were used in their study.

While, Dash et al. [11] performed RR-interval separation from ECG rhythms and P-wave detection, and reported atrial fibrillation and normal sinus rhythm detection rates of 98 % and 93 %, respectively, using the MIT-BIH Atrial Fibrillation Database and the MIT-BIH Arrhythmia Database.

The present study applied a second-order system derivation to ECG signals of atrial fibrillation patients. Our initial study showed that a second-order system could be applied to the characterization and classification of atrial fibrillation [14]. This study reports advance findings of applying second-order system to atrial fibrillation ECGs for MIT-BIH Atrial Fibrillation Database compared to [14] which only reported a patient record, as an early step of characterizing human heart oscillatory behavior.

II. Methodology

A. Data collection

The ECG archive of the MIT-BIH Atrial Fibrillation Database provided by Physiobank met our needs for this study [15]. Data were provided in binary format with 12-bit resolution in the range of ±10 mV. The sampling frequency was 250 Hz, and the typical bandwidth of a recording was approximately 0.1-40 Hz. ECG signals were windowed into 4-second episodes. A total of 2,844 episodes of normal heart rhythm signals and 2,844 episodes of atrial fibrillation signals were used in this study. ECG signals from both Lead I and Lead II were used.

J. Goh and C.T. Lim (eds.), *7th World Congress on Bioengineering 2015*, IFMBE Proceedings 52, DOI: 10.1007/978-3-319-19452-3_6

B. Pre- and post-processing processes

During the pre-processing stage, a Butterworth band-pass filter of 1-30 Hz was chosen [16]. LabVIEW system-design platform from National Instrument was used in this study.

A second order dynamic system is a system's response that can be described by a second order ordinary differential equation (ODE) as in equation (1).

$$\frac{d^2}{dt^2}x(t) + 2.\zeta.\omega\frac{d}{dt}x(t) + \omega^2 x(t) = G\omega^2 u(t) \quad (1)$$

where *x(t)* is the response of the system, *u(t)* is the input to the system, ζ is the damping coefficient; to determine how much the system oscillates as the response decays toward steady state, ω is the natural frequency; to determine how fast the system oscillates during any transient response, and *G* is the gain of the system; to determine the size of steady state response when input settles out at constant value.

According to Al-Dabbas et. al [17], [18], a second order system or second order dynamic system is a reduced-order agent model that describes the oscillatory behaviour of a complex physical and biological system by monitoring the behaviour patterns of the semantic concept. The output pattern of a simple second order dynamic system is defined by three parameters, i.e. natural frequency (ω), damping coefficient (ζ) and forcing input (u) which is shown in equation (2) [18].

$$\omega^{-2}.x'' + 2\zeta\omega^{-1}x' + x = \mu;\ x(0) = x_0;\ x'(0) = x'_0 \quad (2)$$

These three parameters were extracted from ECG signals (represent as $x(t)$) to determine their characteristics for further analysis. By differentiating equation (3) with respect to t for two values of time, the damping coefficient, ζ, and the natural frequency, ω, are shown in (4) and (5), respectively.

$$\omega^{-2}.x''' + 2\zeta\omega^{-1}x'' + x' = 0 \quad (3)$$

$$\zeta = -\left[\frac{\omega^{-2}.x''' + x'}{2\omega^{-1}.x''}\right] \quad (4)$$

$$\omega^2 = \frac{x''.x'''' - (x''')^2}{x'.x''' - (x'')^2} \quad (5)$$

The forcing input, μ, as (3).

III. Results and Discussions

A. Extraction of features - Characterization and Classification of human heart oscillatory behavior

There were three extracted parameters from the second-order system, namely, the damping coefficient, ζ; natural frequency, ω; and forcing input, μ, based on equations (5), (6), and (3), respectively. Twelve parameters/features were examined, including the natural frequency, ω; damping coefficient, ζ; forcing input, μ; ratios of forcing input to natural frequency, μ/ω; ratios of forcing input to damping coefficient, μ/ζ; ratios of natural frequency to damping coefficient, ω/ζ; the derivative of the natural frequency with respect to time, dω/dt; the derivative of the damping coefficient with respect to time, dζ/dt; the derivative of the forcing input with respect to time, dμ/dt; the derivative of the forcing input with respect to the natural frequency, dμ/dω; the derivative of the forcing input with respect to the damping coefficient, dμ/dζ; and the derivative of the natural frequency with respect to the damping coefficient, dω/dζ. Figure 1 shows the distribution for a sample from the MIT-BIH AF dataset (data #07910) of atrial fibrillation (AF) and normal sinus rhythm conditions (N) for 237 episodes of each occurrence. From Figure 1, the threshold values for the natural frequency, ω; forcing input, μ; and damping coefficient, ζ, were determined as 0.120 rad/s, 0.044 s/rad, and 0.00 mV, respectively. The feature values for AF episodes were higher than those for N periods for the three parameters plotted, with only a few AF episodes observed below the threshold values. The oscillation of human heart during atrial fibrillation shows higher oscillation than during normal sinus rhythm for the same patient. The natural frequency, ω, for N was 0.11 rad/s in a previous report [7], whereas in this study, the natural frequency for N was 0.12 rad/s.

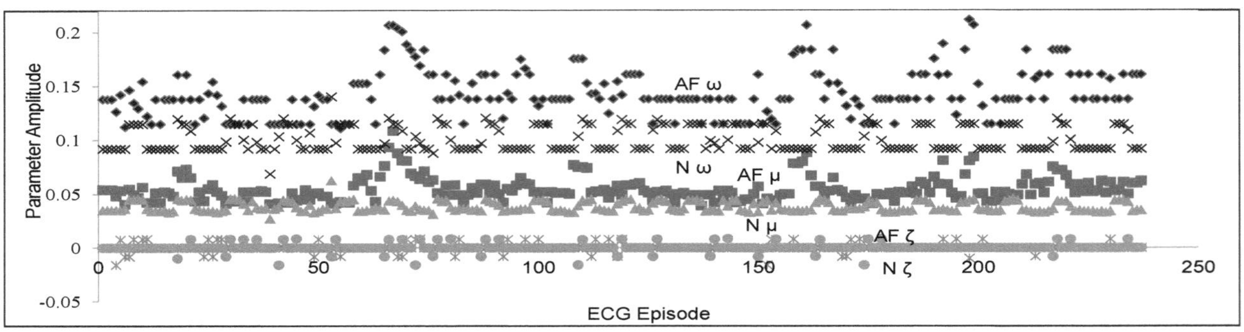

Fig. 1 Distribution of the mean amplitudes of natural frequency, ω (rad/s); forcing input, μ (mV); and damping coefficient, ζ.

B. *Statistical Analysis*

Based on the samples from the MIT-BIH AF dataset that were extracted using the second-order system, the average parameter values associated with the atrial fibrillation (AF) and normal sinus rhythm (N) episodes were summarized. Both ECG signals of Lead I and Lead II were compared between N and AF to observe the significant difference. For classification, the results are shown in Table 1. The statistical t-test for both leads were observed to monitor the more significant lead to extract ECG features based on its oscillatory behavior using second-order system method.

From Table 1, there are four parameters of Lead I provide significant difference in value. The parameters of Lead I which had significant difference ($p < 0.01$) between N and AF are natural frequency, ω (rad/s); ratios of natural frequency to damping coefficient, ω/ζ (rad/s); the derivative of the natural frequency with respect to time, dω/dt (rad/s^2); the derivative of the forcing input with respect to time, dμ/dt (mV/s). According to Table 1, there is a parameter of Lead II provides significant difference in value. The parameter is the ratio of natural frequency to damping coefficient, ω/ζ (rad/s). Other extracted features of Lead II ECG signal could not provide significant difference between N and AF conditions. The damping coefficient, ζ, can determine the rate of oscillation towards steady state. During this study, ζ had no significant differences for both Lead I and Lead II ECG signals. This satisfies our findings, of which ζ had no significant differences between AF and N. In 1978, Pepine et al. had mentioned no significant differences of oscillation ratio between heart failure and non-heart failure group [19]. Therefore, ζ is not suitable for characterizing the oscillatory behaviour of human heart.

Table 1 t-test for Lead I and Lead II ECG Signals

Parameter	Lead I	Lead II
ω	0.0000*	0.2037
ζ	0.8083	0.3489
μ	0.1072	0.8060
μ/ω	0.0211	0.0547
μ/ζ	0.2950	0.0355
ω/ζ	0.0040*	0.0026*
dω/dt	0.0027*	0.0118
dζ/dt	0.3273	0.4149
dμ/dt	0.0070*	0.0403
dμ/dω	0.3658	0.2052
dμ/dζ	0.0322	0.4510
dω/dζ	0.0439	0.0646

*Indicator : * = (p < 0.01)*

To assess the performance of the acquired data, a classification function using sensitivity and specificity parameters was employed using the sample data. Table 2 shows the sensitivity, specificity and accuracy values for samples of Lead I ECG signals. The classification used 2-layer feed-forward back-propagation neural network. Several different combinations of features were analyzed. By using a parameter, i.e. ω, ω/ζ, dω/dt, or dμ/dt each, the accuracies are more than 95 % for each parameters. While using two significant with dω/dt; and ω/ζwith dμ/dt) lessen the accuracy to 70.8 % and 72.9 %. Also, ω/ζ feature could not provide 100 % compared to dω/dt and dμ/dt (Table 3). Despite that, combination of three parameters (ω, dω/dt, dμ/dt) provide

Table 2 Sensitivity and Specificity of N and AF classification using artificial neural networks (ANNs)

Parameter	Condition: Positive (AF) True Positive (TP)	Condition: Positive (AF) False Negative (FN)	Condition: Negative (N) True Negative (TN)	Condition: Negative (N) False Positive (FP)	Specificity (%) $\left[Sp = \frac{TN}{TN+FP}\right]$	Sensitivity (%) $\left[Se = \frac{TP}{TP+FN}\right]$	Accuracy (%)
ω	2607	237	2844	0	100.0	91.7	95.8
ω/ζ	2607	237	2844	0	100.0	91.7	95.8
dω/dt	2844	0	2844	0	100.0	100.0	100.0
dμ/dt	2844	0	2844	0	100.0	100.0	100.0
dω/dt and dμ/dt	5688	0	5451	237	95.8	100.0	97.9
ω and ω/ζ	5688	0	2370	3318	41.7	100.0	70.8
ω and dω/dt	5688	0	5451	237	95.8	100.0	97.9
ω and dμ/dt	5451	237	5451	237	95.8	95.8	95.8
ω/ζand dω/dt	5688	0	2607	3081	45.8	100.0	72.9
ω/ζand dμ/dt	2607	3081	5451	237	95.8	45.8	70.8
ω , ω/ζ and dω/dt	2607	5925	8295	237	97.2	30.6	63.9
ω , dω/dt and dμ/dt	7584	948	8532	0	100.0	88.9	94.4
ω , ω/ζ , dω/dt and dμ/dt	11376	0	2607	8769	22.9	100.0	61.5

accuracy of 94.4 %. Whereas, combining all significant parameters (4 features, which were. ω, ω/ζ, dω/dt and dμ/dt) provide accuracy of 61.5 %.

IV. Conclusions and Future Work

In conclusion, of the 12 parameters extracted from the data in our study, only 4 (ω, ω/ζ, dω/dt, and dμ/dt) can be used to differentiate atrial fibrillation (AF) from normal sinus rhythm ECG signals (N) based on probability values determined from the statistical two-tailed t-test for Lead I. Comparing Lead I and Lead II ECG signals, Lead I is considered better than Lead II for AF and N classification. As revealed in Tables 1 and 2, more parameters (4 features) from Lead I showed significant differences than those obtained from Lead II (1 feature). By using artificial neural networks (ANNs) of 2-layer feed-forward back-propagation, each significant parameter (of the four parameters) provides more than 95 % accuracy. Therefore, the use of a second-order system was found to be effective in the classification of atrial fibrillation.

The oscillatory behavior of the human heart based on a second-order system, hopefully can reveal the onset of heart arrhythmia. Further study is needed to detect the onset of atrial fibrillation based on heart oscillatory behavior to aid patients before suffering a minor stroke.

Acknowledgment

The authors would like to express their appreciation to MOHE and Universiti Teknologi Malaysia for supporting and funding this study under ERGS Grant No. R.J130000.7823.4L062 and Zamalah UTM Scholarship for N.A. Abdul-Kadir.

References

[1] J. B. Shea and S. F. Sears, "A patient's guide to living with atrial fibrillation.," *Circulation*, vol. 117, no. 20, pp. e340–3, May 2008.

[2] S. Nattel, "New ideas about atrial fibrillation 50 years on," *Nature*, vol. 415, no. 6868, pp. 219–226, Jan. 2002.

[3] L. R. Caplan, *Stroke*, 1st ed. Demos New York: American Academy of Neurology, 2006, pp. 4–5.

[4] V. L. Roger, A. S. Go, D. M. Lloyd-Jones et. al, "Heart disease and stroke statistics--2012 update: a report from the American Heart Association.," *Circulation*, vol. 125, no. 1, pp. e2–e220, Jan. 2012.

[5] D. Al-Dabass and M. Ren, "Semantic Mining Dynamics for Games Language Processing," *First Asia Int. Conf. Model. Simul.*, pp. 313–319, Mar. 2007.

[6] D. Al-Dabass, R. Cant, and C. Langensiepen, "Semantics Mining for Opponent Strategy Estimation," in *CGAMES*, 2006.

[7] M. A. Othman, N. M. Safri, I. A. Ghani et al., "A new semantic mining approach for detecting ventricular tachycardia and ventricular fibrillation," *Biomed. Signal Process. Control*, vol. 8, no. 2, pp. 222–227, Mar. 2013.

[8] K. Yodogawa, Y. Seino, T. Ohara et al., "Prediction of atrial fibrillation after ischemic stroke using P-wave signal averaged electrocardiography.," *J. Cardiol.*, vol. 61, no. 1, pp. 49–52, Jan. 2013.

[9] N. D. Asha and K. P. Joseph, "Linear and Nonlinear Analysis of Heart Rate Variability in Ventricular Tachycardia and Supraventricular Tachycardia," *J. Med. Imaging Heal. Informatics*, vol. 3, no. 2, pp. 294–300, Jun. 2013.

[10] M. Stridh and M. Rosenqvist, "Automatic Screening of Atrial Fibrillation in Thumb-ECG Recordings," in *Computing in Cardiology*, 2012, pp. 193–196.

[11] S. Dash, K. H. Chon, S. Lu et al., "Automatic real time detection of atrial fibrillation.," *Ann. Biomed. Eng.*, vol. 37, no. 9, pp. 1701–9, Sep. 2009.

[12] J. P. V. Madeiro, P. C. Cortez, J. A. L. Marques et al., "An innovative approach of QRS segmentation based on first-derivative, Hilbert and Wavelet Transforms," *Med. Eng. Phys.*, vol. 34, no. 9, pp. 1236–1246, Nov. 2012.

[13] P. Langley, M. Dewhurst, L. Y. Di Marco et al., "Accuracy of algorithms for detection of atrial fibrillation from short duration beat interval recordings.," *Med. Eng. Phys.*, vol. 34, no. 10, pp. 1441–1447, Dec. 2012.

[14] N. A. Abdul-Kadir, N. Mat Safri, and M. A. Othman, "Classification of Paroxysmal Atrial Fibrillation using Second Order System," *J. Teknol. (Science Eng.*, vol. 67, no. 3, pp. 57–64, 2014.

[15] G. B. Moody, R. G. Mark, and A. L. Goldberger, "PhysioNet: Physiologic Signals, Time Series and Related Open Source Software for Basic, Clinical, and Applied Research," in *Conference Proceedings: Annual International Conference of the IEEE Engineering in Medicine and Biology Society*, 2011, vol. 2011, pp. 8327–8330.

[16] O. J. Escalona and M. E. Reina, "A Fast and Robust Time-Series Based Decision Rule for Identification of Atrial Fibrillation Arrhythmic Patterns in the ECG," in *Computing in Cardiology*, 2010, pp. 995–998.

[17] D. Al-Dabass, A. Zreiba, D. Evans et al., "Simulation of Three Parameter Estimation Algorithms for Pattern Recognition Architecture," in *4th UK Simulation Society Conference*, 2004, vol. 0, no. 8, pp. 170–176.

[18] D. Al-Dabass, A. Zreiba, D. J. Evans et al., "Parameter Estimation Algorithms for Hierarchical Distributed Systems," *Int. J. Comput. Math.*, vol. 79, no. 1, pp. 65–88, Jan. 2002.

[19] C. J. Pepine, W. W. Nichols, and C. R. Conti, "Aortic Input Impedance in Heart Failure," *Circulation*, vol. 58, pp. 460–465, 1978.

Corresponding author:
Author: N. Mat Safri
Institute: Universiti Teknologi Malaysia
Street: Faculty of Electrical Engineering
City: UTM Johor Bahru
Country: Malaysia
Email: norlaili@fke.utm.my

A Simplified Approach to Identify the Fetal ECG from abdECG and to Measure the fHR

Ridhu H. Nair, J. Rolant Gini, and K.I. Ramachandran

Department of Electronics and Communication Engineering, Amrita School of Engineering, Amrita Vishwa Vidyapeetham, Coimbatore-641112, India

Abstract— **Fetal ECG (fECG) recording aids physicians to diagnose congenital disorders and other anomalies like asphyxia at the early stages of pregnancy. The fECG extraction has been an area of intensive research. Despite the existence of sophisticated and detailed algorithms – based on adaptive filters, independent component analysis (ICA), &c – filtering out the fECG, buried in the noise and mixed up with the maternal ECG (mECG) remains a challenging task. Some residues of mECG are always present in the fECG extracted with all such techniques. A simple algorithm has been developed here to identify the local maxima in the pre-processed abdominal ECG (abdECG) through thresholding; it locates the mECG peaks explicitly. At the outset, the abdECG has been refined by removing the baseline wander and power line interference at a pre-processing stage. With these as pivots the mECG component is eliminated and the fECG of good quality culled out. The fetal heart rate (fHR) and information required to know the condition of fetal heart can be extracted from this fECG effectively. Extraction of these information helps reducing the rate of fetal mortality, and improving the health condition of fetus as well as mother. Performance of the method is better than the conventional adaptive filtering method and the same is proven quantitatively. A processor based realization of the scheme adds to its credibility substantially to ensure its usability in practice.**

Keywords— **fECG, mECG, abdECG, fHR, congenital disorders**

I. INTRODUCTION

A high fetal mortality rate has led to the need of ensuring the well-being of the fetus. Congenital diseases, asphyxia, arrhythmias, and other disorders are the main causes of fetal deaths. Currently arrhythmias are the most common amongst these. Physicians use the fECG waveform to predict the abnormalities in fetal heart during early stages of pregnancy. An early diagnosis can help in effective treatment of disorders. Ultrasound technique is being used to monitor the fetus since a long time. However this might result in false conclusions, especially when the fetus is asleep for a long time; the technique also needs the presence of a well-trained person, clinician, or obstetric nurses. The heart rate variability cannot be clearly assessed using ultrasound technique. Monitoring the fetal heart rate would be useful in clinical predictive medicine. The fECG measurement is carried out by placing electrodes on the scalp of the fetus during labor. This results in a clear waveform, but the method is inconvenient and invasive. A non-invasive method is to measure abdECG which is obtained by placing electrodes at certain locations on the maternal abdomen. This could be performed during any stages of pregnancy and is not limited to recordings during labor [8]. The abdECG is a combination of mECG, fECG, and biological disturbances such as EHG, EMG, and the like. Other noises like baseline wandering and power line interferences also might be present. The fECG being buried in these noises, its extraction is difficult. Placement of the electrodes plays a vital role as the relative fECG strength will be good if the electrodes are close enough to the fetal heart. The details of the fetus like malnourishment, mental illness and other information if diagnosed earlier, could be treated at an early stage. The fetal presentations change from stage to stage during pregnancy. The positions, brow, face, shoulder, breech, and vertex are the various specific fetal presentations. The fetus moves around a lot during the first two trimesters and does not possess a particular presentation. The vertex presentation is the suitable one for birth, commonly known as head down position. The recording of signals from the maternal abdomen is highly influenced by the fetal presentations [8].

Heart being a vital organ, its improper functioning can result in various diseases; these can be diagnosed by analyzing the ECG. Hence the fECG extraction and analysis are important to reduce the fetal mortality rate.

The structure of the rest of the paper is as follows: Literature survey is detailed in Section II; Section III describes the proposed method in detail including the pre-processing stage, mECG peak identification and cancellation, and the fetal peak detection; heart rate measurement is discussed subsequently. It is followed by results and discussions in Section IV. Conclusion is in Section V.

J. Goh and C.T. Lim (eds.), *7th World Congress on Bioengineering 2015*,
IFMBE Proceedings 52, DOI: 10.1007/978-3-319-19452-3_7

II. LITERATURE SURVEY

Various methods are available for fECG extraction, which could be based on multi-channel or single channel data. Techniques involving multi-channels make use of two or more channels to extract the fECG signal. Single channel based methods involve processing of a single abdominal channel for extraction without the aid of information from other channels. Some of the existing methods for the extraction of fECG from abdECG are Blind Source Extraction [4], Wavelet Analysis, Polynomial Networks [5], Neural Networks [2], Adaptive Filtering [6], and the like. The Blind Source Extraction (BSE) techniques, Independent Component Analysis (ICA) [10], Principal Component Analysis (PCA) [9], and Adaptive filtering techniques are multi- channel based methods. The Blind Source Extraction techniques involve extraction of source signals from a mixture, without the knowledge of how signals are being mixed [13]. These make use of all channels and the separation is not effective in most cases. The ICA uses less number of channels compared to BSE, but the selection of channel plays an important role and affects the extraction. The Adaptive Filtering methods require a reference channel; the fECG cannot be recovered completely here. Some mECG components still remain even after the cancellation. The event synchronous noise canceller, which is an extension of adaptive noise canceller is also used for fECG extraction [3]. This method removes the mECG which is a repetitive disturbing signal, by the use of a cyclic template. But the peak detection using this method is difficult and any further processing requires a separate algorithm. Wavelets are used in combination with adaptive filtering algorithms; this involves use of two channels [11]. The use of wavelet analysis and subsequent extraction methods gives better results; however it involves more computational effort. The empirical mode decomposition methods which are being used for de-noising, are not effective and the signals remain noisy even after the technique is employed [15]. Single channel based methods like polynomial networks or singular value decomposition [12] methods are also available. Some single lead based methods involve many stages [14]. Polynomial Networks consider the abdECG to be a combination of mECG and fECG. The results are not of practical use since the background noise is ignored [5]. Some wandering or noise due to uterine contractions in one form or another would always be present. In short, fECG extraction remains an area of intensive research.

A simple approach to identify the fECG of good quality and to infer the fetal heart rate from it, is being discussed here. The method yields good results, even though the algorithm involves less number of steps. This uses a single abdominal channel and the fECG is being extracted without using any other channel information or reference channel. The concept is implemented in MATLAB software and the real signals from DaISy database are used.

III. APPROACH TO fECG IDENTIFICATION

The mECG is the major signal corrupting the fECG extraction from the abdECG. Identification of maternal peaks and cancelling out the mECG signals to cull out the fECG and to measure the FHR is the focus of this work. The proposed approach is explained as follows.

A. PRE-PROCESSING STAGE

The abdECG is a mixture of various signals and noises. The extraction of the desired signal from this noisy signal is difficult. Respiratory movements of the patient, change in the position of instrument, and interaction between electrodes and skin, lead to a drift from the baseline called 'baseline wander' which is a low frequency noise; it limits the utility of ECG signal and affects clinical evaluation. This stage is totally dependent on the abdECG signal available in the database. According to the noise present, necessary filtering needs to be performed. Here signal is corrupted by baseline wandering and is removed using a FIR band pass filter [7]. The filter passes the signal components of frequencies between 5-20Hz and stops the rest. Hence the lower frequencies contributed by the baseline wander are removed. The filtered signal is then normalized for further processing.

B. MATERNAL PEAK IDENTIFICATION AND CANCELLATION

The maternal R peaks are detected using the concepts of thresholding and local maxima. A threshold value is set according to the nature of the signal. Based on thresholding, the region of the signal above this threshold is only considered. This consideration is based on the fact that mECG peaks are of higher amplitude than fECG peaks. In this region, points of local maxima which are the R peaks of the mECG, are found. The mECG is reconstructed around the peaks and cancelled out effectively. The remaining signal consists mainly of the fECG component.

C. FETAL PEAK DETECTION AND FHR MEASUREMENT

The concept of thresholding and peak detection used above, is being adapted according to the extracted information. The threshold values might be modified. This when applied to the fetal signal extracted, the fetal R peaks are obtained. In case of those channels which are noisy, some post processing could be performed before the detection of peaks. This could even be a simple high pass filter to filter out the lower frequencies which might still be present after the extraction. The fetal heart rate is calculated as

$$fHR = \frac{Number\ of\ peaks\ detected*60}{Duration\ of\ signal} \quad (1)$$

The FHR calculated gives physicians a clear idea of the arrhythmias and other abnormalities in the fetus.

IV. Results and discussions

The proposed method is implemented using MATLAB and Fig. 1(a) shows the channel 2 abdECG; (b) preprocessed signal; (c) maternal peaks; and (d) fetal ECG. The fECG peaks detected are indicated in the Fig. 2. A fetal heart rate of 132 bpm (beats per minute) is obtained for channel 2. The normal range of FHR is between 120 to 160 bpm. The FHR obtained here falls within this range in all cases.

In order to validate our method and for further assessment, the proposed approach is implemented for channels 3 and 4 of abdECG. The results are summarized in Table1.

The peaks which are present and are detected are truly diagnosed peaks (TD). Some peaks which detected when they are actually not present are categorized as false positives (FP). When the actual peaks are not being detected, it is considered as false negative (FN) [1]. The accuracy and sensitivity are calculated and tabulated.

Table 1 Accuracy and Sensitivity

Ch. No.	Total peaks	Peaks detected	FP	FN	TD	Acc. (%)	Sens. (%)
2	22	22	0	0	22	100	100
3	21	21	0	0	21	100	100
4	21	22	2	1	20	86.95	95.23

$$Accuracy = \frac{TD}{TD+FP+FN} \times 100 \quad (2)$$

$$Sensitivity = \frac{TD}{TD+FN} \times 100 \quad (3)$$

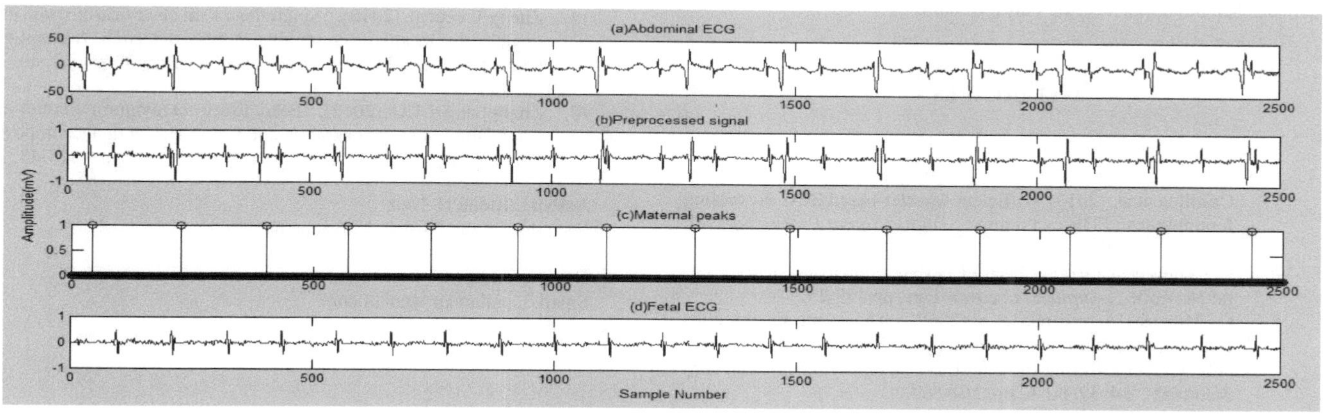

Fig. 1 Identification of maternal peaks and mECG removal

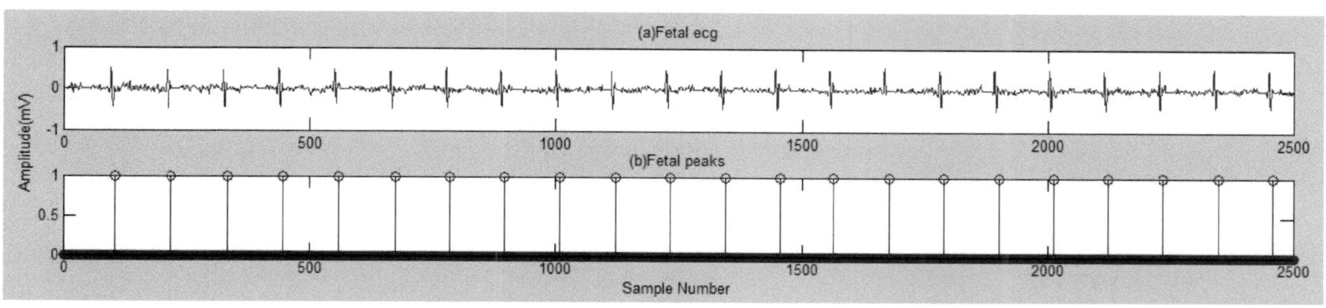

Fig. 2 Fetal ECG and peaks detected

Channel 4 corresponds to a location away from the fetus and hence is noisy. It is reasonable to expect results of comparatively poorer accuracy here. The average accuracy level of 95.65% achieved here is really encouraging; it is conspicuously higher than that achieved by other methods.

V. Conclusion

A simple approach is presented in this paper for fetal ECG extraction. The mECG is identified and cancelled out to get fECG and the heart rate of fetus is measured from it. However this approach might not work in case of twin fetuses. Critical comparison of the results with those from other methods needs to be done. The method also warrants hardware validation using a suitable processor.

Conflict of interest

The authors declare that they have no conflict of interest.

Acknowledgments

We would like to acknowledge the fruitful discussions with Dr. T. R. Padmanabhan of Amrita Vishwa Vidyapeetham.

References

1. Castilloa et al. (2013), 'Efficient wavelet-based ECG processing forsingle-lead FHR extraction.' , *Digital Signal Processing*, vol. 23 no. 6 pp. 1897-1909.
2. G. Camps et al. (2001), 'Fetal ECG Extraction using an FIR Neural Network', *computers in cardiology*, pp.249-252.
3. G. Mihaela Ungureanua et al.(2009), 'The event synchronous canceller algorithm removes maternal ECG from abdominal signals without affecting the fetal ECG', *Comput. in Biology and Medicine* , vol. 39, no. 6, pp. 562-567.
4. Hongjuan Zhang et al. (2009), 'Semi-blind source extraction algorithm for fetal electrocardiogram based on generalized autocorrelations and reference signals', *J. of Comput. and Applied Mathematics*, vol. 223, pp. 409–420.
5. Khaled Assaleh and Hasan Al-Nashash(2005) 'A Novel Technique for the Extraction of Fetal ECG Using Polynomial Networks', *IEEE Trans. Biomed. Eng.* , vol. 52, no.6, pp. 1148-1152.
6. Mahmoud Ahmed Suliman Ali and XiaoPing Zeng (2011), 'An auto detection system for Electrocardiogram of Foetal heart Issues', *Int. J. Comput. Sci. Issues*, vol. 8, no. 2, pp. 40-45.
7. P. Kathirvel et al. (2011), 'An Efficient R-peak Detection Based on New Nonlinear Transformation and First-Order Gaussian Differentiator', *Cardiovascular Engineering and Technology*, vol. 2, no. 4, pp. 408-425.
8. Reza Sameni and Gari D. Clifford (2010), 'Review of Fetal ECG Signal Processing;Issues and Promising Directions', *Open Pacing Electrophysiology* Therapy J. vol.3, no.1, pp.1-30.
9. Reza Sameni et al. (2008), 'Multichannel Electrocardiogram Decomposition Using Periodic Component Analysis', *IEEE transactions on biomedical engineering*, vol. 55, no. 8.
10. Ruben Mart'ın-Clemente et al. (2011), 'Fast Technique for Noninvasive Fetal ECG Extraction', *IEEE transactions on biomedical engineering*, vol. 58, no. 2.
11. Shuicai Wu et al. (2013), 'Research of fetal ECG extraction using wavelet analysis and adaptive filtering', *Comput. in Biology and Medicine* , vol. 43, no.10 pp. 1622-1627.
12. XiaoPing Zeng et al. (2013), 'Fetal ECG Extraction by Combining Single-Channel SVD and Cyclostationarity-Based Blind Source Separation', *Int. J. of signal processing, Image Processing and Pattern Recognition*, Vol. 6, no. 4, pp. 367-376.
13. Yunxia Li andZhang Yi (2008), 'An algorithm for extracting fetal electrocardiogram', *Neurocomputing*, vol. 71, no. 7-9, pp. 1538-1542.
14. Zheng Wei et al. (2010), ' Single-lead fetal electrocardiogram estimation by means of combining R-peak detection, resampling and comb filter', *Medical Engineering & Physics* , vol. 32, no. 7, pp. 708-719..
15. Zhongliang LUO (2012), 'Fetal Electrocardiogram Extraction using Blind Source Separation and Empirical Mode Decomposition', *J. of Computational Inform. Syst.* , vol. 8, pp. 4825-4833.

Author: Ridhu H. Nair
Institute: Amrita Vishwa Vidyapeetham
City: Coimbatore
Country: India
Email: ridhunair@gmail.com

Mathematical Modelling of Blood Perfusion and Oxygen Transport in the Cerebral Microvasculature of Ischemic Stroke

Z.J. Zhou[1,2], Z. Liu[1,2], Y. Cai[1,2], and Z.Y. Li[1,2]

[1] State Key Laboratory of Bioelectronics, Southeast University, Nanjing, China
[2] School of Biological Science and Medical Engineering, Southeast University, Nanjing, China

Abstract— **Most frequently (80%) strokes result from the occlusion of one or several brain vessels and are called ischemic strokes (in the other cases, strokes are hemorrhagic strokes). Measurements of haemodynamics and oxygen delivery on microscopic scales are technically difficult or impossible in many cases, especially during brain activity. One way to study the disease is to establish mathematical models to better understand the dynamic process of blood perfusion and oxygen transport in an ischemic stroke. In this paper, we propose a mathematical modelling system to investigate the haemodynamics and oxygen transport through a 2D cerebral microvascular network during ischemic stroke. The microvessel network is based on the anatomical brain microcirculation structure. The haemodynamic calculation is carried out on the microvessel network by fully coupling the intravascular blood perfusion, the transvascular flow and the interstitial fluid flow. In addition, the compliance of microvessels and blood rheology with hematocritic distribution are also considered. The coupling procedure is based on the iteratively numerical simulation techniques in our previous study for tumour microvessels. The oxygen delivery is described by the time-varying oxygen advection diffusion equation which includes oxygen diffusion and advection in individual microvessel segments, oxygen flux across the vascular wall, and then oxygen diffusion and consumption within the brain tissue. The haemodynamic information and oxygen distribution are investigated under physiological and pathological conditions.**

Keywords— **Ischemic stroke; Blood perfusion; Oxygen transport; Cerebral microvasculature; Mathematical modelling**

I. INTRODUCTION

Stroke is a transient or permanent loss of brain function resulting from a disturbance in the blood supply to the brain, and is the third cause of death and the first cause of disabilities in adults in developed countries. When a brain vessel is occluded, the surrounding tissues receive less oxygen and glucose, which is responsible for the evolution of brain cells towards necrosis. Ischemic stroke might lead to irreversible brain damage and even death if the blood supply is not restored within a short timeframe.Understanding of blood flow and oxygen transport in the brain microcirculation is of the most importance to study the pathophysiological mechanisms and develop new therapeutic strategies of ischemic stroke since no treatments are currently available for most stroke patients.

The complexity of pathophysiological knowledge of microenvironment in brain tissue has generated an increasing interest in mathematical modelling and numerical simulation of blood perfusion, haemodynamics and mass transport in cerebral microvasculature. S. Lorthois *et al.* [1] established an anatomically accurate large human intra-cortex vascular network which can be considered as the union of a random homogeneous capillary mesh and of quasi-fractal trees with a lower cut-off corresponding to the characteristic capillary length. Reichold*et al.* [2] proposed a vascular graph modeling framework that can simulate blood pressure, flow and scalar transport in realistic vascular networks. However, the interstitial fluid flow and transvascular flow, which have important influences on the cerebral microvascular blood perfusion after the damage of the blood brain barrier (BBB) in an ischemic stroke, are excluded in above models. Mathematical modeling of oxygen transport have been developed to simulate the spatial distribution of oxygen levels in cerebral microvessel network. Magnus W Roos [3] proposed a theoretical model of cerebral microischemia, including the acute change in glucose and oxygen transport. Qianqian Fang *et al.* [4] established a discrete model based on the advection-diffusion equation, using a hybrid numerical algorithm and a general work-flow to model the 3D time-varying oxygen transport. This model can be used to simulate a complex anatomical micro vascular network and provides a quantitative and computationally feasible approach for dynamic modeling.

In this paper, we propose a mathematical modelling system to investigate the haemodynamics and oxygen transport through a 2D cerebral microvascular network during ischemic stroke. The haemodynamic calculation is carried out on the microvessel network by fully coupling the intravascular blood perfusion, the transvascular flow and the intersitial fluid flow. The oxygen delivery is described by the time-varying oxygen advection diffusion equation which includes oxygen diffusion and advection in individual microvessel segments, oxygen flux across the vascular wall,

J. Goh and C.T. Lim (eds.), *7th World Congress on Bioengineering 2015*,
IFMBE Proceedings 52, DOI: 10.1007/978-3-319-19452-3_8

and then oxygen diffusion and consumption within the brain tissue. The haemodynamic information and oxygen distribution are investigated under physiological condition and ischemic stroke. Futhermore, the influences of different levels of vessel occlusion on the oxygen transport in tissues are discussed.

II. METHOD

A. 2D cerebral microvascular network

The image data for 2D cerebral microvascular network is obtained from the anatomical thick sections (300μm) of a 60 year old female's brain by confocal laser microscopy in the published paper [5]. Then image processing is carried out to delete unnecessary vessel segments and generate the basic model. Five main vessels are set to be three arteriole with inlet pressure Pin=75mmHg at y=0 and two venules with outlet pressure Pout=15mmHg at y=0. Another venule vessels are also set to be outlet with the same pressure. We classify vessel branches according to the Strahler system, a well-established method for describing stream order. In Strahler's system, leaf segments are assigned Strahler order one. The Strahler order will increase when two vessels with the same Strahler orders join into one vessel. However, two vessels with different Strahler order meeting will not create a vessel with higher order. In our model, there are five Strahler orders to show a brief tree architecture of an arteriolar branching pattern. Strahler orders (N=0, 1, 2, 3, 4), corresponding to the radius of 16,10,7,5, 3.5μm, respectively. The relationship between Strahler order and the vessel radius is based on the experimental equation [6].

B. Haemodynamic analysis

The haemodynamic calculation is carried out on the microvascular network by coupling the intravascular blood flow with the interstitial fluid flow. Briefly, the basic equation for the intravascular blood flow is the flux concentration and incompressible flow at each node. Flow resistance is assumed to follow Poiseuille's law in each vessel segment. The interstitial fluid flow is controlled by Darcy's law. The intravascular and interstitial flow is coupled by the transvascular flow, which is described by Starling's law.

a) Main principles and equations

The main equations for blood flow calculation are as follows:

$$Q_v = \frac{\pi R^4 \Delta P_v}{8\mu \Delta l} \quad (1)$$

$$Q_t = 2\pi R \cdot \Delta l \cdot L_p(P_v - P_i - \sigma_T(\pi_v - \pi_i)) \quad (2)$$

$$Q = Q_v - Q_t \quad (3)$$

Where Q is the flow rate of each vessel segment, which has a value zero at each node of the vessel network due to the assumption of flux conservation and incompressible flow. Q_v is the vascular flow rate without fluid leakage; Q_t is the transvascular flow rate. Δl and R are the mean length and radius of vessel segment. Pv and Pi are the intravascular pressure and the interstitial pressure, respectively. Lp is the hydraulic permeability of the vessel wall. σ_T is the average osmotic reflection coefficient for plasma proteins; π_v and π_i are the colloid osmotic pressure of plasma and interstitial fluid, respectively.

The velocity of intravascular Uv and interstitial flow Ui satisfies

$$U_v = Q/\pi R^2 \quad (4)$$

$$U_i = -K\nabla P_i \quad (5)$$

$$\nabla \cdot U_i = \frac{L_p S}{V}(P_v - P_i - \sigma_T(\pi_v - \pi_i)) \quad (6)$$

Where K is the hydraulic conductivity coefficient of the interstitium; S/V is the surface area per unit volume for transport in the interstitium.

The distribution of red blood cells (RBCs) at a microvascular bifurcation is calculated based on the approach proposed by Pries [7]. The details of blood rheology simulation were described in Wu *et al.* [8].

b) Iterative steps for microcirculation simulation:

1. Set P_{in} and P_{out} of every vascular in the region and initial solutions P_V^o, P_i^o.

2. Solving Pv and relative errors $errP_V = \sum\left|P_{V,(j,k)} - P_{V,(j,k)}^o\right| / \sum P_{V,(j,k)}$ by the iteration computations described above.

3. If there is in the health condition, skip this step. On the contrary, solving Pi and relative errors $errP_i = \sum\left|P_{i,(j,k)} - P_{i,(j,k)}^o\right| / \sum P_{i,(j,k)}$ by the iteration computations described above.

4. According to the given initial R, calculating intravascular hematocrit H and blood viscosity μ, viscosity is obtained according to Pries empirical equation.

5. Compute Uv using Eq. (4).

6.Calculate the maximum error $err = \max(errP_V, errP_i)$. The new set of solutions is fed back into step 2 with $P_V \Rightarrow P_V^o, P_i \Rightarrow P_i^o$. Repeat 2-5 until $err \le 1e^{-6}$, indicating the microcirculation reaches the steady state.

C. Oxygen transport:

We used advection-diffusion equation to describe oxygen transport [4]:

$$\frac{\partial C_O}{\partial t} = \vec{v} \cdot \nabla C_O + \nabla \cdot (D_{O2} \nabla C_O) - OC \qquad (7)$$

Where C_O is oxygen concentration, $\vec{v}$ denotes the intravascular blood flow velocity which is obtained from the haemodynamic simulation, D_{O2} is the diffusion coefficient of oxygen, and OC is the oxygen consumption rate by tissues.

The computational space is separated into three domains to characterize three distinct physiological processes, which are (a) the oxygen convection equation inside the vessel, (b) the oxygen flux across the vessel wall and (c) the free oxygen diffusion in the tissue. Specifically, the Eq. (7) is applied on the different simulation domains as follows:

$$\text{Vessel network: } \frac{\partial C_O}{\partial t} = -\vec{v} \cdot \nabla C_O \qquad (8)$$

$$\text{Tissue: } \frac{\partial C_O}{\partial t} = \nabla \cdot (D_{O2} \nabla C_O) - OC \qquad (9)$$

Oxygen flux across the vessel wall is simplified, we consider the oxygen in the vessel as diffusion source with a coefficient relating to the vessel wall permeability and the concentration difference inside and outside the wall.

The initial condition of oxygen concentration is set to be a relatively high value at the arteriole inlets, and a much lower homogeneous value in the vessels to get a driving force for advection, as for the background, the concentration value is set to be an approximate value of zero. No-flux boundary conditions are used in the simulation field. The central difference scheme is utilized to investigate the oxygen delivery through the cerebral microvessels and the brain tissue.

III. RESULT

A. Result of heamodynamic

The spatial distribution of intravascular blood pressure and flow velocity in the network are shown in Fig1 A&B. High pressure regions evidence the functional territories of the three arterioles whereas low pressure regions evidence the functional territories of venules. Regarding flow maps, high flow segments correspond to the main trunks of arteriolar trees, with significantly decreasing flow in secondary vessels and capillaries. It is noteworthy that zero flow are highly distributed in the capillary segments with highest Strahler order, *i.e.*, smallest vessel diameter.

B. Oxygen transport in physiological condition

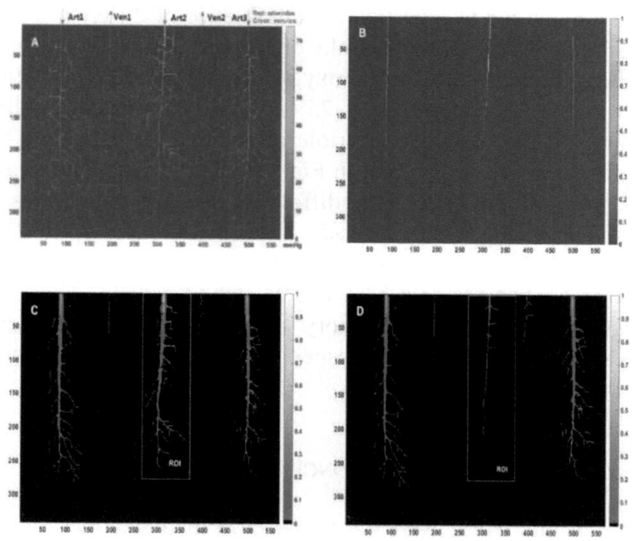

Fig.1 A. Inravascular pressure (Pv) distribution in the anatomical network; B. Inravascular velocity (Uv) distribution; C. oxygen concentration distribution in physiological condition; D. oxygen concentration distribution when the arteriole2 occluded by 75%. The surrounding area of the occluded arteriole is defined to be the region of interest (ROI), shown a white solid line box in C and D.

Oxygen concentration distribution is normalized to be 0 to 1, and shown in Fig.1 C. The arterioles clearly has the highest oxygen concentration compared to the venules and capillaries. The high oxygen around the arterioles drops off rapidly with distance, while the lower oxygen around the capillaries is much more spatially uniform. The mean oxygen concentration versus vessel radius of different vessel orders is shown in Fig.2 A

C. *Oxygen transport during ischemic stroke*

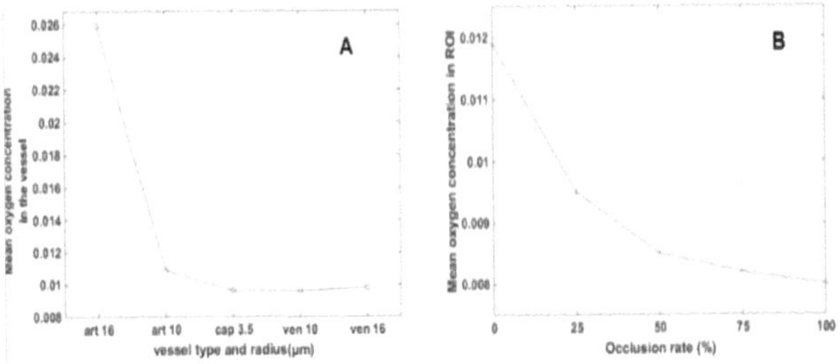

Fig.2 A. The mean oxygen concentration of different vessel type and radius in physiological condition; B. The mean oxygen concentration in different occlusion rate conditions.

Four occlusion rate (25%, 50%, 75%, 100%) of arteriole 2 are applied to investigate the influence of different occlusion levels in ischemic stroke on the whole oxygen delivery. Fig.1 D demonstrated the oxygen concentration distribution when the occlusion rate is 75%. We define the surrounding area of the occluded arteriole as region of interest (ROI), shown as white line box in Fig. 1C & D. The mean oxygen concentration in ROI in different occlusion conditions is shown in Fig.2 B. The tissue oxygen decreases up to 21% when 25% occlusion occurs in ROI. However, 100% occlusion contributes to a 33% decreasing of oxygen. This suggests that the compensatory action caused by collateral circulation is significant especially in the severe occlusion during ischemic stroke.

IV. Conclusions

In this work, we have proposed a mathematical modelling system of haemodynamics and oxygen delivery through the cerebral microvessels. Based on a 2D microvascular network from a published anatomical image, haemodynamic calculation is carried out by fully coupling the intravascular blood flow, the interstitial fluid flow and the transvascular flow. In addition, the compliance of vessel wall and the rheology of blood are included in the coupled model. The haemodynamic information such as intravascular blood velocity is used in the oxygen transport model. The computational space is separated into three domains to characterize three distinct physiological processes, which are (a) the oxygen convection equation inside the vessel, (b) the oxygen flux across the vessel wall and (c) the free oxygen diffusion in the tissue.

The results of haemodynamics and oxygen concentration distribution during ischemic stroke are shown and compared with the normal condition. Furthermore, the influences of different levels of vessel occlusion on the oxygen transport in tissues have been discussed. The sensitivity of certain parameters in the model, such as microvessel density, vessel wall permeability, and the difference of pressure in initial conditions, can be analyzed in the future.

Acknowledgment

This research is supported by the National Basic Research Program of China (973 Program) (No. 2013CB733800), the National Nature Science Foundation of China (No. 11302050, No. 11272091), the Nature Science Foundation of Jiangsu Province (No. BK20130593),

Conflict of Interest

The authors declare that they have no conflict of interest.

References

1. Lorthois S, Cassot F, Lauwers F (2011). Simulation study of brain blood flow regulation by intra-cortical arterioles in an anatomically accurate large human vascular network: part i: methodology and baseline flow.NeuroImage, 54, 2, 1031–1042.
2. Reichold J, Stampanoni M, Lena Keller A, Buck A, Jenny P, Weber B (2009). Vascular graph model to simulate the cerebral blood flow in realistic vascular networks. Journal of Cerebral Blood Flow & Metabolism, 29, 8, 1429–1443.
3. Roos M W (1485). A revised mathematical model of cerebral microischemia. Physiological Measurement, 25, 6, 1485-1493.
4. Fang Q *et al.* (2008). Oxygen advection and diffusion in a three-dimensional vascular anatomical network. Opt Express, 11.
5. CASSOT F, LAUWERS F, FOUARD C, PROHASKA S, LAUWERS-CANCES V (2006). A novel three-dimensional computer-assisted method for a quantitative study of microvascular networks of the human cerebral cortex. Microcirculation, volume 13, 1, 1-18(18).
6. M Fisher C (1969). The arterial lesions underlying lacunes. ActaNeuropathologica, 12, 1, 1-15.
7. Pries A.R., Secomb T.W. AR, P. (2005). Microvascular blood viscosity in vivo and the endothelial surface layer. American Journal of Physiology. Heart and Circulatory Physiology, 289, 6.
8. Wu J, Xu S, Long Q*et al.* (2008). Coupled modeling of blood perfusion in intravascular, interstitial spaces in tumor microvasculature. J Biomech, 41, 5, 996-1004.

Author: Zhi-Yong Li
Institute: School of Biological Science and Medical Engineering
Street: Sipailou 2
City: Nanjing
Country: China
Email: zylicam@gmail.com

The Assessment of Toxicity of Boron Nitride Nanoparticle Using Atomic Forced Microscopy

M.A.I. Rasel, T. Li, T.D. Nguyen, and Y.T. Gu

Mechanical Systems and Asset Managements, Queensland University of Technology, Brisbane, Australia

Abstract— **Nanoparticles have considerably found practical applications in the field of biomedical engineering in recent years. The large surface to volume ratio and the flexibility to deliver in desired positions make them special than any other nanomaterials. However, before being used in human body, the toxicity of Boron nitride nanoparticle (BN NP) must be evaluated. The upake of nanoparticles was confirmed using confocal imaging. Then the stiffness of both normal and cell cultured with BN nanoparticles, were compared using Atomic forced microscopy (AFM). The idea is to observe whether the particles alter the structural integrity of cells by changing their stiffness.**

Keywords— **Nanoparticles, Cell mechanics, Atomic Forced Microscopy, Boron nitride, Cell stiffness.**

I. Introduction

Nanoparticles have gained much popularity in recent years because of their noble properties. Scientists are utilizing the unique properties of nanoparticles in sophisticated biomedical engineering applications like biosensing, drug delivery, molecular imaging, anticancer therapy and so on. Recently Boron nitride caught the attention of researchers when Ciofani el al. reported Boron nitride nanotubes (BNNT) to be cytocompatible [1] [2].

Boron nitride possesses superior physical, chemical and thermal properties which led to many practical applications like industrial tool manufacturing, photocaralysis, electrical devices, lubrication and so on [3] [4] [5] [6]. However, in order to use BN in biomedical applications, their toxicity needs to be evaluated. Although BNNT has been reported to be cytocompatible, BN NP is still to be evaluated. Nanoparticles are popular in biomedical engineering applications due to their structure, shape and size [7] [8].

As nanoparticle is expected to be used *in vivo* for biomedical applications, the viability of targeted biological components (human cells) needs to be evaluated. As an integrated part of human body, cell is subjected to a wide variety of stimulations. The structural integrity of cells enables it to withstand the mechanical stimulations. For proper functioning of the human body, cell structural integrity (mechanical properties) has to be retained. Any deviation can result in the breakdown of physiological functions and may lead to different diseases. For example deterioration of stiffness of Chordrocytes is believed to be the one reason for development and progression of osteoarthritis [9].

Therefore, in this study, the physical response of cells is focused while assessing the toxicity of BN NP cells. The idea is to quantify the stiffness of cells and compare it between normal cells and cells infiltrated with nanoparticle. The aim is to investigate whether there is any variation in cellular stiffness because of cellular uptake of nanoparticle.

Cells were cultured for 24h with BN NP and imaged with confocal microscopy in order to confirm the uptake. Later the stiffness of both normal and cell infiltrated with BN NP were quantified and compared. Atomic Forced Microscopy (AFM) a state-of-the-art experimental facility for nano-imaging and indentation were used for the stiffness quantification. The obtained results are discussed and analyzed.

II. Methods and Materials

A. Preparation of Nanoparticle

The nanoparticles were prepared by adding varying amount of UHQ water and diluting them in different concentrations. The solution was then ultrasonicated for 4h (Misonix Sonicator, 3000). The resultant was then allowed to settle and sterilized by autoclaving at 121° C for 20 mins.

B. Cell Culture

The cells were seeded in six well plates and cultured using Dulbecco's Modified Eagle's Medium (low glucose) (GIBCO, Invitrogen Corporation, Melbourne, Australia) supplemented with 10% fetal bovine serum (FBS) (HyClone, Logan, UT) and 1% penicillin and streptomycin (P/S) (GIBCO, Invitrogen Corporation, Melbourne, Australia).

C. Confocal Imaging

Cells were fixed using 4% paraformaldehyde (Sigma-Aldrich). They were then repeatedly washed with PBS. Cells were then stained with 1:100 of DAPI and Alexa Fluor 568 Phalloidin (GIBCO, Invitrogen Corporation, Melbourne, Australia) for 10-15 minutes. Samples were then

J. Goh and C.T. Lim (eds.), *7th World Congress on Bioengineering 2015*,
IFMBE Proceedings 52, DOI: 10.1007/978-3-319-19452-3_9

washed again and taken for imaging. Nikon A1R confocal, Japan with 40X oil immersion lens was used.

D. Atomic Forced Microscopy

To quantify the young's modulus of each cells, the forced-indentation curve ($F-\delta$) was processed using the modified Hertzian model proposed by Dimitriadis et al. [10]. As colloidal probe cantilevers were used, the relationship between the force (F) and indentation (δ) can be expressed as,

$$F=\frac{4E}{3(1-\nu^2)}R^{\frac{1}{2}}\delta^{\frac{3}{2}}\left[\begin{array}{l}1-\frac{2\alpha_0}{\pi}\chi+\frac{4\alpha_0}{\pi^2}\chi^2-\frac{8}{\pi^3}\left(\alpha_0^3+\frac{4\pi^2}{15}\beta_0\right)\chi^3\\+\frac{16\alpha_0}{\pi^4}\left(\alpha_0^3+\frac{3\pi^2}{5}\beta_0\right)\chi^4\end{array}\right] \quad (1)$$

Here, $\chi=\sqrt{\frac{R\delta}{h}}$, α_0 and β_0 are the functions of Poisson ratio ν, R is the radius of the rigid indenter and E is the young's modulus. For simplicity, the Poisson ratio is assumed to be 0.25 [11].

As the cells have strong attachment with the substrate, α_0 and β_0 can be expressed as,

$$\alpha_0=-\frac{1.2876-1.4678\nu+1.3442\nu^2}{1-\nu} \quad (2)$$

$$\beta_0=\frac{0.6387-1.0277\nu+1.5164\nu^2}{1-\nu} \quad (3)$$

Therefore the relationship between force (F) and indentation (δ) can be expressed as,

$$F=\frac{16E}{9}R^{\frac{1}{2}}\delta^{\frac{3}{2}}\left[1+1.133\chi+1.283\chi^2+0.769\chi^3+0.0975\chi^4\right]$$

(4)

III. Results and Discussion

A. Nanoparticle Characterization

In order confirm the material as boron nitride energy-dispersive X-ray spectroscopy (EDS) was conducted. EDS is an analytical technique for elemental analysis and chemical characterization of materials. The obtained results demonstrated that the material consists of 49% boron and 49% nitrogen. Therefore the material is confirmed as boron nitride (BN).

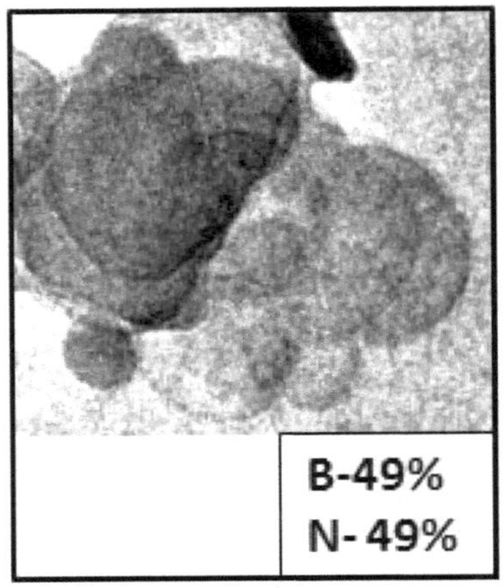

Fig 1 EDS analysis of Boron nitride nanoparticles

B. Nanoparticle Uptake

In order to make sure that BN NPs are taken in by the cells, cells are cultured with BN NP and imaged using confocal. Cells were cultured for 24h with a concentration of 25 µg/ml of BN NP. The obtained images are given below:

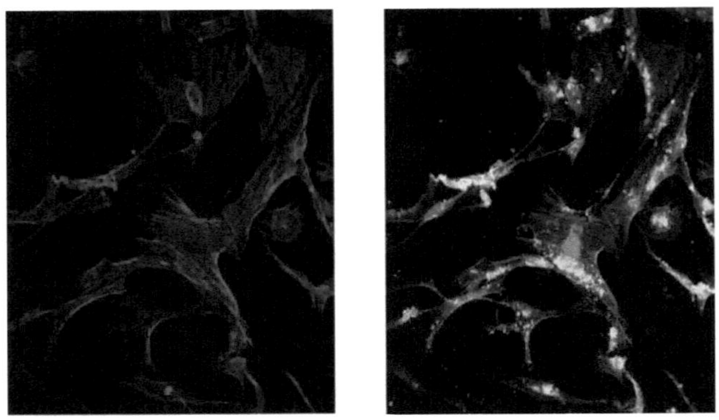

(a) (b)

Fig 2 Confocal Image of the cell cultured with BN NP

As cells are stained with DAPI and Phalloidin, the cell nucleus (blue), the actin filaments (red) and BN NP (green) are clearly seen (Fig 2b). The actin filament network is separated with image processing to provide an idea of the cell shape and nanoparticle localization. It is clear from the figure that, BN NP nanoparticles are successfully taken in by the cells. The green BN NP's are clearly observed inside the cell.

C. Accessing the toxicity

For advanced biomedical applications (like drug delivery) nanomaterials are delivered in the cells in a controlled manner. After going through the cell membrane, the material may cause cellular dysfunctionality depending on it's toxic nature. Therefore, nanomaterials are evaluated before being used *in vivo*. In this study we are trying to evaluate BN NP as a potential drug delivery vehicle. The idea is to investigate whether BN NP physically causes any damage to a single living cell by comparing cell stiffness.

Two sets of cells were cultured and prepared for the experimental study. One set were cultured in normal conditions and the other were cultured with BN NP for 24h with a concentration of 25 μg/ml. After trypsinization, the cells were allowed to settle in petri dish coated with poly-D-lysone (PDL, Sigma-Aldrich). The aim is to let the cells attach with the substrate and maintain their circular shape. Once the cells were ready they were tested with Atomic Forced Microscopy (AFM).

As mentioned earlier, AFM is an experimental technique facilitating high resolution imaging and indentation of nano sized materials (Cells, tissues etc.). The basic principle is to indent materials with a nano tip attached with a movable cantilever. The applied force is calculated from the deflection of cantilever detected by attached sensors. Afterwards, the elastic modulus of sample is quantified from the generated Force (F) indentation (δ) curve.

The AFM used is this study is a Nanosurf FlexAFM (Nanosurf AG, Switzerland) with a mounted Leica DM IRB microscope. The cantilever probe used is the SHOCONG-SiO2-A-5 (AppNano) with a probe diameter of 5 μg/ml and spring constant of 0.224-0.3114.

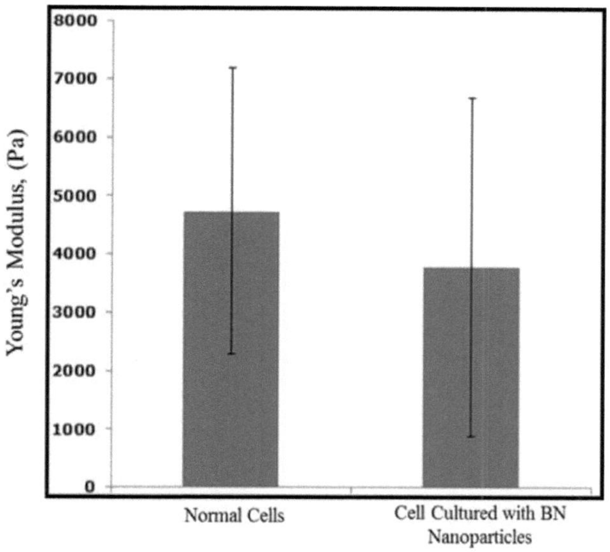

Fig 3 Stiffness comparison between normal and cell cultured with BN NP cells

Fig 3 shows the young moduli of both normal and cell cultured with BN NP. It can be observed that the average stiffness of these two sets of cells are very close (statistically insignificant). Despite uptaking a significant amount of nanoparticles, cell stiffness did not vary much. This is significant, because it indicates, even though nanoparticles are spread in the cell cytoplasm, its structural integrity is not compromised. Mechanically cells are as strong and functioning as normal cells.

IV. Conclusion

This study aimed at assessing the toxicity of BN NP for advanced biomedical use. However, instead of biological testing, the physical property (stiffness) of cells was evaluated. Atomic Forced Microscopy (AFM) was used as the test ring. The cellular uptake of BN NP was confirmed by confocal imaging. BN NPs are clearly observed to be distributed inside the cell.

AFM was successfully used to quantify the stiffness of cells. After comparing the stiffness of both normal and cell infiltrated with BN NP, it can be confirmed that BN NP fails to change the bulk stiffness of cell signifying cells structural integrity is retained. From the physical prospective BN NP does not seem to have any adverse effect on cells.

We believe this is a good approach to assess the toxicity of nanomaterials. However, more in depth studies are necessary to get conclusive evidence that BN NP are actually cytocompatable and can be used *in vivo*. None the less, physically assessing biomaterials for structural compromise is still a handy tool to assess the bio-toxicity of biomaterials.

V. Acknowledgement

We would like to acknowledge the QUT Postgraduate Research Award (QUTPRA) and ARC Future Fellowship Grant (FT100100172) for financially supporting this project.

VI. Reference

1. Ciofani, G., V. Raffa, A. Menciassi and A. Cuschieri (2008). 'Cytocompatibility, interactions, and uptake of polyethylene-imine-coated boron nitride nanotubes by living cells: Confirmation of their potential for biomedical applications', BIOTECHNOL BIOENG, **101**, 850-58.
2. Ciofani, G., S. Danti, D. D'Alessandro, S. Moscato and A. Menciassi (2010). 'Assessing cytotoxicity of boron nitride nanotubes: interference with the MTT assay', BIOCHEM BIOPH RES CO, **394**, 405-11.
3. Lyutaya, M. D., I. G. Chernysh and O. A. Frenkei (1970). 'Chemical properties of nitrides of the AIIIBV type', SOV POWDER METALL+, **9**, 503-08.
4. Lin, L., Z. Li, Y. Zheng and K. Wei (2009). 'Synthesis and application in the CO oxidation conversion reaction of hexagonal boron nitride with high surface area', J AM CERAM SOC, **92**, 1347-49.
5. Wu, J. C. and S.-J. Lin (2008). 'Novel BN supported bi-metal catalyst for oxydehydrogenation of propane', Chem. Eng. J., **140**, 391-97.
6. Sueyoshi, H., N. T. Rochman and S. Kawano (2003). 'Damping capacity and mechanical property of hexagonal boron nitride-dispersed composite steel', J. Alloys Compd., **355**, 120-25.
7. De, M., P. S. Ghosh and V. M. Rotello (2008). 'Applications of nanoparticles in biology', Adv. Mater., **20**, 4225-41.
8. Steichen, S. D., M. Caldorera-Moore and N. A. Peppas (2013). 'A review of current nanoparticle and targeting moieties for the delivery of cancer therapeutics', Eur. J. Pharm. Sci, **48**, 416-27.
9. Trickey, W. R., G. M. Lee and F. Guilak (2000). 'Viscoelastic properties of chondrocytes from normal and osteoarthritic human cartilage', J. Orthop. Res., **18**, 891-98.
10. Dimitriadis, E. K., F. Horkay, J. Maresca, B. Kachar and R. S. Chadwick (2002). 'Determination of elastic moduli of thin layers of soft material using the atomic force microscope', Biophys. J ., **82**, 2798-810.
11. Zhou, E., C. Lim and S. Quek (2005). 'Finite element simulation of the micropipette aspiration of a living cell undergoing large viscoelastic deformation', MECH ADV MATER STRUC, **12**, 501-12.

Author: M.A.I Rasel
Institute: Queensland University of Technology
Street: 2 George Street
City: Brisbane
Country: Australia
Email: m.rasel@qut.edu.au

Fabrication of Cellulose-Gelatin Based Endothelialed Vascular Graft with SMCs/ADSCs Seeding in Bioreactor

Ts-Ching Yu and Ming-Long Yeh*

Medical Engineering Department, National Chen Kung University, Taiwan

Abstract—**Cardiovascular disease (CVD) is a major health issue over the worldwide. The application of tissue engineering technology holds great promise for improving outcomes in CVD patients. Currently, the use of synthetic vascular grafts has several limitations, including thrombogenicity, increased risk of infection, and lack of growth potential. Natural vascular grafts have many advantages including excellent degradation, biocompatibility, but low mechanical property. Therefore, to form a biodegradable and antithrombotic vessel graft as temporary substitute is a problem to overcome in recent researches. The purpose of our study is using natural materials to rapidly form an endothelialized vascular graft in vitro. We utilized cellulose and gelatin to fabricate a composite scaffold. In order to enhance the mechanical property and decrease the degradation rate of scaffold, we added genipin to crosslink cellulose and gelatin. We mixed the cellulose, gelatin, and genipin solution, and injected solution into self-made mold to form tubular hydrogel. At last, we could obtain the tubular scaffold by freeze-drying. With regard to fabrication of vascular graft, the smooth muscle cells (SMCs) extracted from rabbit's carotid arteries would be seeded into scaffold to secret the collagen and elastic fiber, while adipose derived stem cell (ADSCs) would be seeded onto the lumen of scaffold as the source of endothelial cells. The material property of scaffold were analyzed by swelling test, biocompability, and FTIR. Moreover, we observe the SMCs distribution and extracellular matrix (ECM) deposition of vascular graft by histology.**

Keywords— **natural polymer, genipin, SMCs, endothelia cell.**

I. Introduction

Natural vascular grafts have many advantages including excellent degradation, biocompatibility, but low mechanical property. Therefore, to form a biodegradable and antithrombotic vessel graft as temporary substitute is a problem to overcome in recent researches. The purpose of our study is using natural materials to rapidly form an endothelialized vascular graft in vitro.

Cellulose can be considered for biomedical application because of owning biocompatibility and available[1]. Gelatin was chosen in this study because it is a derivative of collagen. Gelatin exhibit anti-antigenicity, high biodegradable rate, and high cell affinity properties[2]. In order to afford cells enough time to proliferate within the scaffold in vivo, we have to reduce the degradable rate of the scaffold through crosslinking. Genipin was used as an ideal crosslinking agent, which has less cytotoxicity compared to traditional crosslinking agent[3]. The cellulose/ gelatin hydrogel are nontoxic, inexpensive, and own high surface-to-volume ratio of pores, which is good for cell to adhesion and proliferation.

Studies have shown that smooth muscle cells (SMCs) could secrete the type I collagen that mimicking the extracellular matrix. Rabbit SMCs are seeded by perfusion of a cell suspension from the lumen (inner diameter 2 mm) through the wall of a tubular scaffold.

For endothelialization, it's reported the ability of adipose-derived stem cells(ADSCs) to differentiate into cells functional features of endothelial cells when treated with the shear stress[4]. Mechanical stimulation of ADSCs seeded in vascular tissue engineering scaffolds promotes the endothelial layer formation. Moreover, various growth factors were used to promote biomaterial endothelialization.

In our study, cellulose and gelatin were fabricated as a tubular scaffold, and seeded with SMCs to mimic the natural vessel environment.

II. Materials and Methods

Type B gelatin (pharmaceutical grade, pH 5.64, pI 4.9) were used. Cellulose(Degree of polymerization < 350, Particle Size: < 1% +60 mesh, Alfa Aesar). Genipin (Molecular Weight 226.23, ≥98% (HPLC), powder, Sigma) All other chemicals used in this work are of analytical grade.

Scaffold fabrication

Type B gelatin were swollen in deionized water at room temperature and then dissolved at 37°C under agitation to obtain 50 wt%(w/w) solutions. Add 1g cellulose into 3.5ml deionized water, then, add 1ml gelatin solution under agitation. The gelatin/cellulose solution was cross-linked by genipin, and injected into designed mold. A tubular rod is as the mold(outer diameter: 4mm, internal diameter: 2mm; length: 5cm), After injecting, the mold was rotated at room temperature for 10 hours.

Swelling property

* Corresponding author.

J. Goh and C.T. Lim (eds.), *7th World Congress on Bioengineering 2015*,
IFMBE Proceedings 52, DOI: 10.1007/978-3-319-19452-3_10

The water absorption capacities of the scaffolds were determined by swelling them in phosphate buffered saline (PBS) at room temperature. The freeze-dried scaffolds were placed in the PBS solution for 5 h. The swelling ratio of the scaffold was calculated from the equation:

$$\textbf{Swelling Ratio (\%)} = \frac{Wt - W0}{W0} \times 100\% \tag{1}$$

Smooth muscle cell extraction

SMCs were isolated from rabbit aortas. After removing endothelium, adventitia, fat and connective tissue, were dissected and incubated in a sterile conical tube containing an enzymatic dissociation buffer under agitation on an orbital shaker (60 rpm) for 90 min at 37◦C. This buffer contains 1.0 mg collagenase in 10 ml DMEM medium. Following the complete dissolution of the matrix, centrifuged at 1000 × g for 5 min.

Cell seeding

The novel cell seeding technique is named swell-rotating method, which utilizes the water absorption property of gelatin/cellulose hydrogel. The cell would be penetrated into scaffold wall during the scaffold absorbed the medium. Prepare the medium containing ($1.5x10^5$) cells. The freeze-dried scaffolds were immersed in medium and shaked for 1 hour at 37◦C. Then, the scaffold was maintained in the medium for 4 hour in order to assist cells to adhesion. The scaffold was put into 4% formalin a half hour and underwent frozen cut. For observing the cell distribution, the cells would be stained DAPI before seeding.

Cell viability

ADSCs were seeded onto gelatin/cellulose composite sheet scaffold for 1, 3, 5, 6 days. Prepare the culture medium of the cells and composite sheet in 96-well plates containing a final volume of 100 µl/well. An optional set of wells can be prepared with medium only for background subtraction. Add 20 µl MTS solution to each well. Incubate 1 hour at 37°C. Record absorbance at 490 nm.

FTIR

The FTIR spectra of freeze-dried gelatin, and composite scaffold in absorption mode in the range of 4000~650 cm^{-1}. 32 scans were performed to establish accuracy.

III. Results and discussions

Scaffold fabrication

The gelatin/cellulose composite scaffold contains elasticity, and appropriate mechanical strength properties. The scaffold exhibit deep blue color because of the crosslink between –NH group on the gelatin and genipin(Figure 1).

Cell viability

All samples biocompatibility was determined by MTS assay. By 5th culture of ADSCs and cross-linked scaffold shows good cell compatibility. Day 1 cell viability is up to 92%, and day 6 reaches about nearly 80%. It shows that gelatin/cellulose cross-linked scaffold afford a nontoxic environment for cell to live(Figure 2).

Swelling test

The swelling ratios at 5th hour were investigated since it was the least time for the initially cell attachment[5]. The results showed that the weight of swollen scaffolds were two times than the dried scaffolds, as presented in Table 1. Gelatin is widely known for its hydrophibic, which allows gelatin to absorb water up to 5 times of its dry weight. It shows that crosslinking may influence the absorption ability of the material.

Cell seeding

It's obviously to see that seeding ADSCs into the three-dimensional composite scaffold by swell-rotating method is an efficient way. Histologic analysis showed that many cells adhered to the scaffold within short incubation time, and adhered more uniformly around the scaffold lumen. Figure 3a and 3b showed the whole cell distribution of the scaffold, figure 3c and 3d showed the cell accumulation within the pore. Because of auto-fluorescence, cellulose exhibits red color. The swell-rotating method increases cell seeding efficiency. Compared to other cell seeding methods, it has higher cell attachment ratio, and more uniform cell distribution in the scaffold.

FTIR

The figure 4a shows the FTIR spectra of the dried gelatin films in the 900-1800 cm^{-1} finger-point region of wave numbers. According to previous studies [6.7], the absorption peaks at 1690-1760, 1500-1600, 1340-1470, 1180-1360, 1050-1300 cm^{-1} were attributed to C=O, C=C, C-H, C-N, and C-O, respectively. The results demonstrated that the FTIR spectra of dried composite films corresponded to gelatin cross-linked with genipin. Conformation of the cellulose chains and their strong packing depends on intermolecular and intramolecular H-bond. As the free –OH group peak appears at 3650-3590 cm^{-1}. Figure 4b shows that the broader peak in the 3600-3200 cm^{-1} . It's implies that the H-bond of scaffold decreases, and cellulose indeed cross-linked with genipin.

IV. Conclusions

The gelatin/cellulose composite scaffold was stabilized by crosslinking with genipin. The scaffold exhibits good swelling ability enough to be utilized for cell seeding, and excellent biological environment for cell proliferation. The swell-rotating method causes to uniform and dense cell attachment. Implanting the scaffold into rabbit artery is the next step, and I will observe thrombogenicity by angiography.

V. Figures and Table

Figure 1. The genipin cross-linked gelatin/cellulose composite scaffold.

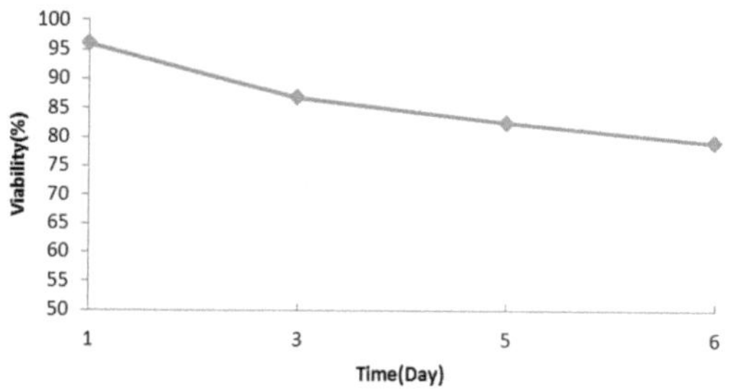

Figure 2. MTS assay cell viability that cultured for 1, 3, 5, 6 days.

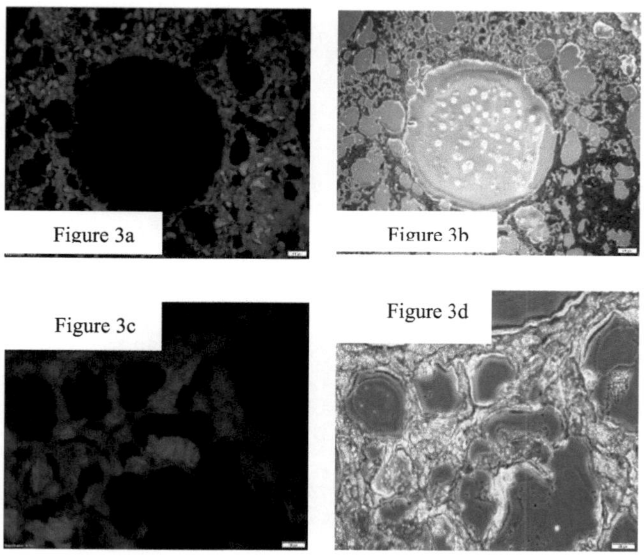

Figure 3. Using swell-rotating cell seeding method. (a) The whole section of scaffold in fluorescency. (b) The whole section of scaffold in white light. (c) The cells accumulation in the pores of scaffold in fluorescency under 200X. (d) The cells accumulation in the pores of scaffold in white light under 200X.

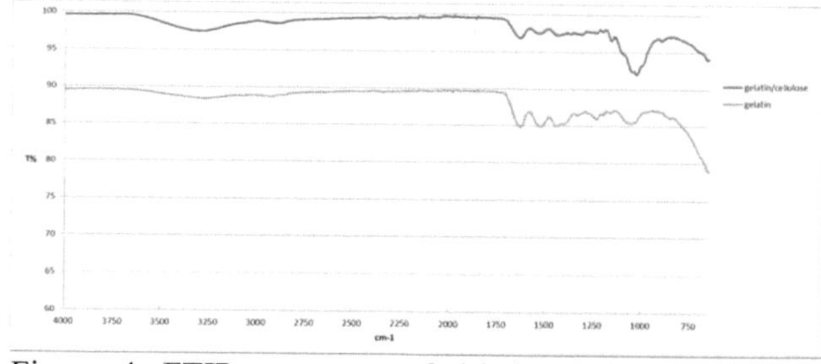

Figure 4. FTIR spectrum of dried gelatin powder and genipin cross-linked gelatin/cellulose powder with a scanning range of 4000-650 cm^{-1}.

	dried weight(g)	*wet weight(g)*	*swelling ratio(%)*
gelatin	0.52	2.95	467.3
gelatin/ cellulose	0.28	0.86	211.4

Table 1. The swelling ratio of gelatin and gelatin/cellulose composite.

Acknowledgment

This work was supported in part by the National Chen Kung University, Biomedical Engineering Department.

Conflict of Interest

The authors declare that they have no conflict of interest.

References

1. John Rojas, Eduardo Azevedo.(2011) Functionalization And Crosslinking of Microcrystalline Cellulose in Aqueous Media: A Safe And Economic Approach. IJPSR. Volume 8, Issue 1, Article-006
2. Hutson, Che B. et al.(2011) Synthesis and Characterization of Tunable Poly(Ethylene Glycol): Gelatin Methacrylate Composite Hydrogels. Tissue Engineering Part A 17.13-14 : 1713-1723.
3. Balamurugan Manickam, Rajesh Sreedharan et al.(2014) Genipin – The Natural Water Soluble Cross-linking Agent and Its Importance in the Modified Drug Delivery Systems: An Overview. Current Drug Delivery, 139-145
4. Francesca Colazzo, Fahad Alrashed et al.(2014) Shear stress and VEGF enhance endothelial differentiation of human adipose-derived stem cells. Growth Factors; 32(5): 139–149

5. Juthamas Ratanavarporn, Siriporn Damrongsakkul. et al.(2006) Comparison of Gelatin and Collagen Scaffolds for Fibroblast Cell Culture. JOM, Vol.16 No.1 pp31-36
6. Bai-Shuan Liu, Chun-Hsu Yao et al.(2008) Effect of Drying Method on the Characteristics of Genipin Cross-linked Gelatin Films. JMBE. 29(1): 29-38
7. Shih-Ta Chang, Li-Chen Chen et al.(2012) Nano-biomaterials application: Morphology and physical properties of bacterial cellulose/gelatin composites via crosslinking. Food Hydrocolloids; 27 ,137-144

Three Dimensional Panorama Visualization for Endoscopic Video

Atul Kumar[1,2], Yen-Yu Wang[1,2,4], Kai-Che Liu[1,2], Wan-Chi Hung[1,2], Shih-Wei Huang[2], Wen-Nung Lie[3], and Ching-Chun Huang[3]

[1] IRCAD-Taiwan
[2] Chang Bing Show Chwan Memorial Hospital, Taiwan
[3] National Chung Cheng University, Taiwan
[4] National Changhua University of Education-Taiwan

Abstract—**Introduction: Endoscopy is widely used in the surgical world. Reduced surgical injury during endoscopic surgery makes the patient's life easy. However, the endoscopic surgery becomes a challenge for the novice surgeons because of the narrow field of view and the lack of 3D perception in the 2D endoscope image. Such limitations of the endoscope may be addressed with a 3D panorama created with the endoscopic 2D images. Few studies have been reported to create a 3D panorama from endoscopic images. However, to our knowledge, no study has reported a 3D panorama system which uses only a single camera image from the conventional endoscope to make a 3D shape of the organs, and further stitch those 3D shapes to create a 3D panorama of the surgical scene. Such system would enable surgeons to have an extended field of view of the surgical scene with a 3D perception.**

Method: Images from the endoscopic surgery video were used in this study. A shape from shading (SfS) algorithm was applied to create a 3D shape of the organs in the images. Characteristic feature points were identified on the images using a feature detection algorithm (SURF). The matching feature points in the consecutive images were found using a feature matching algorithm (BRIEF). An iterative closest point (ICP) algorithm was applied to stitch 3D shapes from the consecutive images to synthesize the 3D panorama.

Results: The method was applied on 100 consecutive video frames from an endoscopic video of a patient. The root mean square error for the registration of the consecutive image feature points was <4 mm. The 3D panorama from the method helps in visualizing a larger area of the surgical anatomy. An improved version of the method may be applied to a real time video from the endoscope.

Keywords— **Panorama · Mosaicing · 3D surface reconstruction.**

I. Introduction

Endoscopic surgery has revolutionized the surgical world. However, it has several limitations creating challenges to the surgeons.

A conventional (single camera) endoscope has limited field of view which can be increased using techniques such as mounting a fish eye lens, and mosaicing techniques. Mosaicing is a technique for stitching the images together to create a larger scene image. The images are acquired from different views with a partial overlap in the images. A panorama image is created by stitching the overlapping images together and making a larger image.

A conventional (single camera) endoscopic image (2D image) does not provide the depth perception of a scene in the image. However, there are several ways of calculating or retrieving the depth map of a scene using its 2D image, which include methods of shape from shading and depth from linear perspective [20]. A 3D structure can be reconstructed by using the depth map with the 2D color image. The 3D structures (surfaces) from overlapping images can be stitched together to form a 3D panorama of the scene.

A panorama using the 3D structures (surfaces) would not only increase the field of view, would also be more informative and suggest the relationship among the surgical anatomies. Moreover, such panorama can be useful in retrieving the complete 3D structure of an organ. Several studies have reported for 2D panorama using conventional endoscope [3],[4],[12],[15],[19] and few studies have reported 3D panorama (reconstruction) using stereo or conventional endoscope [9],[11],[13],[16].

The current study describes a technique which creates a 3D panorama of the surgical anatomy with the 2D endoscopic images and its depth information calculated from the 2D image. To our knowledge, no such study has been reported.

II. Method

The method used in the study includes, acquisition of video image using endoscope, applying shape from shading algorithm to each of the video frames to make a 3D structure of the surgical anatomy, features matching between consecutive video frames and aligning 3D structure using the matched feature points. A brief description of the algorithms used in the method is included in the respective section.

A. 3D surface reconstruction

Shape from shading: A shape from shading algorithm (*SfS*) [18] was applied to each of the image frames of the

J. Goh and C.T. Lim (eds.), *7th World Congress on Bioengineering 2015*, IFMBE Proceedings 52, DOI: 10.1007/978-3-319-19452-3_11

video sequence for calculating the depth map of the scene in the image. The depth map was further used in reconstructing a 3D structure of the surgical anatomy in the image frame. The *SfS* algorithm finds the depth u(x) for each pixel coordinate $\mathbf{x} = (x, y)$ by solving the Hamiltonian equation (Eq. 1)

$$H(\mathbf{x}, \nabla v) = I(\mathbf{x})\frac{1}{\rho}\sqrt{(v_x^2 + v_y^2) + J(\mathbf{x}, \nabla v)^2} \cdot Q(\mathbf{x})^{\frac{3}{2}} \quad (1)$$

where $H(\mathbf{x}, \nabla v)$ is the Hamiltonian, $J(\mathbf{x}, \nabla v) = \frac{v_x(x+\alpha)+ v_y(y+\beta)+1}{(f+\gamma)}$ and $Q(\mathbf{x}) = (x + \alpha)^2 + (y + \beta)^2 + (f + \gamma)^2$, $v = \ln u$, *I(*$\mathbf{x}$*)* is the intensity at $\mathbf{x}$, α, β and γ are the camera position with respect to the light (in our method camera and light were considered at the same point), v_x, v_y are the partial derivatives of v with respect to the x and y and ρ is the surface albedo (1 in our method). Depth map for one of the image frame is show in Figure 1.

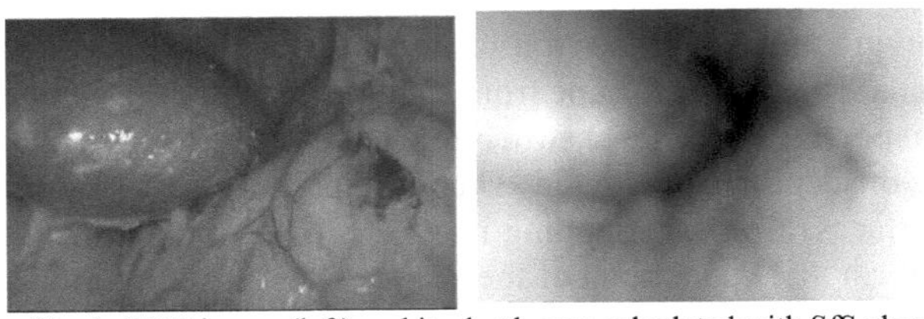

Fig. 1: RGB image (left) and its depth map calculated with SfS algorithm (right)

B. Triangulation and color mapping

The pixel coordinates in a 2D image and its corresponding depth calculated with *SfS* provided the x, y and z point coordinates of each pixel. A Delaunay triangulation technique [8] was used to make a 3D surface from the point coordinates. The points on the surface were mapped with the color of the respective pixel in 2D image. The triangle surface color was interpolated using the bilinear interpolation method [1]. The example surface without color mapping are shown in Figure 2.

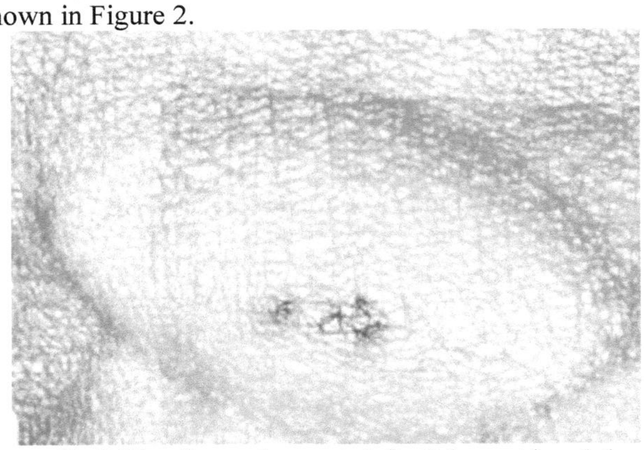

Fig. 2:3D surface mesh generated after Delaunay triangulation

C. Feature detection and matching

Unique feature points on the image frames were detected using the speeded-up robust feature (SURF) detection algorithm [2]. The SURF algorithm uses a Hessian detector where the Hessian is defined as the $\mathcal{H}(\mathbf{x}, \sigma)$ in x at different scale factor σ

$$\mathcal{H}(\mathbf{x}, \sigma) = \begin{bmatrix} L_{xx}(\mathbf{x}, \sigma) & L_{xy}(\mathbf{x}, \sigma) \\ L_{xy}(\mathbf{x}, \sigma) & L_{yy}(\mathbf{x}, \sigma) \end{bmatrix} \quad (2)$$

where $L_{xx}(\mathbf{x}, \sigma)$ is the convolution of the second order Gaussian derivative $\frac{\partial^2}{\partial x^2} g(\sigma)$ in the image *I* at point $\mathbf{x}$, and similar relationship is to $L_{xy}(\mathbf{x}, \sigma)$ and $L_{yy}(\mathbf{x}, \sigma)$ [28]. The detected feature points were assigned a descriptor known as Binary Robust Independent Elementary Features (BRIEF) which is based upon the intensity differences [7] among the pixels near the feature points. An image patch around the feature point is tested with τ (Eq. 3) to create a descriptor of that patch. In a patch p, τ is defined as

$$\tau(\mathrm{p}; x, y) := \begin{cases} 1 \text{ if } \mathrm{p}(x) < p(y) \\ 0 \quad \text{otherwise} \end{cases} \quad (3)$$

where $\mathrm{p}(\mathbf{x})$ is the pixel intensity at $\mathbf{x}$. The descriptor for the patch is defined as

$$f_{n_d}(p) := \sum_{1 \le i \le n_d} 2^{i-1}\, \tau(p; x, y) \quad (4)$$

where n_d is the number of binary tests and in our method n_d=128. The feature points (Figure 3) with matching descriptors were then identified with a nearest neighbor search using Hamming distance measurements in the descriptor space [29].

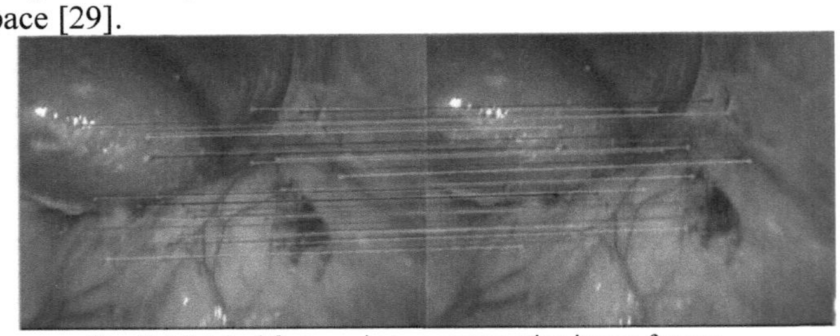

Fig. 3: Matched features in two consecutive image frames

D. Stitching the 3D surface

The pixel position of the points of correspondence (between two consecutive images calculated by the feature matching) provided their 2D coordinates (X,Y). Their Z coordinates were retrieved from their respective 3D surfaces. A rigid body transformation matrix between these matched points was calculated using ICP algorithm [5]. The matrix was used to stitch the 3D surfaces.

III. Results

A recorded video of the laparoscopic surgery, acquired with Karl Storz® laparoscope system (single camera system) (frame size: 720 x 480) was used in this study. The laparoscope camera intrinsic parameters were calculated before the surgery using a camera calibration method proposed by Zhang et.al. [21].

A customized software using libraries of VTK [14] and OpenCV [6] in C++ was prepared for applying the algorithms and to display the 3D panorama. The current method was applied to 100 consecutive frames of the video (with Intel CoreTM i7 960 @ 3.20 GHz, 6.00RAM 64 bit Windows 7, NVidia TESLA C2075) to visualize the 3D panorama (Figure 4). Root mean square error for the alignment between two consecutive frames after ICP was between 1 to 4mm.

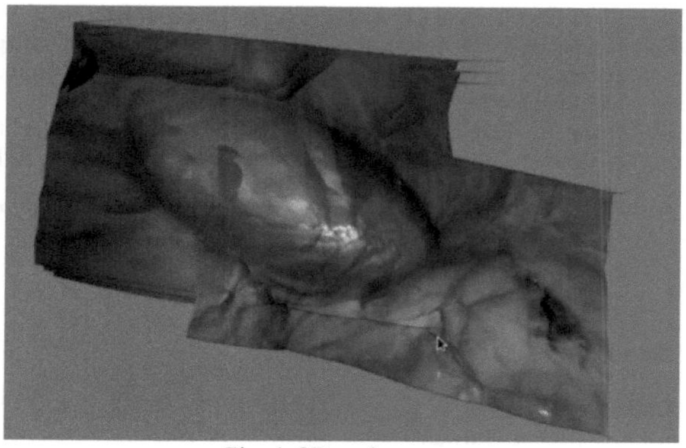

Fig. 4: 3D surface panorama

IV. Discussion

The current work describes a method to create a 3D panorama using images from a single camera endoscope system. The method was applied on 100 consecutive frames of laparoscope video. An improved version of the system will be useful for most of the conventional endoscope systems where surgeons need to visualize a 3D structure of large part of the surgical area. The algorithms with good computational performance [17] were selected for the current system.

Previous studies which describe 3D surface reconstruction from endoscope images have used techniques such as structure from motion (*SfM*) while the current study used shape from shading (*SfS*) method. Compared to the *SfS*, the *SfM* provides a sparse set of 3D points making it difficult to reconstruct a smooth surface of the object. There are several algorithms currently available for the shape-from-shading [10, 20], however, we used the one presented by Visentini-Scarzanella et.al. which is among the recent algorithms for the *SfS* problem and the depth map calculated with the algorithm is more accurate than those from other algorithms [18].

For the feature detection and the feature points matching SURF and BRIEF were used, respectively. The SURF is faster than the many other available algorithms for feature detection [2] and with BRIEF it is promising for the near real-time applications [7].

The presented system would be useful when the surgeons need to visualize 3D shape of the surgical anatomy and an extended view of the surgical field to know the spatial relationship of important structures.

The current method has several limitations such as: (i) it assumes that the organs have moved linearly; (ii) it needs detectable features in every frame of the video; (iii) the specular areas of the endoscope may not have correct depth map; (iv) the cumulative error in the registration (stitching) after some duration of the video was not corrected in this system. In our future work, we would be applying non-linear registration method, the algorithms at GPU level and may include the bundle adjustment in the method.

V. Conclusion

A method to create a 3D panorama of the surgical anatomy using single camera endoscope video is reported. The method can be useful for surgeons in visualizing the extended view and the 3D structures of surgical anatomy. It can be improved further by including non-linear registration and bundle adjustment techniques.

Acknowledgment

The study was supported by the National Science Council (Grant application number:103WHK5400012) and Chang Bing Show Chwan Memorial Hospital (Grant application Number:RA 14003)..

Conflict of Interest

Conflict of interest: Atul Kumar, Yen-Yu Wang, Kai-Che Liu, Ching-Chun Huang,Wen-Nung Lie, Wan-Chi Hung andShih-Wei Huang declare that they have no conflict of interest.

References

1. Allebach, J., Wong, P.W.: Edge-directed interpolation. In: Image Processing, 1996. Proceedings., International Conference on, vol. 3, pp. 707–710. IEEE (1996)
2. Bay, H., Tuytelaars, T., Van Gool, L.: Surf: Speeded up robust features. In: Computer Vision–ECCV 2006, pp. 404–417. Springer (2006)
3. Behrens, A., Bommes, M., Stehle, T., Gross, S., Leonhardt, S., Aach, T.: A multi-threaded mosaicking algorithm for fast image composition of fluorescence bladder images. In: SPIE Medical Imaging, p. 76252S. International Society for Optics and Photonics (2010)
4. Bergen, T., Wittenberg, T., Munzenmayer, C., Chen, C.C.G., Hager, G.D.: A graph-based approach for local and global panorama imaging in cystoscopy. In: SPIE Medical Imaging, p. 86711K. International Society for Optics and Photonics (2013)
5. Besl, P.J., McKay, N.D.: Method for registration of 3-d shapes. In: Robotics-DL tentative, pp. 586–606. International Society for Optics and Photonics (1992)
6. Bradski, G.: Dr. Dobb's Journal of Software Tools
7. Calonder, M., Lepetit, V., Strecha, C., Fua, P.: Brief: Binary robust independent elementary features. In: Computer Vision–ECCV 2010, pp. 778–792. Springer (2010)
8. De Berg, M., Van Kreveld, M., Overmars, M., Schwarzkopf, O.C.: Computational geometry. Springer (2000)
9. Duan, M., Xu, R., Ohya, J.: 3d surface reconstruction based on image stitching from gastric endoscopic video sequence. In: SPIE Optical Engineering+ Applications, p. 88561I. International Society for Optics and Photonics (2013)
10. Durou, J.D., Falcone, M., Sagona, M.: Numerical methods for shape-from-shading: A new survey with benchmarks. Computer Vision and Image Understanding 109(1), 22–43 (2008)
11. Ishii, T., Zenbutsu, S., Nakaguchi, T., Sekine, M., Naya, Y., Igarashi, T.: Novel points of view for endoscopy: Panoramized intraluminal opened image and 3d shape reconstruction. Journal of Medical Imaging and Health Informatics 1(1), 13–20 (2011)
12. Konen, W., Breiderhoff, B., Scholz, M.: Real-time image mosaic for endoscopic video sequences. In: Bildverarbeitung fur die Medizin 2007, pp. 298–302. Springer (2007)
13. Mountney, P., Yang, G.Z.: Dynamic view expansion for minimally invasive surgery using simultaneous localization and mapping. In: Engineering in Medicine and Biology Society, 2009. EMBC 2009. Annual International Conference of the IEEE, pp. 1184–1187. IEEE (2009)
14. Schroeder, W., Martin, K., Lorensen, B.: An Object-Oriented Approach To 3D Graphics. Prentice hall (2013)
15. Soper, T.D., Chandler, J.E., Porter, M.P., Seibel, E.J.: Constructing spherical panoramas of a bladder phantom from endoscopic video using bundle adjustment. In: SPIE Medical Imaging, p. 796417. International Society for Optics and Photonics (2011)
16. Soper, T.D., Porter, M.P., Seibel, E.J.: Surface mosaics of the bladder reconstructed from endoscopic video for automated surveillance. Biomedical Engineering, IEEE Transactions on 59(6), 1670–1680 (2012)
17. Vemuri, A.S., Liu, K.C., Ho, Y., Wu, H.S., Ku, M.C.: Endoscopic video mosaicing: application to surgery and diagnostics. In: Living Imaging Workshop, December, pp. 1–2 (2011)
18. Visentini-Scarzanella, M., Stoyanov, D., Yang, G.Z.: Metric depth recovery from monocular images using shape-from-shading and specularities. In: Image Processing (ICIP), 2012 19th IEEE International Conference on, pp. 25–28. IEEE (2012)
19. Yamauchi, Y.: Non-optical expansion of field-of-view of the rigid endoscope. In: World Congress on Medical Physics and Biomedical Engineering 2006, pp. 4184–4186. Springer (2007)
20. Zhang, R., Tsai, P.S., Cryer, J.E., Shah, M.: Shape-from-shading: a survey. Pattern Analysis and Machine Intelligence, IEEE Transactions on 21(8), 690–706 (1999)
21. Zhang, Z.: A flexible new technique for camera calibration. Pattern Analysis and Machine Intelligence, IEEE Transactions on 22(11), 1330–1334 (2000)

The address of the corresponding author:

Author: Atul Kumar
Institute: IRCAD-Taiwan
Chang Bing Show Chwan Memorial Hospital, Taiwan
Street: No 6-1 Lugong Road, Lukang Township
City: Changhua
Country: Taiwan
Email: sharmaatul11@gmail.com

Salmonella Detection on Microfluidic CD Using Loop Mediated Isothermal Amplification

Abkar Ahmed Sayad[1,2], Fatimah Ibrahim[1,2], Mas S. Mohktar[1,2], and Kwai Lin Thong[2,3]

[1] Department of Biomedical Engineering, Faculty of Engineering, University of Malaya, 50603 Kuala Lumpur, Malaysia
abkar8819@gmail.com, mas_dayana@um.edu.my
[2] Centre for Innovation in Medical Engineering (CIME), Faculty of Engineering,
University of Malaya, 50603 Kuala Lumpur, Malaysia
Fatimah@um.edu.my
[3] Microbiology Unit, Institute of Biological Sciences, Faculty of Science, University of Malaya 50603 Kuala Lumpur, Malaysia
thongkl@um.edu.my

Abstract— ***Salmonella* is a major foodborne pathogen threat and an economic burden for the food business, individuals and society. Therefore, low-cost, portable, and easy-to-use devices for *Salmonella* detection are required to prevent foodborne outbreaks and ensure food safety. This paper describes *Salmonella* detection on centrifugal microfluidic compact disc (CD) using Loop Mediated Isothermal Amplification (LAMP) technique. LAMP amplification and *Salmonella* detection are simultaneously integrated on this developed centrifugal microfluidic device. A cheap hot air gun was used to produce a temperature for the DNA amplification resulting in system miniaturization and low cost. The whole process from reagent preparation, *Salmonella* DNA amplification and detection was automatically performed on the microfluidic CD and SYBR Green I was used for detection process. Our system offers a rapid and automated diagnostic platform. Therefore, reducing the usage of manual and expensive operated equipment, that subsequently could reduce manpower, shortened time-to-result, and lower the cost.**

Keywords— **foodborne; microfluidics; lab-on-a-disc; detection; *Salmonella***

I. Introduction

The genus *Salmonella*, a member of the family Enterobacteriaceae, is a Gram-negative facultative anaerobe, rod-shaped, non-spore forming and motile. *Salmonella* causes many human diseases such as, chronic asymptomatic carriage, enteric fever, bacteremia, and gastroenteritis. Particularly, salmonellosis causes the morbidity and mortality worldwide, with 21 million cases of typhoid fever and 1.3 million cases of gastroenteritis estimated per year [1]. Common outbreaks of Salmonellosis are frequently related to eggs, fruits, vegetables, poultry and raw milk. Therefore, food contamination by *Salmonella* has become a global health concern and failure to detect this foodborne pathogen would cause a significant losses to the food industry and health sectors. Therefore, a rapid detection method is needed to overcome food contamination outbreaks. Several conventional methods for detecting and identifying *Salmonella* have been reported. These methods involve direct cell culture and colony counting, which are time consuming, require expensive equipment and multiple steps [2]. Therefore, PCR-based methods have been introduced to provide more specific and sensitive detection as compared to culture method. PCR assays have become widely used for the detection of *Salmonella.* Even though, PCR assays can allow rapid detection within 24 h, but there are various limitations, such a, the need for trained personnel, high risk of contamination, low detection limit, and expensive equipment/reagents such as thermal cycling [3,4]. Notomi *et al.* have developed a novel method called loop-mediated isothermal amplification (LAMP) [5]. This method introduced a simple, rapid, specific, and cost-effective nucleic acid amplification technique. In addition, LAMP is a nucleic acid amplification method that amplifies DNA/RNA under isothermal conditions (constant temperature) utilizing a strand displacement reaction which eliminate the need to change the temperature to denature and re-anneal DNA strands. Furthermore, LAMP shows high sensitivity and specificity using a set of six primers and a *Bst* DNA polymerase [5,6].

Point-of-care (POC) devices are capable of rapidly identifying *Salmonella*, resulting in immediate and accurate results. Lab-on-a-chip (LOC) is one of the POC devices which provides this alternative platform for *Salmonella* detection [7]. The process automation, reduction of manually operated personal and equipment, shortened time-to-result, minimize volumes, cheap, miniaturized size and precise volume control are the key factors that make LOC devices powerful tools in *Salmonella* detection [8,9]. Therefore, several studies for the detection of *Salmonella* has been reported. The Group of Rashid Bashir have introduced a silicon chip for the multiplexed detection of foodborne pathogens including *Salmonella* using LAMP [10]. Another study reported by Hyeong Kim *et al.* have shown that detection of *Salmonella* can be performed on a microfluidic CD using recombinase isothermal amplification (RPA) [11]. Therefore, lab-on-a-disc platforms (microfluidic CDs) are considered to be a better tool than lab-on-a-chip platforms

J. Goh and C.T. Lim (eds.), *7th World Congress on Bioengineering 2015*,
IFMBE Proceedings 52, DOI: 10.1007/978-3-319-19452-3_12

because they eliminate the need for tubing and external pumping devices, as the centrifugal force is enough to control the liquids [12-14].

In this paper, we present a technique for *Salmonella* detection on a microfluidic CD using loop mediated isothermal amplification (LAMP). This technique demonstrates a fully automated centrifugal functions for *Salmonella* detection such as, pumping, mixing, and metering the LAMP reagents. DNA amplification is performed on the developed microfluidic CD and a SYBR Green 1 is used for detection. Thus, reducing of manually operated equipment and manpower, immediate results, low cost, portable and miniaturized size will make our microfluidic CD beneficial especially in developing countries.

II. EXPERIMENTAL SETUP.

A. LAMP Assay preparation

DNA from *Salmonella* Enteritidis was prepared by using boiling method. A 2.5 μL (~50 ng/μL) of crude DNA was used as template for the LAMP assays. A set of 6 primers were used to specifically target *Salmonella* DNA. The primers set consists of 2 outer primers (F3 and B3), 2 inner primers (FIP and BIP) and 2 loop primers (LF and LB). Briefly, the LAMP reaction was performed in a total volume of 25 μL composed of 12.5 μl of 2X reaction mix, 8U of *Bst* DNA polymerase (Eiken Chemical Co., Ltd., Tokyo, Japan), 40 pmol of each primer FIP and BIP, 5 pmol of each primer F3 and B3, 20 pmol of each loop primer LF and LB, 2.5 μl (~50 ng/μL) and 2 μL deionized water. The LAMP reaction was set at approximately 63 °C for 60 min prior to the inactivation of *Bst* polymerase for 2 min at 80 °C.

B. Microfluidic CD Fabrication

The microfluidic CD was designed using a computer aided design software (AutoCAD). The Microfluidic CD device consists of top and bottom layers made of polymethyl methacrylate (PMMA) and middle layer made of pressure-sensitive adhesive (PSA) material as shown in fig 1a. The combination of channels and chambers on PMMA material were engraved using a computerized numerical machine (CNC). The channels and chambers on the PSA layer were cut using a cutter plotter machine. All 3 layers were aligned, laminated and pressed-bound together using a custom made system to completely seal the microfluidic CD device. Additional steps for a complete sealing of specific regions such as to prevent liquid loss by evaporation was accomplished by application of epoxy resins such as a Norland UV curable adhesives. Fig 1b shows a portion of the microfluidic CD operations, loading chambers, mixing chambers, metering chambers, sealing chambers and amplification (detection) chambers.

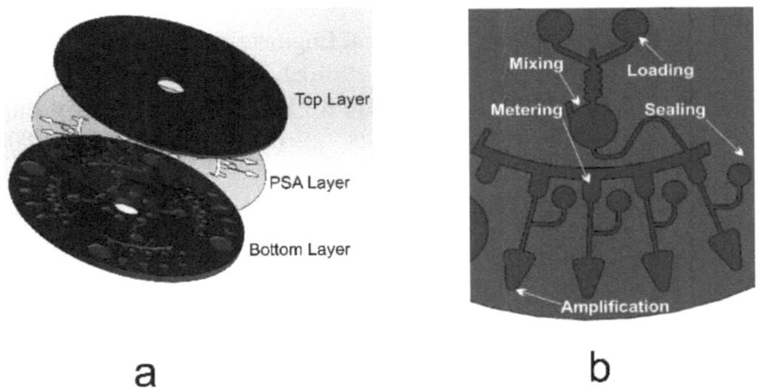

Fig.1 Schematics of *Salmonella* detection on microfluidic CD. 1a is the top, PSA and bottom layer of the microfluidic CD. 1b top view of a part of the microfluidic CD showing the microfluidic operations; loading chambers, mixing, metering, sealing and amplification chambers.

C. LAMP on Microfluidic CD

Once the microfluidic CD device was fabricated and ready to use, a customized CD spin test system was used for testing. The microfluidic operations for *Salmonella* detection starts with loading the LAMP reagents to the loading chambers, loading the Norland UV curable adhesives to sealing chambers and loading *Salmonella* DNA to amplification chambers (see fig 2a). The LAMP reagents were pumped to the mixing chamber in order to mix as shown in fig 2b. Then, the LAMP reagents were transferred to metering chambers and filled chambers 1-3 (see fig 2c). Subsequently, chamber number 4 was filled and the rest of LAMP reagents flowed to the wasting chamber number 5 (see fig 2d & 2e). Then the LAMP reagents was transferred to the amplification chambers where DNA amplification took a place as shown in fig 2f. The additional sealing steps was applied by exposing Norland UV curable adhesives to the UV light source in order for the Norland UV curable adhesives to solidify and prevent evaporation of the LAMP assay. Once the microfluidic CD was completely sealed, the DNA amplification was begun. The hot air gun was set and turned on and the DNA amplification was started. The DNA amplification took about 60 min at approximately 63°C then, the temperature was raised to 80 °C for 2 min for inactivation of *Bst* polymerase in order to stop the DNA amplification. The microfluidic CD was left and cool down for 8 min. Then, 1 μL of SYBR Green I was added to each amplification (detection) chamber for detection process.

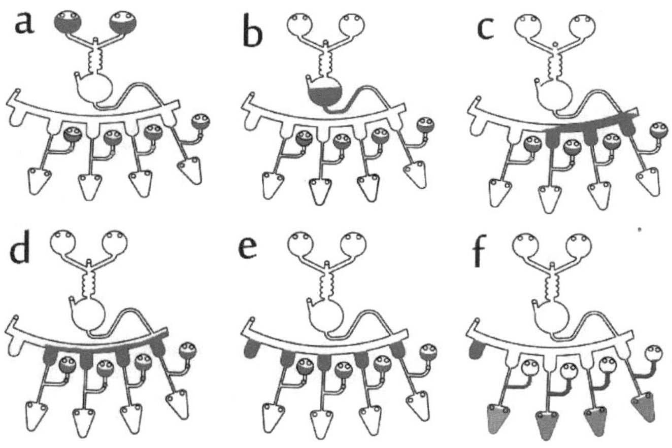

Fig.2 Results of LAMP amplification after adding 1 μL of SYBR Green I. The green color determines a positive result, while an orange color indicates a negative result.

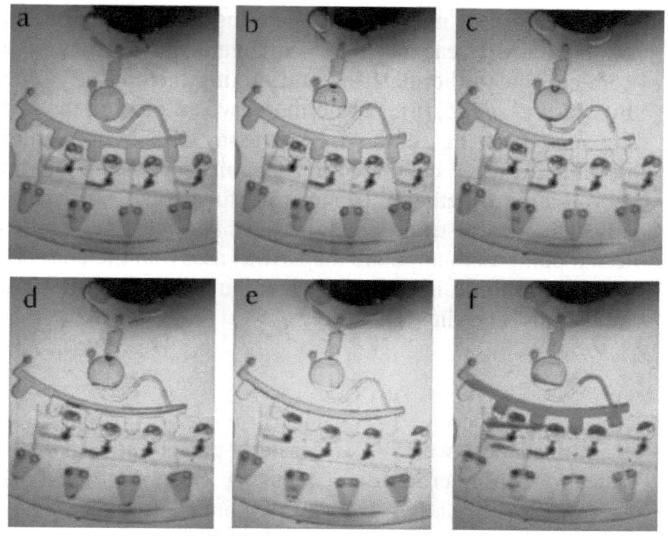

Fig 3.Images showing the results of the microfluidic operations. 3a loading LAMP reagents. 3b mixing the LAMP reagents. 3c metering the LAMP reagents to chambers1-3. 3e filling chamber number 4. 3e rest of the LAMP reagents flow to wasting chamber number 5. 3f transferred the LAMP reagents to amplification chambers, sealing the channels with UV curable adhesive and performing DNA amplification

III. Results and Discussion

A. Salmonella Detection on Microfluidic CD

Fig 3 shows the results of first steps in *Salmonella* detection on microfluidic CD, (refer to fig 2 because the LAMP assay is transparent and is hard to see the process in real images). The steps involved are mixing and metering of the LAMP reagents. The LAMP reagents was first pumped from loading chambers into mixing chamber (see fig 3a & 3b). Then, the LAMP reagents were pumped and filled the first 3 metering chambers (see fig 3c). Subsequently, chamber number 4 was filled after as shown in fig 3d. After all 4 chambers were equally filled with LAMP reagents the rest flowed into wasting chamber as shown in fig 3f. The LAMP reagents did not flow yet to the amplification chambers because the very narrow channel that is designed to only let the LAMP reagents flow through it when increasing the rpm speed to 1500 rpm. Then, the channels that connect the metering chambers with amplification chambers were sealed in order to prevent the LAMP assay evaporation during amplification. Afterward, the hot air gun was set and turned on and the DNA amplification was started. The DNA amplification took 60 min at approximately 63°C and then, the temperature was increased to 80°C to stop the amplification. A 1 μL of SYBR Green I was added to each chamber for visualization of the results (see fig 4)

It can be seen from the results in fig 4 that, the SYBR Green I in chamber 1-3 changed from orange color to green color because the chambers contain DNA *Salmonella*. Hence, chamber 1-3 indicate positive results (*Salmonella* detected) as SYBR Green 1 color changed. On the other hand, chamber number 4 have no SYBR Green 1 color change as it contains deionized water (as negative control). Therefore, chamber number 4 indicate negative result (no *Salmonella* detected).

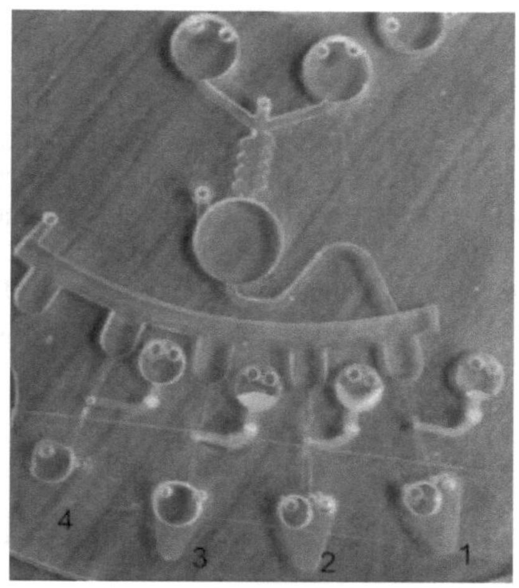

Fig.4 Results of LAMP amplification after adding 1 μL of SYBR Green I. The green color determines a positive result, while an orange color indicates a negative result.

We present a microfluidic metering and mixing process of the LAMP reagents that has several advantages when compared with the metering and mixing method performed by Mark *et al.*[15,16]. Furthermore, we automated the mixing and metering of the LAMP assay which bring more advantages when compared to *Salmonella* conventional and lab-on-a-chip methods. Using centrifugal platform is more advantageous of using chips because we don't need an external pump to manipulate the LAMP assay on the chip. The centrifugal force is enough to manipulate the LAMP assay on the microfluidic CD.

IV. CONCLUSIONS

In this study, we have successfully performed *Salmonella* detection on microfluidic CD using Loop Mediated Isothermal Amplification (LAMP). We have introduced a point of care diagnostic tool as, the LAMP assay preparation, mixing of the LAMP reagents, metering of the LAMP reagents, DNA amplification and detection, can be integrated into a single microfluidic CD. DNA extraction was performed off-chip and the automated process of *Salmonella* detection on the microfluidic CD took approximately 65 min, which is much shorter compared to conventional methods (3-4 h). Visualization of the final detection results was simply done by monitoring the color change of the added SYBR Green 1 to each amplification chamber. The economy and usefulness of this technique (microfluidic CD device) was attributed to the automated processes (Pumping, mixing, metering of the LAMP reagents) performed on the microfluidic CD and to the heating block (hot air gun) used for DNA amplification.

ACKNOWLEDGMENT

This research is financially supported by University of Malaya (UM) research grant, Program RakanPenyelidikan UM (Project No: CG003-2013), Fundamental Research Grant Scheme (FRGS: FP042-2013B), MOSTI Science Fund (Project No: GA013-2013). Fatimah Ibrahim would like to acknowledge Sultan Iskandar Johore foundation for supporting and funding the one off special equipment in the microfluidic research.

CONFLICT OF INTEREST

"The authors declare that they have no conflict of interest".

REFERENCES

1. Crump, J.A.; Mintz, E.D. Global trends in typhoid and paratyphoid fever. Clinical Infectious Diseases 2010, 50, 241-246.
2. Olsen, J.; Aabo, S.; Rasmussen, O.; Rossen, L. Oligonucleotide probes specific for the genus *salmonella* and for salm. Typhimurium. Letters in applied microbiology 1995, 20, 160-163.
3. Mandal, P.; Biswas, A.; Choi, K.; Pal, U. Methods for rapid detection of foodborne pathogens: An overview. American Journal of Food Technology 2011, 6, 87-102.
4. Wang, Y.; Ye, Z.; Ying, Y. New trends in impedimetric biosensors for the detection of foodborne pathogenic bacteria. Sensors 2012, 12, 3449-3471.
5. Notomi, T.; Okayama, H.; Masubuchi, H.; Yonekawa, T.; Watanabe, K.; Amino, N.; Hase, T. Loop-mediated isothermal amplification of DNA. Nucleic acids research 2000, **28**, e63-e63.
6. Zhang, G.; Brown, E.W.; González-Escalona, N. Comparison of real-time pcr, reverse transcriptase real-time pcr, loop-mediated isothermal amplification, and the fda conventional microbiological method for the detection of *salmonella* spp. In produce. Applied and environmental microbiology 2011, 77, 6495-6501.
7. Foudeh, A.M.; Didar, T.F.; Veres, T.; Tabrizian, M. Microfluidic designs and techniques using lab-on-a-chip devices for pathogen detection for point-of-care diagnostics. Lab on a Chip 2012, 12, 3249-3266.
8. Haeberle, S.; Zengerle, R. Microfluidic platforms for lab-on-a-chip applications. Lab on a Chip 2007, 7, 1094-1110.
9. Lutz, S.; Weber, P.; Focke, M.; Faltin, B.; Hoffmann, J.; Müller, C.; Mark, D.; Roth, G.; Munday, P.; Armes, N. Microfluidic lab-on-a-foil for nucleic acid analysis based on isothermal recombinase polymerase amplification (rpa). Lab on a Chip 2010, 10, 887-893.
10. Duarte, C.; Salm, E.; Dorvel, B.; Reddy Jr, B.; Bashir, R. On-chip parallel detection of foodborne pathogens using loop-mediated isothermal amplification. Biomedical microdevices 2013, 15, 821-830.
11. Kim, T.-H.; Park, J.; Kim, C.-J.; Cho, Y.-K. Fully integrated lab-on-a-disc for nucleic acid analysis of food-borne pathogens. Analytical chemistry 2014, 86, 3841-3848.
12. Gorkin, R.; Park, J.; Siegrist, J.; Amasia, M.; Lee, B.S.; Park, J.-M.; Kim, J.; Kim, H.; Madou, M.; Cho, Y.-K. Centrifugal microfluidics for biomedical applications. Lab on a Chip 2010, 10, 1758-1773.
13. Park, J.-M.; Cho, Y.-K.; Lee, B.-S.; Lee, J.-G.; Ko, C. Multifunctional microvalves control by optical illumination on nanoheaters and its application in centrifugal microfluidic devices. Lab on a Chip 2007, 7, 557-564.
14. Steigert, J.; Grumann, M.; Brenner, T.; Riegger, L.; Harter, J.; Zengerle, R.; Ducrée, J. Fully integrated whole blood testing by real-time absorption measurement on a centrifugal platform. Lab on a Chip 2006, 6, 1040-1044.
15. Mark, D.; Metz, T.; Haeberle, S.; Lutz, S.; Ducrée, J.; Zengerle, R.; von Stetten, F. Centrifugo-pneumatic valve for metering of highly wetting liquids on centrifugal microfluidic platforms. Lab on a Chip 2009, 9, 3599-3603.
16. Mark, D.; Weber, P.; Lutz, S.; Focke, M.; Zengerle, R.; von Stetten, F. Aliquoting on the centrifugal microfluidic platform based on centrifugo-pneumatic valves. Microfluidics and Nanofluidics 2011, 10, 1279-1288.

Author to whom correspondence should be addressed; Fatimah Ibrahim.
Department of Biomedical Engineering, Faculty of Engineering, University of Malaya, 50603 Kuala Lumpur,
Mail: fatimah@um.edu.my;
Tel.:+603-7967-6818; Fax: +603-7967-6878.

A Method of Measuring Corneal Young's Modulus

Huei-Jyun Cao[1], Chun-Ju Huang[1], Po-Jen Shih[2], I-Jong Wang[3], and Jia-Yush Yen[1]

[1] Department of Mechanical Engineering, National Taiwan University
[2] Department of Civil and Environmental Engineering, National University of Kaohsiung, Kaohsiung, Taiwan
[3] Department of Ophthalmology, College of Medicine, National Taiwan University

Abstract-**Eye is very important to human, where there is a problem with the eye, it is always desirable to seek medical care early on; however, delay still exists due to miscalculation and examination uncertainties. This paper thus tries to provide a more accurate non-intrusive measurement of the cornea properties for doctor's reference.**

One of the properties that is very sensitive to the condition of the eye is the Young's modulus of the cornea. The cornea material exhibits different behavior in patients with different conditions. This paper proposes a mathematical model for describing the mechanical properties of the cornea. By comparing the model behavior with the non-intrusive measurement data taken off the Oculus Corvis® ST non-contact tonometer, it is possible to deduce the corneal Young's modulus.

Sensitive to the condition of the eye is the Young's modulus of the cornea. The cornea material exhibits different behavior in patients with different conditions. This paper proposes a mathematical model for describing the mechanical properties of the cornea. By comparing the model behavior with the non-intrusive measurement data taken off the Oculus Corvis® ST non-contact tonometer, it is possible to deduce the corneal Young's modulus.

Keywords— **cornea, Young's modulus, Corvis® ST, tonometer, intraocular pressure**

I. Introduction

The new corneal visualization scheimpflug technology (Corvis® ST), an advanced non-contact-tonometer, takes real time images of the profile of corneal deformation response during the air puff measuring by a ultra-high-speed scheimpflug camera (around 140 frames in 0.031 Sec) [1]. Nowadays, the most important medical application of the Corvis® ST is on the intraocular pressure (IOP) measurement. The main purpose of this paper is to develop a dynamic model to analyze the biomechanical properties of the cornea.

In this paper, we demonstrate a new model for the visualization scheimpflug technology in order to add some mechanism knowledge to ophthalmology, and 10 patients' data are applied to develop a preliminary database. To study the dynamics of the entire eyeball during the air-puff emits the cornea, our model is developed from the fundamentally spherical wave function and applied to the external force. In this model, the block mass is assumed to be canceled and replaced with the spherical diaphragm with constant density, and the elasticity of the cornea is replaced by the geometric stiffness composed of parameters: thickness, radius, Young's modulus, Poisson's ratio, IOP and densities. In addition, the estimated material properties may help quantify details of symptoms of the diseases in future.

II. The forced dynamic model

The eye is made up of the three layers and the three transparent structures as shown in Figure 1. The outermost layer is composed of the cornea and sclera, the middle layer is consists of the choroid, ciliary body, and iris, and the innermost is the retina. Within these coats are the aqueous humor (about 0.310 microliter), the vitreous body (about 4 microliter), and the lens (about 0.22 microliter). The aqueous humor is a clear fluid, and the vitreous body is a clear jelly which composing of 98-99% water and occupying two third of the entire volume of the eye [2].

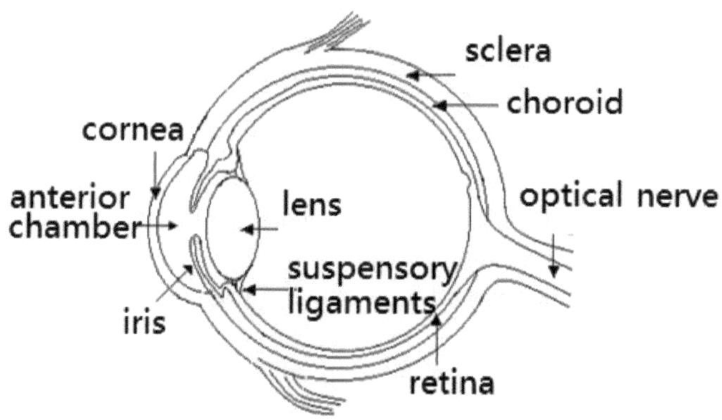

Figure 1.The structure of the eyeball

As shown in Figure 2, this model simplifies the eyeball into the three parts: the air outside the eyeball (the oblique and rectus muscles are not concerned), the diaphragm of the eyeball (including cornea, sclera and retina), and the fluid inside the eyeball (including vitreous humor, lens, and aqueous humor.)

J. Goh and C.T. Lim (eds.), *7th World Congress on Bioengineering 2015*, IFMBE Proceedings 52, DOI: 10.1007/978-3-319-19452-3_13

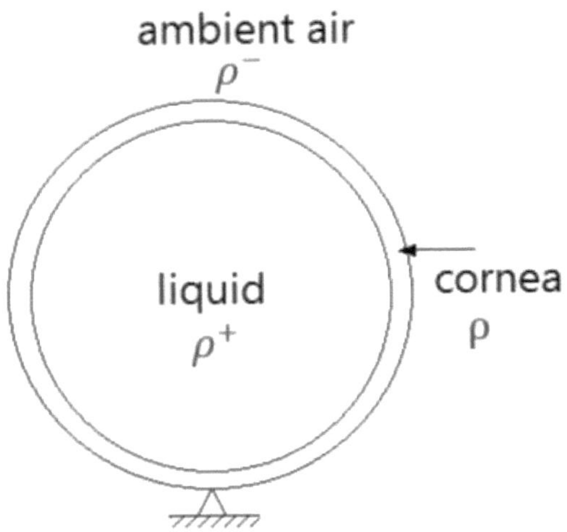

Figure 2 The dynamic model of the eyeball

First, the Euler's equation for the ambient air and the fluid inside the eyeball is

$$\begin{cases} \dfrac{\partial \mathbf{v}}{\partial t} + \mathbf{v} \cdot \nabla \mathbf{v} = -\dfrac{1}{\rho} \nabla p \\ \nabla \cdot \mathbf{v} = 0 \end{cases} \tag{1}$$

where v_i is the covariant component of velocity fields, ρ is the density, p is the pressure, and we neglect the fluid viscosity.

The solution of p^+ inside the eyeball is given by

$$p^+ = \sum_{l=0}^{\infty} \sum_{m=l}^{-l} i \rho^+ R^2 \omega^2 A^+_{lm} Y_{lm}(\theta, \phi) (\frac{r}{R})^l e^{i\omega t} \tag{2}$$

with undetermined constants $A^{\cdot}_{lm}$, and where the angular frequency □, R is the radius of eyeball, and Y_{lm} are the spherical surface harmonics [3]. Note that the superscript '+' denotes the domain inside the eyeball.

Furthermore, the solution of p^- outside the eyeball is

$$p^- = \sum_{l=0}^{\infty} \sum_{m=l}^{-l} i \rho^- R^2 \omega^2 A^-_{lm} Y_{lm}(\theta, \phi) (\frac{r}{R})^{-l-1} e^{i\omega t} \tag{3}$$

Here the air and fluid are assumed to be incompressible, so the velocity component on the surface becomes

$$\begin{aligned} v_R(r, \theta, \phi; t) &= \sum_{l=0}^{\infty} \sum_{m=l}^{-l} -\rho^- R^2 \omega A^+_{lm} Y_{lm}(\theta, \phi) e^{i\omega t} \\ &= \sum_{l=0}^{\infty} \sum_{m=l}^{-l} -\rho^- R^2 \omega A^-_{lm} Y_{lm}(\theta, \phi) e^{i\omega t} \end{aligned} \tag{4}$$

The modified Helmholtz function as shown in Eq. (5) helps derive the displacement function on the stretched diaphragm.

$$\frac{(E+T) \cdot t^3}{12(1-\nu^2)} \nabla^4 w - T \cdot t \nabla^2 w - [p] + \rho \cdot t \frac{\partial^2 w}{\partial t^2} = 0 \tag{5}$$

in which E is the Young's modulus, T is the tension on the diaphragm, t is the thickness of the diaphragm, ν is the Poisson's ratio, ρ is the density of the diaphragm, and $[p]$ is the pressure difference between the two sides of the diaphragm, i.e. $[p] = p^+ - p^-$.

We wish to solve the wave equation for the cornea as Eq. (6)

$$\frac{1}{c^2} \frac{\partial^2 w(\theta, \phi; t)}{\partial t^2} - \nabla^2 w(\theta, \phi; t) = f(\theta, \phi; t) \tag{6}$$

Where c is the frequency-independent, constant phase velocity, f is the forcing term and w is the deformation of the cornea.

We assume the simple harmonic motion is the solution of this equation,

$$w(\theta, \phi; t) = \sum_{l=0}^{\infty} \sum_{m=l}^{-l} w_{lm}(\theta, \phi; t) = \sum_{l=0}^{\infty} \sum_{m=l}^{-l} -i B_{lm} Y_{lm}(\theta, \phi) \cdot e^{i\omega t} \tag{7}$$

where B_{lm} are the undetermined constants.

With Eq. (7), the modified pressure difference term $[p]$ at the diaphragm lead to

$$[p]_{lm} = -i R \omega^2 B_{lm} Y_{lm}(\theta, \phi) e^{i\omega t} (\frac{\rho^+}{l} + \frac{\rho^-}{l+1}) \tag{8}$$

Substitute Eq. (8) into Eq. (5) as the boundary condition and obtain the modified wave equation:

$$\begin{aligned} &[\frac{(E+T) \cdot t^3}{12(1-\nu^2)} \frac{l^2 (l+1)^2}{R^4} + \frac{l(l+1) T \cdot t}{R^2}] w_{lm} + \\ &[\rho \cdot t + R(\frac{\rho^+}{l} + \frac{\rho^-}{l+1})] \frac{\partial^2 w_{lm}}{\partial t^2} = 0 \end{aligned} \tag{9}$$

The solution of the Eq. (9) is

$$w(\theta, \phi; t) = \sum_{l=0}^{\infty} \sum_{m=l}^{-l} \frac{1}{M_l} \int_0^t \frac{\sin \omega_l (t-\tau)}{\omega_l} F_{lm} \, d\tau \cdot Y_{lm}(\theta, \phi) \tag{10}$$

Here, F_{lm} is the coefficient in the series expansion of the external force as the Eq. (11).

$$F_{lm} = \frac{\int_0^{2\pi} \int_0^{\pi} F(\theta, \phi; t) Y_{lm}(\theta, \phi) \sin \theta \, d\theta \, d\phi}{\int_0^{2\pi} \int_0^{\pi} Y^2_{lm}(\theta, \phi) \sin \theta \, d\theta \, d\phi} \tag{11}$$

III. NUMARICAL SIMULATION AND CLINICAL STUDY

The external force in the Corvis ST is the air-puff generated from a nozzle jet as shown in Figure 3, and we assume the pressure distribution deduced from the air-puff is a three-dimensional Gaussian distribution. The time series of the pressure is shown in the Figure 4.

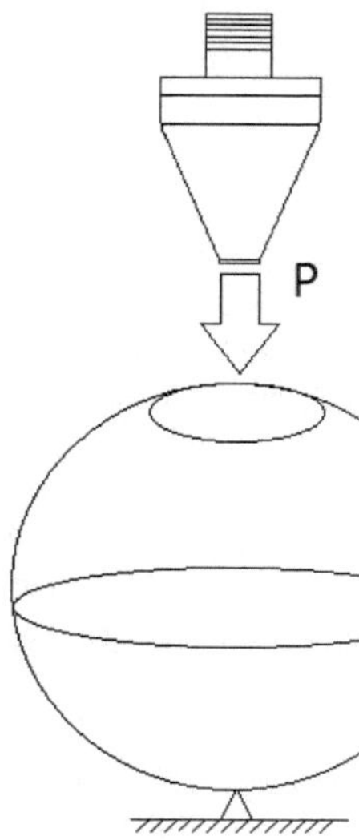

Figure 3 A model of the eyeball subjected by an air impulse

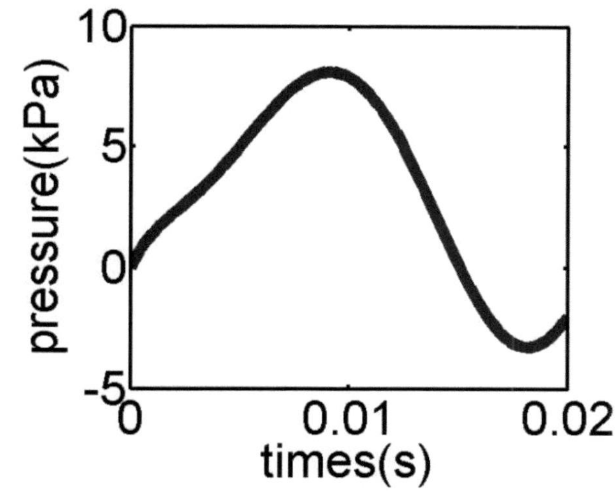

Figure 4 The time series of the pressure

In the clinical study, there were 10 patients joined this study in National Taiwan University Hospital. The study was approved by the Ethics Committees of the National Taiwan University Hospital, Taipei, Taiwan and followed the tenets of the Declaration of Helsinki.

The values of IOPs and the thicknesses could be measured from the Corvis® ST. During the air-puff, the high-speed Scheimpflug camera recorded the movements of the cornea. It gets 140 images during 32 msec. Figure 4 shows the displacements of the cornea of one patient at time 0.000, 9.561, 17.425, and 20.302 msec.

Figure 4 The corneal images taken from one patient by Corvis® ST.

Except of the IOP and the thickness, the other material properties of the eyeball required in Eq. (9) are shown in Table 1. .

Table 1 Material properties of the eyeball

Properties	Symbols	Value
Radius	R	12 mm
Thickness	T	540 μm
Poisson's ratio	ν	0.49
Density of air	ρ^+	1.204 kg/m^3
Density of the fluid inside eyeball	ρ^-	1000 kg/m^3
Density of sclera	ρ	1444 kg/m^3

Then we applied the optimization program of Matlab on the corneal deformation to estimate the Young's moduli of the cornea E.

Figure 5 shows one patient's corneal displacement time series measured from Corvis® ST, and the same time series obtained from our model are also shown. The two curves are not perfectly close, because the setup of the boundary conditions is still a big issue in the analysis.

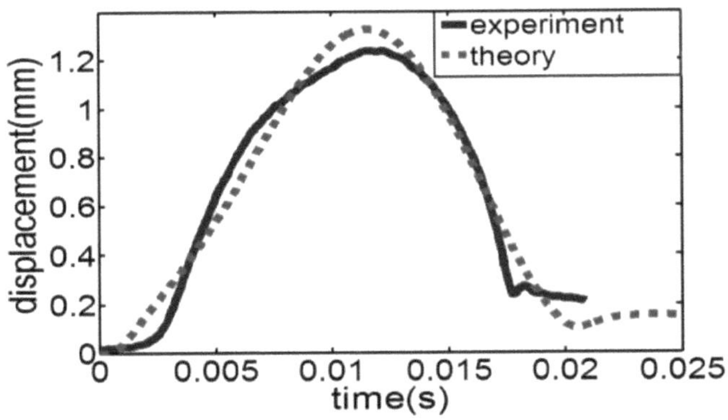

Figure 5 The displacement time series of the cornea obtained from the experiment and the theory.

Table 2 shows the results of Young's moduli among the 10 patients. These values satisfy the acceptable ranges 0.1~10MPa from the tests of the human eyes (in vitro).

IV. Conclusion

This paper proposes a simulation model which helps to estimate instantaneously the eyeball properties during the IOP measuring by the Corvis® ST. In vivo experiments, the images taken from the 10 patients help develop the data base of the Young's moduli and the damping coefficients. The average Young's modulus is 0.2985 MPa, which is in the acceptable range of other studies [4] [5] [6]. This model help to measure human corneal mechanical properties in vivo and quantify the quality of the symptom.

Table 2 The 10 patients' information

Patient No.	Age	Thickness [μm]	IOP [mmHg]	Young's modulus [MPa]
1	38	525	12	0.205
2	33	501	13	0.1301
3	27	565	14	0.7325
4	26	529	15	0.2909
5	49	518	16	0.1347
6	24	577	16.5	0.5288
7	35	547	15	0.2651
8	27	651	17	0.1022
9	51	584	15.5	0.026
10	25	506	13	0.5701
Average	33.5	550.3	14.7	0.2985

Acknowledgment

The authors wish to thank National Science Council of Taiwan, for financial support under contract NSC-103-2228-E-002- 001.

References

1. J. X. Hong, J. J. Xu, A. J. Wei, . S. X. Deng, X. H. Cui, X. B. Yu and X. H. Sun, "A New Tonometer-The Corvis ST Tonometer: Clinical Comparison with Noncontact and Goldmann Applanation Tonometers," Investigative Ophthalmology & Visual Science, vol. 54, no. 1, pp. 659-665, Jan 2013.
2. K. E. Swindle and N. Ravi, "Recent advances in polymeric vitreous substitutes," Expert Review of Ophthalmology, vol. 2, no. 2, pp. 255-265, 2007.
3. J. W. S. Rayleigh, The theory of sound, vol. 1, London: Macmillan, 1926.
4. S. L. Y. Woo, A. S. Kobayashi, W. A. Schlegel and C. Lawrence, "Nonlinear material properties of intact cornea and sclera," Experimental Eye Research, vol. 14, no. 1, pp. 29-39, 1972.
5. A. Elsheikh, D. F. Wang and D. Pye, "Determination of the modulus of elasticity of the human cornea," Journal of Refractive Surgery, pp. 808-818, Oct 2007.
6. K. E. Hamilton and D. C. Pye, "Young's modulus in normal corners and the effect on applanation tonometry," Optometry and Vision Science, vol. 85, no. 6, pp. 445-450, Jun 2008.

Author: Jia-Yush Yen
Institute: Dept. of ME, NTU
Street: No. 1, Sec. 4, Roosevelt Road
City: Taipei
Country: Taiwan
Email: jyen@ntu.edu.tw

Convex Hull Based Detection of Overlapping Red Blood Cells in Peripheral Blood Smear Images

Feminna Sheeba[1], Robinson Thamburaj[2], Joy John Mammen[3], Mohan Kumar[1], and Vansant Rangslang[1]

[1] Department of Computer Science, Madras Christian College, Chennai, India
[2] Department of Mathematics, Madras Christian College, Chennai, India
[3] Department of Transfusion Medicine & Immunohaematology, Christian Medical College, Vellore, India

Abstract— **The Segmentation of Red Blood Cells (RBCs) in blood smear images to obtain their count is often the first step in the diagnosis of various pathological conditions. Although several procedures have been devised for this task, a majority of them suffer from performance degradation due to the overlapping of cells. Various researches have been carried out to split these overlapping cells. The proposed paper aims at suggesting two algorithms to find the concavity points in the overlapping RBCs' contours. In the first approach, the dip points are obtained by analyzing the concave regions, obtained by finding out the Euclidean distance of all points in the overlapping cell to their convex hull. In the second approach, dip point identification is based only on the convex hull of the overlapping cell. The contours of the concave regions are analyzed from the perspective of the centroid. These two strategies were compared with the approach used in an earlier work, which also addressed the splitting of overlapping RBCs, by identifying the dip points using curve fitting and smoothing of the contours. The two approaches proposed in this paper are quite efficient in terms of accuracy and the time taken to achieve results. The specificity of the first approach was 90% and that of the second approach was 94%, meaning that the two new methods are far more advanced than the earlier work for which the specificity was only 75%.**

Keywords— **Segmentation, convex hull, clumped RBCs, dip points.**

I. INTRODUCTION

The usual diagnostic approach in the study of blood disorders is to examine the blood film and obtain a total blood count. Obtaining the differential count of Red Blood Cells (RBCs) is vital in order to obtain the total blood count, RBCs in normal peripheral blood are circular and fairly uniform in size. They have a zone of central pallor about one third the size of the RBC. Though most of the RBCs are singular, few of them are found to be overlapping or clumped in groups. Various researches are carried out in splitting of overlapping cells. If intensity based segmentation was done. Overlapping objects may not be split [1] [2]. Hence a marker controlled watershed Transform may be used which split overlapping cells for which the regional minima were clearly identified [3]. Gametocytes touching RBCs were defected using distance transform of RBCs [4]. Various researches were carried out in segmentation using level set methods to detect overlapping objects. The contour of each cell is obtained using a level set algorithm based on an interactive model [5]. A shape-based approach is proposed to do curve evolution for the segmentation of medical images containing known object types [6]. A study to cluster the nuclei seen in confocal microscopy images was done using a clump-splitting algorithm

One recent common method used for segmenting overlapping or clumped cells is by doing their concavity analysis. A novel nonparametric concavity point analysis-based method for splitting clumps of convex objects in binary images is presented in [7]. The method is based on finding concavity point-pairs by using a variable- size rectangular window. Results obtained with images that have clumps of biological cells show that the method gives accurate results.

In peripheral blood smear images two or three RBCs overlap in various forms resulting into one or more concavities. In some of the overlapping cells, the gradient values in the overlapped region do not show remarkable difference compared to the other areas of the cell. Hence splitting them according to the concavity and convexity of the overlapped cells is more appropriate.

In one of the earlier works of the primary author, Watershed Transform was used to split such overlapping RBCs [8]. The Watershed Transform was able to split almost all the overlapping RBCs except for a few whose regional minima were not clearly identified. Such overlapped cells were split by obtaining dip points using concavity analysis [8]. The proposed work aims at suggesting two more algorithms to find the concavity points in the overlapping RBCs' contours. The first method obtains the dip points by finding out the Euclidean distance of all points in the contour of the overlapping cell to the centroid of the object. In the second approach, the dip points are identified by using the distance transform of all pixels within the convex hull of the overlapping cell. The work also compares both the approaches with the earlier work of the author, which uses

J. Goh and C.T. Lim (eds.), *7th World Congress on Bioengineering 2015*,
IFMBE Proceedings 52, DOI: 10.1007/978-3-319-19452-3_14

curve fitting and smoothing of the contours in order to find the dip points.

The paper is organized with Methodology in Section II, Results and Findings in Section III and Conclusion in Section IV of the paper.

II. Methodology

A. Image Acquisition

The Images for the study were obtained using a digital camera attached to a compound microscope. These images are TIFF images of resolution 1024 x 1024. 150 such images have been used for the study.

B. First Approach

In the first approach, the convex hull of the overlapping cell is obtained. It is understood that the dip points may be found in the region where any two points in the convex hull are far apart. This is shown in Fig. 1. Therefore the Euclidean distance of all the points in the border of the overlapping cell between these two points to the centroid of the cell are computed. It is found that the point which has the minimum distance from the centroid is the dip point.

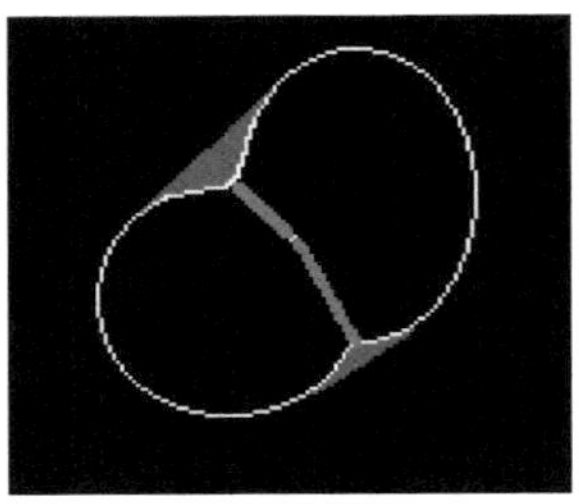

Fig. 1 Dip Points by First Approach

C. Second Approach

In the second approach, the Distance Transform of the region within the convex hull of the overlapping cell is obtained, resulting in the distance of the various pixels in the overlapping cell to the border of the convex hull. The points having minimum distance are marked, which are none other than the dip points of the overlapping cell.

III. Results and Findings

A prototype application was developed for testing the algorithms of both the approaches discussed. The input for the application was peripheral blood smear images. The overlapping cells that were not split by the Watershed Transform were identified as in [8]. The overlapping cells identified were tested for splitting by both the approaches. The step-by-step results are shown in Table 1. The results of these two approaches were compared with the results in [8]. It is seen that the approach in [8] used curve fitting and smoothing, which is a time consuming process. This was rectified in our approaches and hence the methods were made quite simple.

Table 1

1 Dip Point		**2 Dip Points**		**3 Dip Points**	
App 1	App 2	App 1	App 2	App 1	App 2
Contour	Contour	Contour	Contour	Contour	Contour
Centroid	Mask	Centroid	Mask	Centroid	Mask
Convex Hull Pts	Dist. Transfm	Convex Hull Pts	Dist. Transfm	Convex Hull Pts	Dist. Transfm
1 Dip Point	Concave Regions	2 Dip Points	Concave Regions	3 Dip Points	Concave Regions
Split Cells	1 Dip Point	Split Cells	2 Dip Points	Split Cells	3Dip Points

IV. Conclusion

The two approaches proposed in this paper are quite efficient in terms of accuracy and the time taken to achieve results. The specificity of the first approach was 90% and that of the second approach was 94%, meaning that the two new methods are for more advanced than the earlier work for which the specificity was only 75%.

Acknowledgment

The authors would like to thank the Department of Transfusion Medicine and Immunohaematology, Christian Medical College, Vellore, India for providing them with sample images used for the study.

References

1. Prasad, A.S., Latha, K.S., Rao, S.K.: Separation and counting of blood cells using geometrical features and distance transformed watershed. International Journal of Engineering and Innovative Technology (IJEIT), (2013) 3(2)
2. Feminna, S., Robinson, T., Nagar, A.K., Mammen, J.J.: Segmentation of peripheral blood smear images using tissue-like P-Systems. IJNCR–BICTA 2011 Special Issue 3(1), (2012) 16–27
3. Feminna, S., Robinson T., J.J. Mammen , H.M.T. Thomas and Atulya K. Nagar.: White Blood Cell Segmentation and Watermarking. In: Proceedings of the IASTED International Symposia Imaging and Signal Processing in Healthcare and Technology, ISPHT 2011,Washington DC, USA (2011).
4. Feminna, S., Robinson T., Joy John Mammen and Atulya K. Nagar, Detection of plasmodium falciparum in peripheral blood smear images. In: Proceedings of BICTA 2012, 202 (2013) 289–298
5. Qi, X., et al.: Robust segmentation of overlapping cells in histopathology specimens using parallel seed detection and repulsive level set. IEEE Transactions on Biomedical Engineering, 59(3) (2012) 754–765
6. Tsai, A., et al.: A shape-based approach to the segmentation of medical imagery using level sets. IEEE Transactions on Medical Imaging, 22(2) (2003) 137
7. LaTorre, A., et al.: Segmentation of neuronal nuclei based on clump splitting and a two-step binarization of images. Expert Syst. Appl, 40(16) (2013) 6521–6530
8. Feminna, S., Robinson T., J.J. Mammen, H.M.T. Thomas and Atulya K. Nagar.: Splitting of overlapping cells in peripheral blood smear images by concavity analysis. Combinatorial image analysis lecture notes in computer science DOI: 10.1007/978-3-319-07148-0_21 Volume 8466, (2014) 238-249
9. Gonzalez, Woods, Digital Image Processing 3ed (Prentice Hall 2008)
10. [13]Gonzalez, Woods,Eddins, Digital Image Processing using Matlab (Pearson Education 2009)

Detection of Overlapping Tuberculosis Bacilli in Sputum Smear Images

Feminna Sheeba[1], Robinson Thamburaj[2], Joy John Mammen[3], R. Nithish[1], and S. Karthick[1]

[1] Department of Computer Science, Madras Christian College, Chennai, India
[2] Department of Mathematics, Madras Christian College, Chennai, India
[3] Department of Transfusion Medicine & Immunohaematology, Christian Medical College, Vellore, India

Abstract— **Tuberculosis (TB) is a common and lethal infectious disease caused by a germ (bacterium) called Mycobacterium tuberculosis. Early diagnosis of the disease is one of the primary challenges in curtailing its spread and is a critical step in the TB-Control Program worldwide. Among the most common methods used in the diagnosis of TB is the manual microscopic examination of a ZN-stained sputum smear which is a time-consuming and error-prone process. The diagnosis crucially depends on the number of viable or dormant mycobacteria in the sputum, which are seen as red colored rod-shaped objects in the smear under a microscope. This also means that the mycobacteria have to be detected accurately in order to arrive at the correct count, the accuracy of which could be affected when there are overlapping bacilli in the images. The use of Image Analysis in the detection of the mycobacteria will introduce a paradigm shift. The proposed work identifies such overlapping mycobacteria and uses techniques to total them accurately, which is an extension of an earlier work focusing only on segmentation of the tiny organisms. Normal bacilli are just 2-4 micrometers in length and 0.2-0.5 um in width. All the organisms that fall above their average size or show a variation in the ratio of the major-to-minor axis are identified to be overlapping mycobacteria, which are then used for further analysis. The count of mycobacteria that overlap is computed by obtaining the branch points in the skeleton of the overlapping object. The dataset used in the research consisted of eighty images, which were tested using a prototype application that achieved a success rate of 70%.**

Keywords— **Tuberculosis, Mycobacterium, Overlapping Bacilli, Branch Points.**

I. INTRODUCTION

Tuberculosis (TB) is a common and lethal infectious disease caused by a bacterium called Mycobacterium tuberculosis, which spread through the lymph nodes and blood stream to any organ in one's body especially to the lungs. Classically, there is a correlation between the presence of Acid Fast Bacilli (AFB) in sputum specimens and the method of tuberculosis diagnosis. The sensitivity of both smear and culture positivity mainly depends on the number of viable or dormant mycobacteria in the sputum. The bacillary load in the collected sputum depends on the severity of the disease and the process of sputum collection which is influenced by the time of collection The volume, quality and also the time of collection of sputum specimens are important for the increase in the TB case detection rate from suspected TB cases by both smear and culture methods especially in TB prevalent countries. Diagnosis of active TB relies on radiology microbiological culture of body fluids, tuberculin skin test as well as microscopic examination. Microscopically, an acid fast stain of sputum smear will show the organisms as slender red rods. Using technology to assist in screening of slides by image analysis will introduce a paradigm shift in the current scenario. As the diagnosis depends on the number of viable or dormant mycobacteria in the sputum, detection of the presence of the number of tuberculosis is bacilli in sputum smear images are highly essential. In order to detect the tuberculosis bacilli, appropriate segmentation techniques are used. As the bacilli show red under a microscope, color segmentation or K-means clustering have to be performed as a preliminary step [1] [2]. Watershed Transform is another good segmentation techniques which also segments overlapping cells [3] [4]. As Watershed Transform may not be able to segment overlapping cells, whose regional minima are not clearly seen. Hence splitting of overlapping cells are tried based on concavity analysis [5] and level set methods [6]. Automated, multi-stage, color-based Bayesian segmentation is used to identify TB objects [7]. Modified Watershed segmentation approach is used to seg-

J. Goh and C.T. Lim (eds.), *7th World Congress on Bioengineering 2015*,
IFMBE Proceedings 52, DOI: 10.1007/978-3-319-19452-3_15

ment the TB affected areas of the lungs seen through X-Ray of the patients [8]. In the recent years shape modeling is used to identify shapes of mycobacterium using invariant shape features and chromatic channel thresholding [9]. One such method is identifying the bacilli by extracting Heuristic acknowledges extracted from the bacilli shape contour [10]. Artificial neural network is also used to identify the bacilli in sputum smear images [11].

After acquiring a basic knowledge on need for segmentation of TB bacilli, as a study was conducted by the primary author to segment the TB bacilli using color segmentation techniques [12]. This involved microscopic examination of ZN stained sputum smear images under a microscope, in which the rod shaped bacilli are seen as red in color. The manual microscopic examination of the sputum smear is a time consuming error prone process. Therefore there is a need for automation in detection of the mycobacteria A count of the bacilli was computed by conducting few tests on this study. However it is seen that few bacilli overlap and hence are counted as a single one. This degrades the count of the bacilli. Therefore, the proposed work aims at detecting such overlapping bacilli by obtaining the branch points in the skeleton of the overlapping objects.

II. Methodology

A. Image Acquisition

The images for this study are obtained using a digital camera DFC280 attached to a compound microscope. It uses a choice of 1.3 megapixel standard resolution or 2.9 megapixel high resolution. The image is digitized with a 10-bit AD conversion that features a 700:1 dynamic range. The TWAIN interface is used to transfer images to the Leica IM50 Image Manager. The images acquired are 24 bit colored tiff images with a resolution size of 1280x1024 pixels.

B. Image Pre-processing

The RGB images are first converted to HSV color models and then converted to gray scale images. The images acquired may have poor illumination. Hence the contrast of the images is improved by equalizing the histogram of the image. Background elimination is done as the images may have artifacts in the background.

C. Detection of bacilli

Most of the bacilli do not overlap and hence they can be easily detected as discussed in [12]. These overlapping objects are found as single blobs after segmentation and hence they are counted only once. These overlapping bacilli are differentiated from the ones that are single based on their area and ratio of major to minor axis and such blobs are skeletonized. The branch points of each of the skeletons are obtained and their count gives the number of bacilli that overlap.

III. Results and Findings

Fifty sputum smear images were used for the study and one such image is shown in Fig. 1. It was first converted into a binary image and necessary algorithms were applied to segment the tuberculosis bacilli, which are shown in Fig. 2. If Fig.2 is carefully observed, 8 non-overlapped bacilli and 2 overlapped bacilli are seen. The overlapped bacilli were identified by using few geometric measures as discussed. Non-overlapped bacilli are eliminated leaving behind only the overlapped bacilli. This is seen in Fig. 3. The skeletons of the overlapping cells were obtained. The number of overlapping bacilli N are computed as N is equal to the number of branch points.

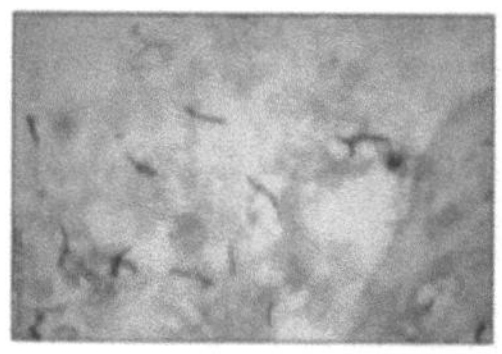

Fig. 1 Source image

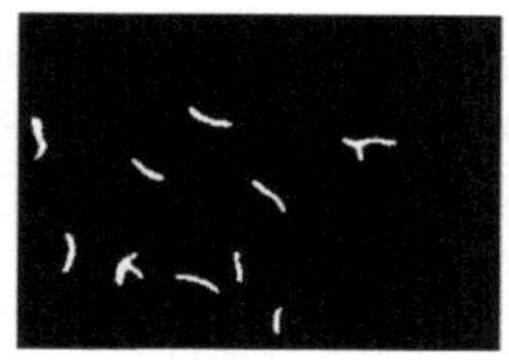

Fig. 2 Segmented TB bacilli

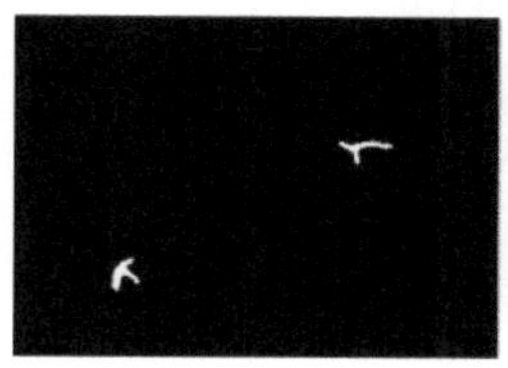

Fig. 3 Overlapping bacilli

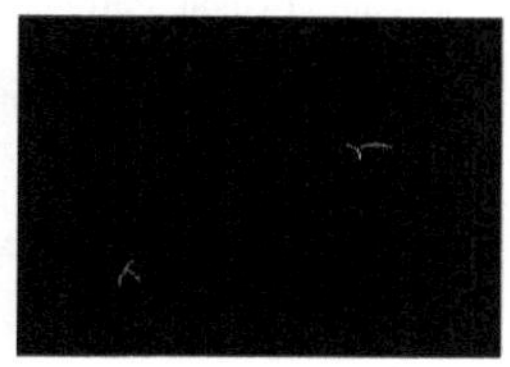

Fig.4 Skeletons of the bacilli

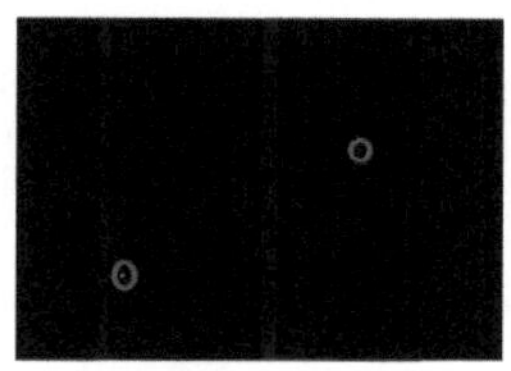

Fig 5 Branch points

IV. References

1. Joy J. Mammen, Maqlin P., FeminnaSheeba, T. Robinson,Making Malarial Diagnosis More Reliable: Using Image Analysis for Identification of Plasmodium Falciparum Gameotcytes, Abstract, J Pathol Inform. 2011; 2: 43e, s19-20.

2. Feminna, S., Robinson T., Joy John Mammen and Atulya K. Nagar, Detection of plasmodium falciparum in peripheral blood smear images. In: Proceedings of BICTA 2012, 202 (2013) 289–298.

3. Feminna, S., Robinson T., J.J. Mammen , H.M.T. Thomas and Atulya K. Nagar.: White Blood Cell Segmentation and Watermarking. In: Proceedings of the IASTED International Symposia Imaging and Signal Processing in Healthcare and Technology, ISPHT 2011,Washington DC, USA (2011) .

4. Hassan Khajehpour et al, Detection and Segmentation of Erythrocytes in Blood Smear Images Using a Line Operator and Watershed Algorithm, Journal of Medical Signals and Sensors, Vol 3, Issue 3, 2013 .

5. Feminna, S., Robinson T., J.J. Mammen, H.M.T. Thomas and Atulya K. Nagar.: Splitting of overlapping cells in peripheral blood smear images by concavity analysis. Combinatorial image analysis lecture notes in computer science DOI: 10.1007/978-3-31907148-0_21 Volume 8466, (2014) 238-249 .

6. Tsai, A., et al.: A shape-based approach to the segmentation of medical imagery using level sets. IEEE Transactions on Medical Imaging, 22(2) (2003) 137

7. P. Sadaphal, J. Rao, G. W. Comstock, M. F. Beg, Image processing techniques for identifying Mycobacterium tuberculosis in Ziehl-Neelsen stains, The International Journal of Tuberculosis and Lung Disease. INT J TUBERC LUNG DIS 12(5):579–582, 2008.

8. Automated Detection of Early Lung Cancer and Tuberculosis Based on XRay Image Analysis, Proceedings of the 6th WSEAS International Conference on Signal, Speech and Image Processing, Lisbon, Portugal, 2006, pp.110 – 115.

9. Automatic identification of Mycobacterium tuberculosis by Gaussian mixture models. Journal of Microscopy, Vol. 223, Pt 2 August 2006, pp. 120–132 accepted 2006. Journal compilation 2006 The Royal Microscopical Society.

10. Automatic identification techniques of tuberculosis bacteria, Instituto de Optica (CSIC), Serrano 121 28006, Madrid, Spain b CICESE, Division de Fisica Aplicada, Dept. de Optica, Carretera Tijuana-Ensenada, B.C Mexico.

11. The Automated Identification of Tubercle Bacilli using Image Processing and Neural Computing Techniques, Proceedings of the 8th International Conference on Artificial Neural Networks, vol 2, pp 797-802 Skövde, Sweden, 1998.

12. Feminna Sheeba*, Robinson Thamburaj, Joy Sarojini Michael, P Maqlin and Joy John Mammen Segmentation of sputum smear images for detection of tuberculosis bacilli BMC Infected diseases 2012 12(Suppl 1):O14 doi:10.1186/1471-2334-12-S1-O14

13. Gonzalez, Woods, Digital Image Processing 3ed (Prentice Hall 2008)

14. Gonzalez, Woods,Eddins, Digital Image Processing using Matlab (Pearson Education 2009)

Morphology Based Detection of Abnormal Red Blood Cells in Peripheral Blood Smear Images

S. Kulasekaran[1], Feminna Sheeba[2], Joy John Mammen[3], B. Saivigneshu[2], and S. Mohankumar[2]

[1] Healthcare Technology Innovation Centre, Chennai, India
[2] Department of Computer Science, Madras Christian College, Chennai, India
[3] Department of Transfusion Medicine & Immunohaematology, Christian Medical College, Vellore, India

Abstract— **Red blood cells are the most abundant type of blood cells in the human body, delivering oxygen to body tissues.The count of these vital cells is often the first step done in analyzing a patient's pathological condition. Normal RBC's are biconcave in shape with a central pale area, and any deviation in size, shape, volume, structure or color represents an abnormal cell. Such abnormalities are detected by viewing the blood-smear images through a microscope, a time consuming and error-prone method. This process can be automated by analyzing the individual cells in a peripheral blood smear image and segmenting the cells using appropriate segmentation techniques. The proposed study aims at Morphology-based detection of abnormal red blood cells in peripheral blood smear images, based on their size and shapes. Abnormalities such as Anisocytosis, Macrocytosis and Microcytosis are detected based on the size of the RBCs. Variations in the shape of RBCs couldindicate various abnormalities. Convex hull based detection of speculated RBCs, is carried out in Acanthrocytosis and Echinocytosis. The condition Eliptocytosis, where some of the RBCs turn elliptical is detected using Houghman Transform. In the abnormality called Rouleaux the RBCs appear as stack of coins, which are detected by applying a watershed algorithm to individual stacks and counting the number of cells in the stack. Sickle cell anemia is another common condition in people, where few RBCs are sickle or crescent shaped and this shape is determined using the roundness factor. Codocytes resemble a bull's eye, and can be identified by examining if the segmented RBCs have rounded areas within the cell. Dacrocytes are tear drop RBCs, which can be detected by analyzing the extreme points of the cell. The experiment was conducted for fifty images and the success rate achieved was 80%.**

Keywords— **Red blood cells, Morphology, Houghman Transform, Watershed, Extreme points.**

I. Introduction

A peripheral blood smear image has a rich representation of White Blood Cells (WBCs), Red Blood Cells (RBCs) and Platelets. Red blood cells are the most abundant type of blood cells delivering oxygen to body tissues, the count of which is often the first step in analyzing a patient's pathological condition. Due to various pathological disorders, certain RBCs change their size, shape, volume, structure or color and become an abnormal cell. Such abnormalities are detected by viewing the blood-smear images through a microscope, which is a time consuming and error-prone process. Thus there is a need for automation ofanalysis of individual red cells in the blood smear image and segmentation of such cells using appropriate techniques. There are various segmentation techniques that are used to segment all types of cells in the blood smear images. These vary from simple intensity based segmentation to fuzzy, neural network or Tissue-like P Systems based segmentation. Intensity based segmentation was carried out to segment White Blood Cells in the blood smear images [1]. A Marker Controlled Watershed Transform was used to segment the nuclei of WBCs [2] and a Distance Transform based Watershed Transform [3] was used to segment the RBCs. Splitting of overlapping cells was possible to a greater extent when Watershed Transform was used [4] . In segmentation methods based on region growing, the homogeneity criterion is achieved by using the gray value and standard deviation of the regions. Certain segmentation methods use the snake algorithm [5] and few others used Tissue-like P Systems [6]. Fuzzy logic can also be applied to segment WBCs [7] [8].Appropriate segmentation techniques were also used to segment abnormal RBCs based on their size and shapes. Feature extraction was used to decide the type of abnormal RBCs [9].

The proposed study aims at a Morphology-based detection of abnormal red blood cells in peripheral blood smear images, based on their sizes and shapes. The rest of the paper is organized as Background in section II, Methodology in section III, Results and Findings in section IV and Conclusion in section V.

J. Goh and C.T. Lim (eds.), *7th World Congress on Bioengineering 2015*,
IFMBE Proceedings 52, DOI: 10.1007/978-3-319-19452-3_16

II. BACKGROUND

The four aspects in evaluating if a red blood cell is normal or abnormal are its size, color, shape and inclusion bodies. Normal RBCs are pink to reddish in color, 6-8 microns in diameter and are round and biconcave disc shaped with high surface to volume ratio, helping them to be flexible and bendable as they pass through capillaries and allowing them to absorb greater amounts of oxygen during their circulation throughout the blood. Any variations in these characteristics of RBCs are an abnormality. Few of such abnormal RBCs based on their size and shape are discussed below:

- *Anisocytosis*: is a variation in size of RBCs, which may be an indication of anemia [10].
- *Microcytosis:* is the abnormality where the size of the red blood cells becomes less than 6 micrometers and has less amount of hemoglobin. This is due to iron deficiency, defective hemoglobin, porphyrin synthesis, Thalassaemia, Sideroblastic anemia, or Lead poisoning [10].
- *Marocytosis*: which are larger than normal RBCs, is due to the occurrence of immature red blood cells, vitamin B12 or folate deficiency and have a blue-grey coloration [10].
- *Ovalocytes:* are oval shaped red blood cells, which are due to Thalassaemia major and is hereditary.
- *Elliptocytosis:* is a disorder passed down through families in which the red blood cells are abnormally shaped. Elliptocytosis affects about 1 in every 2,500 people of northern European heritage. It is more common in people of African and Mediterranean descent. The symptoms arefatigue, shortness of breath and yellow skin and eyes (jaundice) [10].
- *Codocytes*: are nucleated RBCs, which are immature red blood cells, the presence of which indicate accelerated erythropoiesis or severe bone marrow stress in an adult [10].
- *Acanthocytes and Echinocytosis:* are spheroid RBCs with a few large spiny projections. This can be seen in patients with alcoholic cirrhosis, who have inability to fully absorb dietary fats through the intestines or severe liver disease [10].
- *Dacrocytes:* are RBCs with the shape of a tear drop.These abnormal cells are due to severe iron deficiency, Thalassemia major, bone marrow fibrosis orMegaloblastic anemia [10].
- In *rouleax formation*, the RBCs are arranged in the form of a coin stack. It is due to the increase in the blood concentration of fibrinogen, globulin and paraprotein [10].
- In *sickle anemia* the RBCs are sickle shaped. This is a Hb-S disease [10] [11].

In order to analyze these abnormal cells and diagnose various disorders segmentation of such cells is necessary.

III. METHODLOGY

A. Image Pre-processing

The test images are first converted to binary images using Otsu's thresholding. The objects that are touching the borders are eliminated and holes filled in them. Morphological operations like dilation and erosion are applied on to all the blobs in order to smoothen them and strengthen their borders [12][13]. The WBCs are eliminated based on their sizes as they are visibly larger than RBCs. Detection of abnormal cells are discussed in the following sub-section.

B. Detection of Abnormal Red Blood Cells

Microcytes and Polychronasisis are detected by analyzing the size of the RBCS.The elliptical and oval shaped RBCs, which are Elliptocytes and Ovalocytes respectively are identified using Houghman Transform [14]. It is seen in codocytes, which are nucleated RBCs, after segmentation the holes in those cells are visible even after pre-processing. Acanthocytes & Echinocytosis are identified by generating a convex hull for each of the RBCS and analyzing the distance between the points in the convex hulls. In order to detect the cells with the rouleax formation, a distance transform based watershed algorithm is applied to the individual stacks so that they split. The split cells are then counted and if the count is above a particular value it is inferred that such RBCs have Rouleax formation. In sickle anemia the RBCs are sickle shaped. These sickle shaped cells are detected using the roundness factor of blobs. The tear dropped RBCs called dacrocytes are detected by obtaining the extreme points of each of the cell.

IV. RESULTS AND FINDINGS

A prototype application was developed using MATLAB for testingthe images used for the study. 50 blood smear images with various RBC abnormalities were tested. The images were first converted to grey images and necessary algorithms were applied to them and the results were obtained. A sample set of images, which were used for testing, their pre-processed binary images and the resultant segmented images of the abnormal cells are shown in Fig. 1, Fig. 2, Fig.3 and Fig.4.

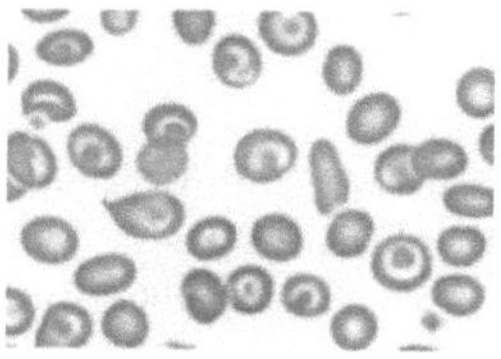

(a) Image with Bull's eye

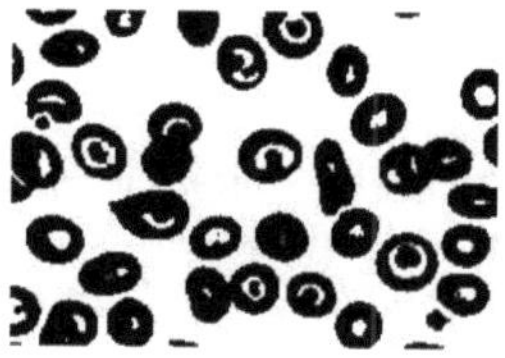

(b) Bull' eye - binary image .

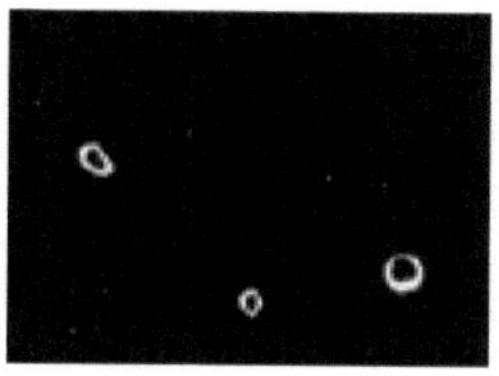

(c) Resultant image – Bulls Eye

Fig. 1.

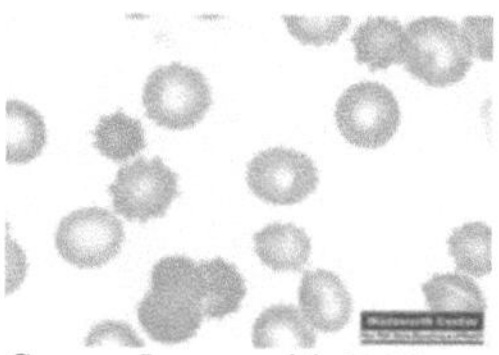

(a) Source Image with Echinocytes

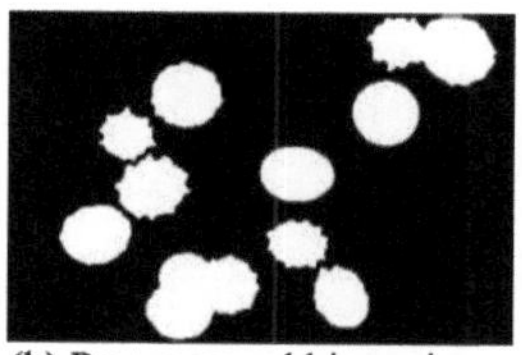

(b) Preprocessed binary image with Echinocytes

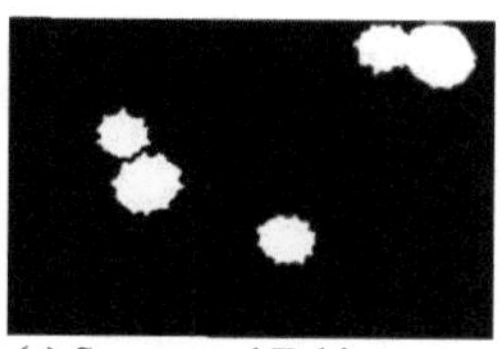

(c) Segmented Echinocytes

Fig. 2

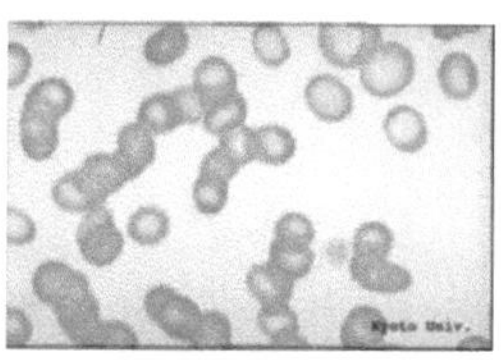

(a) Source Image with RouleauxFomation

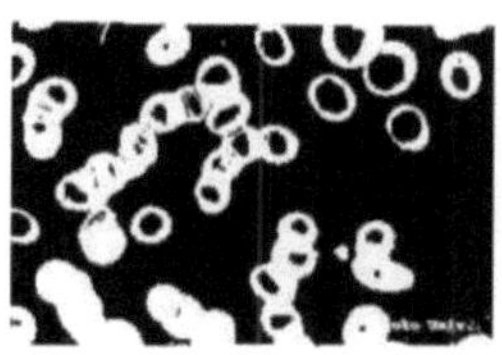

(b) Binary image

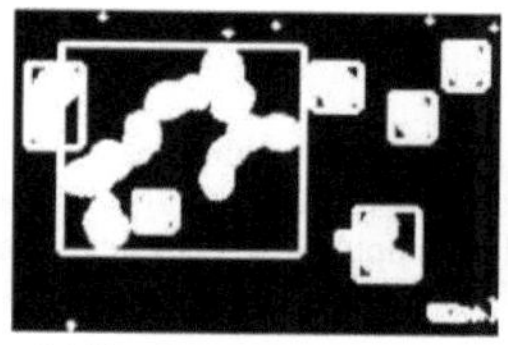

(c)The bounding box of all objects

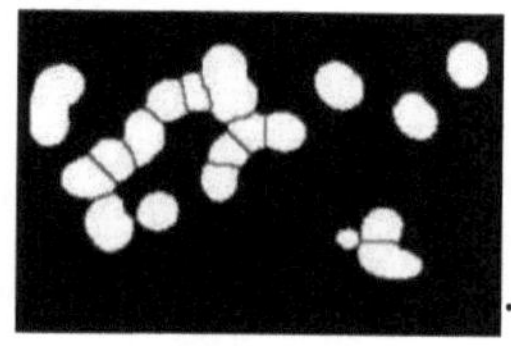

(d) After Watershed Transform

Fig. 3.

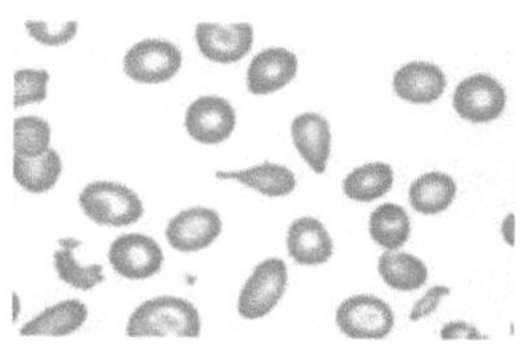

(a) Source Image with Tear drops

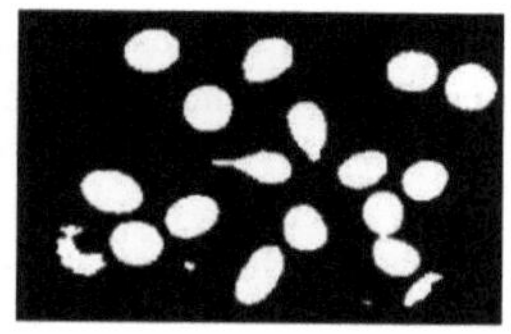

(b) Binary image

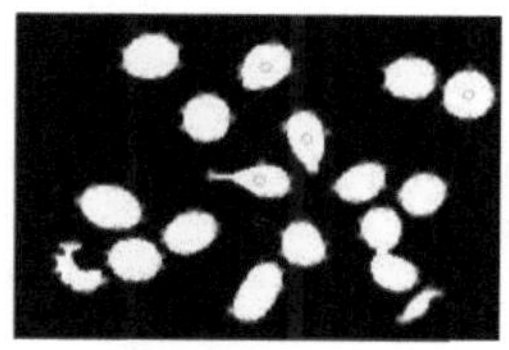

(c) Identified Tear drops using Extrema

Fig. 4

V. Conclusions

The abnormal RBCs were segmented based on their sizes and shapes. The success rate of the study was 80%. This is due to segmentation using geometric measures. Hence the future work involves segmentation by shape modeling.

VI. Acknowledgment

The authors would like to thank the Department of Transfusion Medicine and Immunohaematology, Christian Medical College, Vellore, India for providing them with sample images used for the study.

References

1. Feminna, S., H.M.T. Thomas, And J.J. Mammen, Segmentation And Reversible Watermarking Of Peripheral Blood Smear Images. Proc. Ieee Conference On Bio Inspired Computing: Theories And Applications 2010. 2: 1373-6
2. FeminnaSheeba, Robinson Thamburaj, Joy J. Mammen, Hannah M. Thomas, Atulya K. Nagar, White Blood Cell Segmentation And Reversible Watermarking, Proc. The IASTED International Symposia on Imaging and Signal Processing in Healthcare and Technology (ISPHT 2011), pp. 63-68
3. Feminna, S., Robinson T., J.J. Mammen, H.M.T. Thomas and Atulya K. Nagar.: Splitting of overlapping cells in peripheral blood smear images by concavity analysis. Combinatorial image analysis lecture notes in computer science DOI: 10.1007/978-3-319-07148-0_21 Volume 8466, (2014) 238-249
4. Hassan Khajehpour et al, Detection and Segmentation of Erythrocytes in Blood Smear Images Using a Line Operator and Watershed Algorithm, Journal of Medical Signals and Sensors, Vol 3, Issue 3, 2013
5. Ongun, G., et al., An Automated Differential Blood Count System. Proc. of IEEE EMBC 2001.
6. Feminna, S., Robinson, T., Nagar, A.K., Mammen, J.J.: Segmentation of peripheral blood smear images using tissue-like P-Systems. IJNCR–BICTA 2011 Special Issue 3(1), (2012) 16–27
7. NiponTheera-Umpon, Patch-Based White Blood Cell Nucleus Segmentation using Fuzzy Clustering. ECTI Transactions on El. Eng. and Electronics Communications, 2005. 3: 15-19.
8. Chinwarapath, S., et al., A Modified Fuzzy Clustering for White Blood Cell Segmentation. Proc. of the 3rd Int. Symposium on Biomedical Engineering ISBME 2008,: 356-359.
9. Jameela Ali et al, Isolation and Classification of Red Blood Cells in Anemic Microscopic Images, International Journal of Medical, Health, Biomedical and Pharmaceutical Engineering Vol:8, No:10, 2014
10. Shagana. J.A.,Diagnostic Cells in the Peripheral Blood Smear, Journal of Pharmaceutical Sciences and Research,Vol. 6(4), 2014, pp. 213-216, ISSN:0975-1459
11. Aruna N.S et al, Edge Detection of Sickle Cells in Red Blood Cells, International Journal of Computer Science and Information Technologies (IJCSIT), Vol. 5 (3) , 2014, pp. 4140-4144, ISSN:0975-964
12. Gonzalez, Woods, Digital Image Processing 3ed (Prentice Hall 2008)
13. Gonzalez, Woods,Eddins, Digital Image Processing using Matlab (Pearson Education 2009
14. Chung-Fang Chien, Yu-Che Cheng, and Ta-Te Lin*, Department of Bio-Industrial Mechatronics Engineering, National Taiwan University,No. 1, Sec. 4, Roosevelt Road, Taipei 106, Taiwan. *Corresponding author: m456@ntu.edu.tw,Robust ellipse detection based on hierarchical image pyramid and Hough transform, Vol. 28, No. 4 / April 2011 / J. Opt. Soc. Am. A 581

Medical Image Encryption Using Block-Based Scrambling and Discrete Wavelet Transform

S. Kulasekaran[1], Feminna Sheeba[2], B. Saivigneshu[2], C. Dayalan[2], and P. Cyril Rex[2]

[1] Healthcare Technology Innovation Centre, Chennai, India
[2] Department of Computer Science, Madras Christian College, Chennai, India

Abstract— **In today's age of manifold advances in the field of medical imaging, a significant amount of sensitive and personal information related to patients is being transmitted electronically via images. With the advent of e-Health and Telemedicine in the vast field of medicine, there is a need to guarantee the authenticity and validity of the images being exchanged. The much-need security of medical images imposes the conditions of confidentiality, reliability and availability, and these can be attained by various Image Authentication methods and one such authentication is Image Encryption. The proposed work aims at an Image Encryption technique, which is a combination of Tiling, Scrambling and Image Transformation and Encryption of the image. The proposed architecture for encryption and decryption of a medical image is using a symmetric key, which gives the size of the tiles and the hash code of the image. The encryption algorithm divides the image into tiles of arbitrary size, scramble them using a scrambling technique and transform the scrambled image using Discrete Wavelet Transform (DWT). The hash code in the key is used to find out if tampering has taken place during transmission of the medical image.**

Keywords— **Red blood cells, Morphology, Houghman Transformation, watershed, extreme points.**

I. Introduction

The recent advances in the field of medical imaging and transmission of images across networks, include transmission of significant amount of sensitive and personal information of patients as images, With the advent of e-health and telemedicine, there is a need for authentication and validation of images being exchanged across networks. Moreover such medical images have to be confidential, reliable and securely available which can be attained with image authentication. Image authentication verifies the originality of an image by detecting malicious manipulations. This can be attained by various ways including encryption, cryptography, digital signatures and watermarking.

A method is suggested which can distinguish malicious manipulations from JPEG lousy compression regardless of how high the compression ratio is [1]. An image scrambling encryption algorithm which makes use of one-dimensional chaos system for shuffling the pixel bits with the bit-plain of size M×8N is suggested in [2]. Detailed cryptanalytic results suggest that the image scrambling scheme can only be used to realize perceptual encryption [3]. Cryptography enables significant information to be stored or transmitted over non-secure networks, so that only authorized recipients can read it. Message authentication techniques are used in image integrity and authentication systems. Hash functions, private or public key systems and digital signatures are also used. The problem of a digital signature is that a conventional signature is physically attached to the signed document and that a conventional signature is authenticated by comparing it with a certified one [4]. In the recent times image authentication is achieved by the technique of watermarking. A spatial domain watermarking where the embedding is done in the non-region of interest is suggested in [5]. In order to gain robustness against attacks a frequency domain watermarking was suggested in [6]. In order to keep the images in perfect condition without any loss of information, the original image should be recovered upon the extraction of the embedded watermark [7] [8][9].

The proposed work aims at authentication of images, where the image is split into equal sized blocks, scrambling of the blocks, transforming the image and reversing the process at the receiving end. A key is generated during the watermarking process, which is used as a secret key for reversing.

II. Methodology

The source image is first split into various blocks or tiles with an arbitrary size. Scrambling algorithms which involves interchanging the first and third quadrants in each block is applied. This scrambled image is then transformed using the discrete wavelet transform. This transformed image is sent across the networks. In this process, a key is generated. This key is used to restore the original image in the receiving end. This is done by inversing the process by

J. Goh and C.T. Lim (eds.), *7th World Congress on Bioengineering 2015*,
IFMBE Proceedings 52, DOI: 10.1007/978-3-319-19452-3_17

using the secret key that was generated during transmission. The images when transmitted in networks may be tampered by unauthorized users. This tampering can be detected by obtaining the hash value of the image and matching it with the one that is in the key.

III. Results and Findings

A prototype application was developed do test the image authentication method. 50 images were used for testing. One of the source images used is shown in Fig 1. The image was converted to a square image by adding black pixels to the width and height of the image as shown in Fig.2. The image was divided into various blocks of size 16 x 16. Each of the blocks was scrambled by interchanging the first and fourth quadrants. This is shown in Fig.3. The scrambled image was then transformed using DWT transformation [6] [10][11][12], which is shown in Fig. 4. A key was generated with the block size and the hash code of the image.

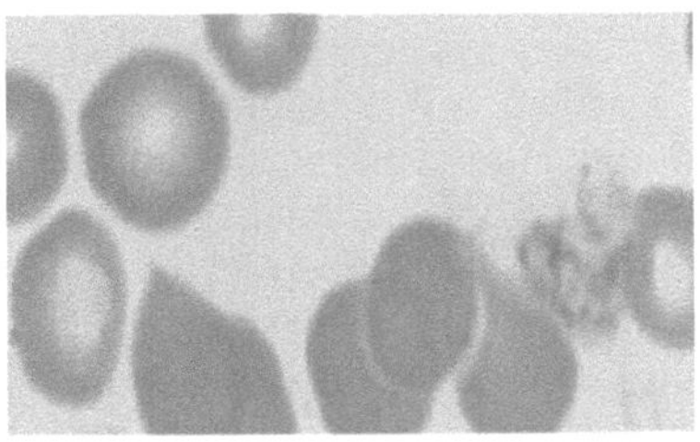

Fig.1 Source image

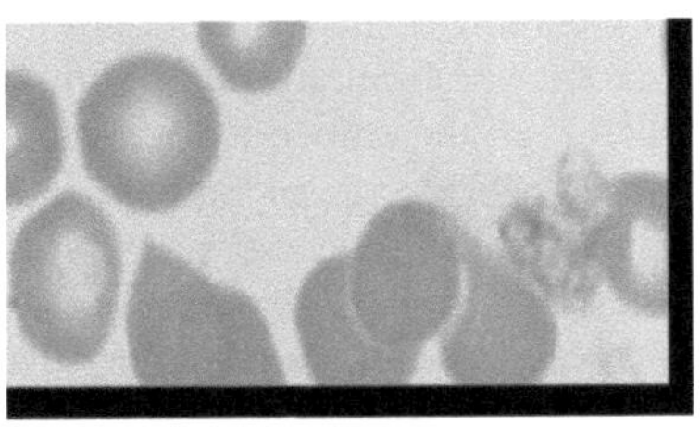

Fig.2 Square Image

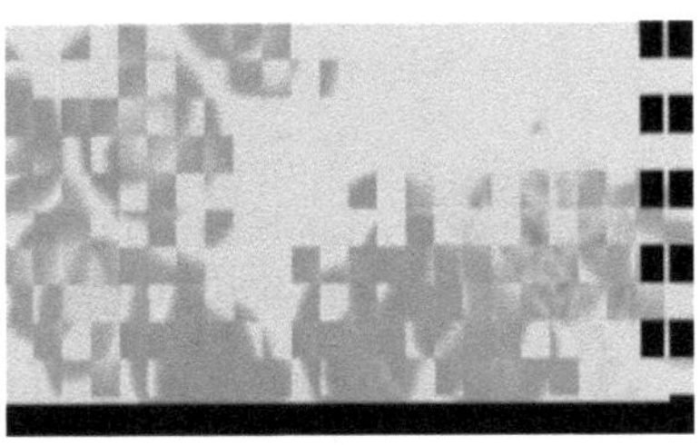

Fig.3 Scrambled Image

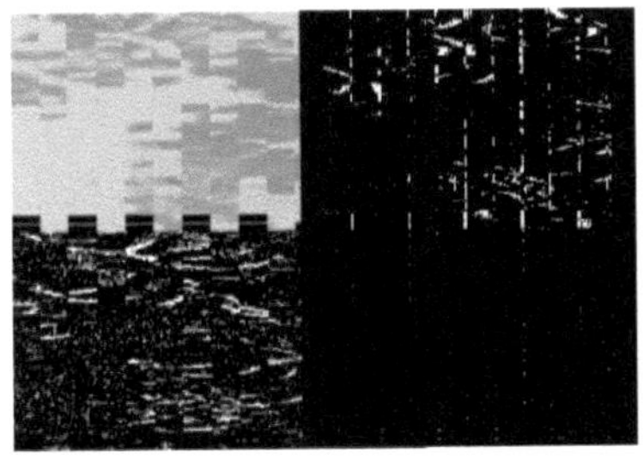

Fig.4 DWT Transformed Image.

In the receiving the image was decrypted using the secret key. Inverse DWT was applied to get the scrambled image, which is shown in Fig. 5. This image was then unscrambled which is shown in Fig 6. The hash code of the unscrambled was calculated and matched with the hash value in the Key. It was found that both were the same meaning that the image was not tampered.

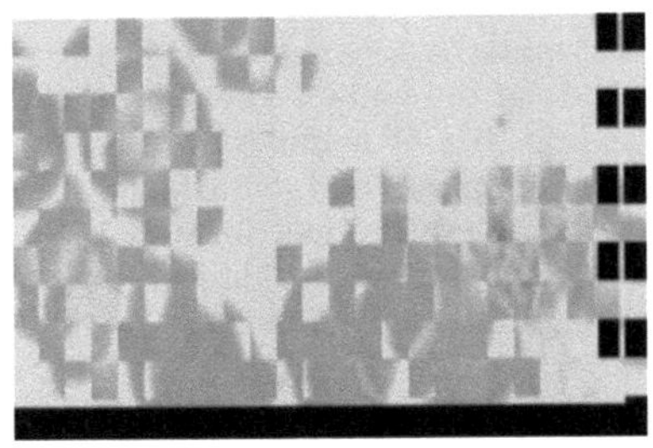

Fig.5 Scrambled Image

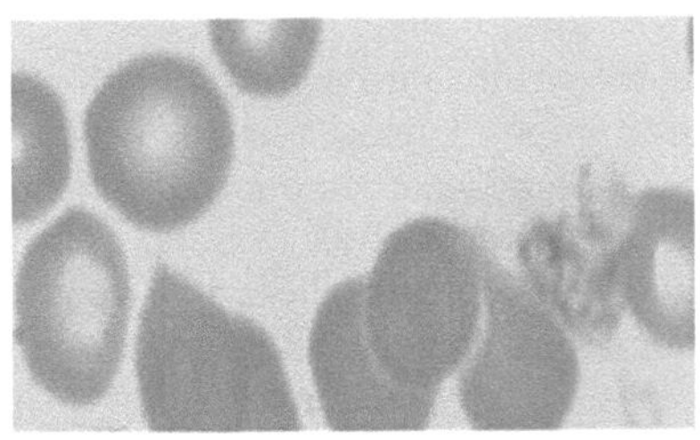

Fig.6 Unscrambled Image

IV. Conclusion

Image authentication is the need of the hour as images are sent across the networks for taking second opinion from clinicians. Also the images have to be stored for future study. Hence tampering of images can be found out using the image authentication method discussed. The method gave expected results which could be easily incorporated in applications involving telemedicine and e-health.

REFERENCE

1. Jianfeng Lua , Meng Wanga , Junping Daib,Qianru Huanga , Li Lia,* and Chin-Chen Changc d, Multiple Watermark Scheme based on DWT-DCT Quantization for Medical Images Journal of Information Hiding and Multimedia Signal Processing 2015 ISSN 2073-4212 Ubiquitous International Volume 6, Number 3, May 2015

2. Liang Zhao, Avishek Adhikari, Di Xiao, Kouichi Sakurai, Cryptanalysis on an Image Scrambling Encryption Scheme Based on Pixel Bit, Conference: Digital Watermarking - 9th International Workshop, IWDW 2010, Seoul, Korea, October 1-3, 2010, DOI: 10.1007/978-3-642-18405-5_5

3. Shujun li, chengqing li, kwok-tung lo, guanrong chen Cryptanalysis of an image scrambling scheme without bandwidth expansion, Circuits and Systems for video technology IEEE transactions, volume:18, issue: 3, pp. 338 – 349, 2008, doi: 10.1109/tcsvt.2008.918116

4. Adil Haouzia & Rita Noumeir, Methods for image authentication: a survey Multimed Tools Appl (2008) 39:1–46, Springer Science + Business Media, LLC 2007, DOI 10.1007/s11042-007-0154-3

5. Feminna, S., H.M.T. Thomas, and J.J. Mammen, Segmentation and Reversible Watermarking of Peripheral Blood Smear Images. Proc. IEEE Conference on Bio Inspired Computing: Theories and Applications 2010. 2: 1373-6.

6. Feminna, S., Robinson T., J.J. Mammen , H.M.T. Thomas and Atulya K. Nagar.: White Blood Cell Segmentation and Watermarking. In: Proceedings of the IASTED International Symposia Imaging and Signal Processing in Healthcare and Technology, ISPHT 2011,Washington DC, USA (2011).

7. M.Sreerama Murty, D.Veeraiah, 3A.Srinivas Rao, Digital Signature and Watermark Methods For Image Authentication using Cryptography Analysis, Signal & Image Processing : An International Journal (SIPIJ) Vol.2, No.2, June 2011

8. Imen Fourati Kallel, Mohamed Salim Bouhlel, Jean-Christophe Lapayre, "Improved Tian's Method for Medical Image Reversible Watermarking", GVIP Journal, Volume 7, Issue 2, August 2007.

9. Feminna, S., Robinson T., Joy John Mammen and Atulya K. Nagar, Detection of plasmodium falciparum in peripheral blood smear images. In: Proceedings of BICTA 2012, 202 (2013) 289–298

10. FeminnaSheeba,Robinson Thamburaj,Atulya K. Nagar, Image Authentication using Reversible Watermarking, Proc. *Conference* on Mathematics in Engineering & Business Management (*ICMEB 2012*) Volume 2. pp.408-410

11. Gonzalez, Woods, Digital Image Processing 3ed (Prentice Hall 2008)

12. .Gonzalez, Woods,Eddins, Digital Image Processing using Matlab (Pearson Education 2009)

Retinal Vasculature Segmentation in Smartphone Ophthalmoscope Images

S. Saranya Devi[1], K.I. Ramachandran[1], and Ashish Sharma[2]

[1] Amrita School of Engineering, Amrita Vishwa Vidyapeetham, Coimbatore, India
[2] Consultant Retina and Head Research, Lotus Eye Hospital and Institute, Coimbatore, India

Abstract— **Retinal imaging system assists ophthalmologists to diagnose the diseases and to monitor the treatment processes. Conventionally, fundus retinal images are obtained from expensive systems like fluorescein angiography and fundus photography but these systems are large tabletop units and can only be handled by trained technicians. Hence, this study reports a low cost, compact and user friendly smartphone ophthalmoscope to perform indirect ophthalmoscopy. By using this system, initial and periodic screening of retina (both center and periphery regions) becomes easier. Traditionally, retinal diseases are diagnosed by manual observations of fundus images and it is a time consuming process. So, automatic retinal disease diagnosing systems are introduced by extracting the essential features of the fundus retinal images. One of the most essential features of the retina is the blood vessels as its morphological changes helps in diagnosing the retinal diseases. Hence, in this study blood vessels are extracted from smartphone ophthalmoscope (SO) images using level set method to develop an automatic retinal disease diagnosing systems for ophthalmologists. The performance of the retinal vasculature segmentation algorithm is compared and analyzed on DRIVE database of retinal images and on smartphone ophthalmoscope images using the measures like sensitivity, specificity and accuracy level.**

Keywords— **Smartphone ophthalmoscope, K-means, Total variation filtering, Bottom-hat transformation, Level set technique**

I. INTRODUCTION

Retina is the light sensitive layer of an eye which helps in vision. Retina can be imaged using fundus photography and the images from this system provides complete information about the features like optic disk, macula, fovea and blood vessels and thus it helps in identifying the retinal diseases. By extracting these features early stages of the retinal diseases can be identified. To diagnose the early stages of the disease, periodic screening of the retina is required. As the cost of imaging the retina becomes costlier using the available retinal imaging systems, periodic screening of the retina becomes difficult by common people. Hence, in this study to reduce the cost of imaging the retina a low cost smartphone ophthalmoscope is introduced after referring to the research conducted by Haddock et al. [1] and Myung et al. [2]. The traditional way of diagnosing the diseases by physician using the fundus images consumes time and involves human errors. This ended up in developing an automatic disease diagnosing system. This system extracts the essential features of the retina and helps in identifying the morphological changes developed in them. Hence, this study reports that the blood vessels are extracted to diagnose the severity of the retinal diseases. This can be done by identifying the morphological changes of blood vessels like vessel shape and length. The retinal image acquired from the SO is affected by noises and hence requires noise filters to remove them. In literature, it is reported that the non-linear filters are better denoising techniques than linear filters as they preserves edge details. Hence, a popular non-linear total variation denoising technique is selected for this study and observed that the noises are smoothened out in the flat regions and preserves fine edges even under low signal to noise ratio [4]. Recently, to segment the blood vessels level set methods are used. In this study, a level set method [7] which considers the objects region information to extract the blood vessels is selected and tested for its performance.

II. MATERIALS AND METHODS

In this section, we present the smartphone ophthalmoscope system setup and elaborate on the retinal vasculature segmentation algorithm.

A. System Design

In this study, smartphone ophthalmoscope is developed using a smartphone in conjunction with a 20D condensing lens. Here, smartphone iPhone 4s is used to capture and store the retinal images. And the 20D condenser lens is used for obtaining 20° field of view to image the retina. It is being proved that the smartphone ophthalmoscope can be used as an initial screening tool for diagnosing the retinal diseases [3]. This system uses the coaxial light source of the iPhone and captures the digital image of the retina. This system basically works like an indirect ophthalmoscope. Fig. 2 (a) represents the setup developed to image the retina.

J. Goh and C.T. Lim (eds.), *7th World Congress on Bioengineering 2015*,
IFMBE Proceedings 52, DOI: 10.1007/978-3-319-19452-3_18

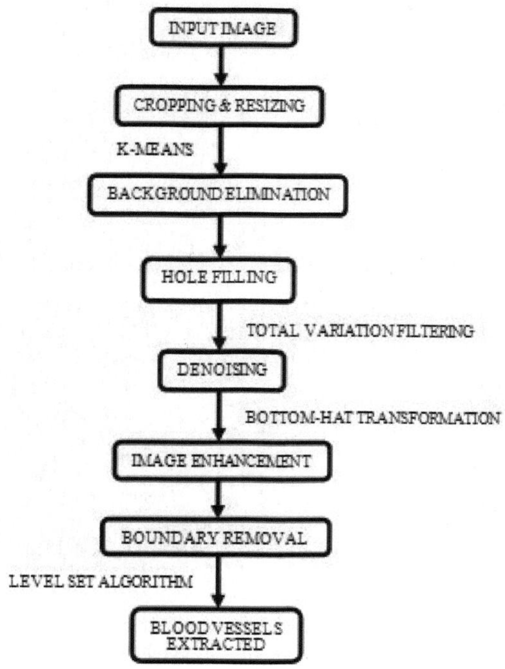

Fig. 1 Process flow to segment blood vessels

Here, approximately the distance between the lens and the smartphone is fixed to 11.5 cm and it can be varied if necessary. Always, the distance between the object and the lens is kept greater than 5 cm to obtain magnified fundus images. Imaging the retina using this setup is possible only by dilating the pupil of the subjects.

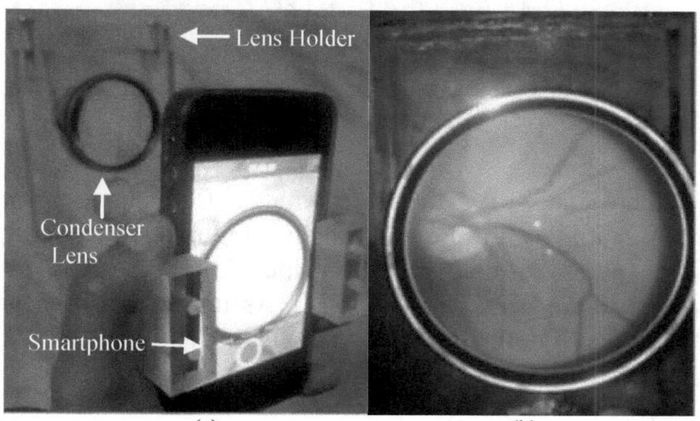

(a) (b)

Fig. 2 (a) Smartphone ophthalmoscope setup (b) A smartphone ophthalmoscope retinal image

B. Retinal Vasculature Segmentation Algorithm

The images acquired from this system are processed in MATLAB R2013a software to segment the blood vessels. Fig. 1 gives the process flow to segment the blood vessels from the database and from the smartphone ophthalmoscope images (SO).

Preprocessing Stage: Preprocessing steps are performed to improve the quality of the fundus retinal images. Thus, in this study the images acquired are cropped and resized to half the size of the original image to reduce the computation. Then, the unwanted background details were removed using k-means algorithm [5]. Here, Fig. 3 shows the background eliminated images. After performing this technique holes are being introduced in the images. Therefore, these holes are filled by using hole filling operation. The resulting image after this process might be affected with noises and hence the total variation filtering technique is used to remove these noises. The denoised image is then enhanced by applying the bottom-hat morphological transformation to enhance the blood vessels alone.

Background Elimination: K-means algorithm is used for background elimination. It is one of the unsupervised clustering techniques used to group the given random data's into different clusters.

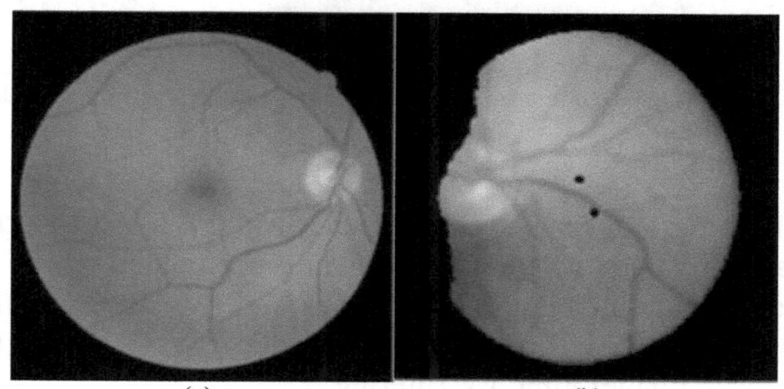

(a) (b)

Fig. 3 Background eliminated images (a) Database image & (b) Smartphone ophthalmoscope image (holes might be generated in this step)

This algorithm extracts only the colour information (RGB). Before performing k-means clustering technique, given RGB image is transformed into L*a*b* color space such that only the colour information are grouped together and helps in the clustering process. After this process, clusters are randomly initialized. Then the data sets are labeled to the closer cluster. Finally, index is calculated for each cluster. Every data set is labeled with its cluster index. And the last two steps are repeated until there is no more move of the data points to different clusters.

Total Variation Minimization: One of the most popular non-linear denoising techniques is the total variation (TV) denoising filter. TV is based on the principle that images with high energy details have high total variation and this variation of the image can be reduced by finding the closeness to the original image [4].

$$J(u) = R(u) + \frac{\lambda}{2} ||Tu_{ij} - y||_2^2 \quad (1)$$
$$R(u) = \sum_{1 \le i,j \le N} |\nabla u_{i,j}| \quad (2)$$

Here, y is the given noisy image; T is the linear operator and u_{ij} is the unknown image. So, to get denoised image u_{ij} both the TV-norm i.e. R (u) and the fidelity term have to be minimized. But, when both the terms are minimized simultaneously, proper denoising cannot be achieved. Hence, by adjusting the regularization parameter, λ both the TV-norm and fidelity term can be minimized such that the resulting denoised image is as close as that of the original image.

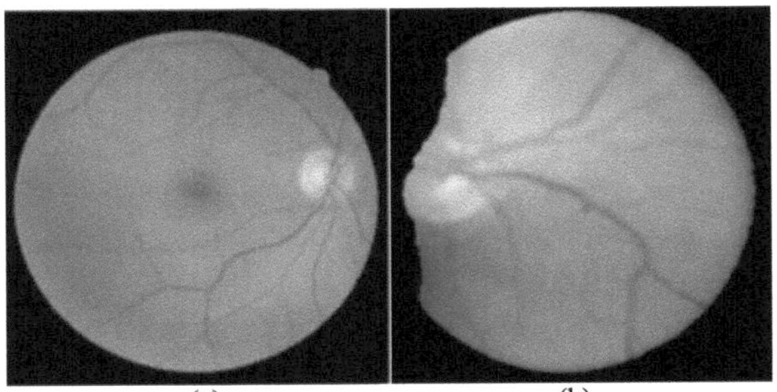
(a) (b)

Fig. 4 Denoised images (a) Database image & (b) Smartphone ophthalmoscope image

In order to verify the efficiency of the TV algorithm quantitatively, PSNR values are computed for 12 smartphone ophthalmoscope images and it is observed that the average PSNR value as 39.83 dB. Fig. 4 shows the denoised images after TV filtering.

Image Enhancement: To extract the blood vessels effectively, bottom-hat transformation is performed. This helps in enhancing the blood vessels from the other details in an image as they are the darkest region in an image. After performing bottom-hat transformation [6], only the green channel is extracted as it contains almost all the blood vessel details in an image. Fig. 5 shows the enhanced database and smartphone ophthalmoscope images after performing bottom-hat transformation and green plane extraction.

Retinal Vasculature Segmentation Algorithm: In recent years, segmentation is performed using level set methods. This technique was first introduced by Sethian et.al and then improvised by Chan et al. [7]. This method segments (Fig. 6 segmented images) an object by finding minimization of energy. The energy equation is defined by this algorithm is,

$$E(C, c_1, c_2) = \mu\, Length(C) + \nu\, Area(inside(C)) + \lambda_1 \int_{inside\,(C)} |u_0(x,y) - c_1|^2\, dxdy + \lambda_2 \int_{outside\,(C)} |u_0(x,y) - c|^2\, dxdy \quad (3)$$

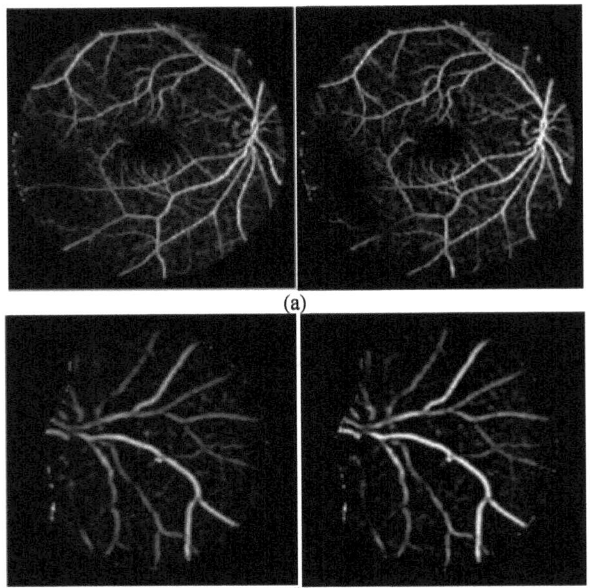
(a)
(b)

Fig. 5 Bottom-hat transformed image & green plane extracted image (a) Database & (b) Smartphone ophthalmoscope image

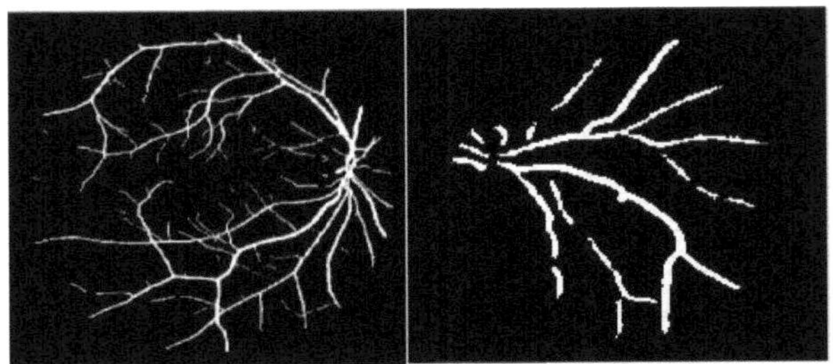
(a) (b)

Fig. 6 Shows, the segmented blood vessels after applying Level set method in (a) Database image & (b) SO image.

Where, $u_0(x,y)$ is the given image, C is the evolving curve and c_1, c_2 are the constants representing the average pixel values inside C and outside C. $\mu \ge 0, \nu \ge 0$ and $\lambda_1, \lambda_2 > 0$ are fixed parameters [7]. Assuming that the level set function ϕ defines the evolving curve C.

$$C = \phi(x,y) = 0$$
$$inside(C) = \phi(x,y) > 0$$
$$Outside(C) = \phi(x,y) < 0 \quad (4)$$

By using the known Heaviside function, H and Dirac measure, δ_0 energy equation is rewritten as [7],

$$E(\phi, c_1, c_2) = \mu \int_{\Omega} \delta_0(\phi(x,y))\, |\nabla\phi(x,y)| dxdy + \nu \int_{\Omega} H(\phi(x,y)) dxdy + \lambda_1 \int_{\Omega} |u_0(x,y) - c_1|^2\ H(\phi(x,y)) dxdy + \lambda_2 \int_{\Omega} |u_0(x,y) - c_2|^2\ H(1 - \phi(x,y)) dxdy \quad (5)$$

III. Results

The performance of the proposed algorithm is tested with DRIVE database. To obtain the statistical quality metrics such as sensitivity (Sen), specificity (Spe) and accuracy (Ac) the segmented blood vessels using the proposed algorithm are compared with the ground truth of the respective fundus images in the DRIVE database.

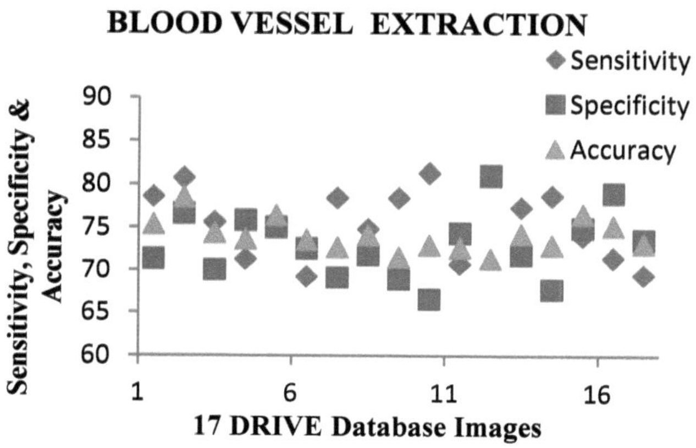

Fig. 7 Statistical quality metrics for 17 DRIVE database images

TP refers to a pixel labeled as vessel by both the proposed algorithm and the ground truth, while TN refers to a pixel that is considered to be non-vessel by both.

$$Sen = \left(\frac{TP}{TP + FN}\right) X\ 100, \qquad Spe = \left(\frac{TN}{TN + FP}\right) X\ 100,$$

$$Ac = \left(\frac{TP + TN}{TP + FP + TN + FN}\right) X\ 100 \quad (7)$$

FN refers to pixels of vessels in the ground truth and missed by the proposed algorithm, and FP refers to pixels falsely considered by the proposed algorithm as vessel. Statistical metric results for 17 DRIVE database images are reported in the Fig. 7. This figure reports that the 17 DRIVE database images has the sensitivity range from 69% to 81%, specificity range from 66% to 76% and accuracy range from 71% to 78%. The average accuracy of the algorithm is 74%.

IV. Conclusions

In this study, smartphone ophthalmoscope is developed and used to image the center and periphery region of the retina. This work also reports that the level set algorithm is used to extract the blood vessels automatically both in database and SO images. Presently, we are trying with different blood vessels enhancement techniques to improve the accuracy of the algorithm.

Acknowledgment

I would like to thank Dr. N. Madhu Mohan, Department of ECE for his patience, guidance and enlightment.

Conflict of Interest

There are no conflicts of interest.

References

1. L. J. Haddock, D. Y. Kim, and S. Mukai (2013), 'Simple, inexpensive technique for high-quality smartphone fundus photography in human and animal eyes,' *J. Ophthalmol.*, vol. 2013, pp. 1–6.
2. D. Myung, A. Jais, L. He, M. S. Blumenkranz, and R. T. Chang (2014), '3D Printed Smartphone Indirect Lens Adapter for Rapid, High Quality Retinal Imaging,' *J. Mob. Technol. Med.*, vol. 3, no. 1, pp. 9–15.
3. Andrea Russo, Francesco Morescalchi, Ciro Costagliola, Luisa Delcassi and Francesco Semeraro (2014), 'Comparison of Smartphone Ophthalmoscopy With Slit-Lamp Biomicroscopy for Grading Diabetic Retinopathy,' *American J. of Ophthalmol.*, vol. 159, no. 2, pp. 360-364.
4. M. J. Bioucas-dias, A. M. Figueiredo, and P. J. Oliveira (2004), 'Total Variation-Based Image Deconvolution: A Majorization-Minimization Approach,' *in IEEE International Conference on Acoustics, Speech and Signal Processing*, no. 2, pp. 1–5.
5. S. R. Dubey, P. Dixit, N. Singh, and J. P. Gupta (2013), 'Infected Fruit Part Detection using K-Means Clustering Segmentation Technique,' *Int. J. Interact. Multimed. Artif. Intell.*, vol. 2, no. 2, pp. 65-72.
6. A. F. M. Hani, L. I. Izhar, and H. a Nugroho (2009), 'Analysis of Foveal Avascular Zone in Color Fundus Image for Grading of Diabetic Retinopathy,' *Int. J. Recent Trends Eng.*, vol. 2, no. 6, pp. 101–104.
7. T. F. Chan and L. a. Vese (2001), 'Active contours without edges,' *IEEE Trans. Image Process.*, vol. 10, no. 2, pp. 266–277.

Author: Dr. K. I Ramachandran
Institute: Amrita Vishwa Vidyapeetham
Street: Ettimadai
City: Coimbatore
Country: Tamil Nadu
Email: ki_ram@cb.amrita.edu

Monitoring Hemodynamic Changes in Brain Infarct Area of Rats Using Diffuse Optical Imaging

Kang-Yu Chu, Chun-Wei Wu, and Jia-Jin Jason Chen

Department of Biomedical Engineering, National Cheng Kung University, Tainan, Taiwan

Abstract—**Stroke is a heterogeneous syndrome caused by various diseases resulting from disruption of cerebral blood flow (CBF) and brain tissue necrosis. Ischemic stroke is the major type of this syndrome induced by brain infarction. The infarct area will accompany with cerebral edema which could change the optical properties of the brain tissue. Near infrared spectroscopy (NIRS) was commonly used in measuring the concentration of the main chromophores in the blood – oxy- and deoxy- hemoglobin for acquiring the hemodynamic or metabolic change in the tissue. In our previous studies, we have shown the hemodynamic activity of stroke rat using frequency domain NIRS (FDNIRS). To precisely locate the ischemic infarct region within whole brain, we develop a non-invasive, diffuse optical imaging (DOI) system based on NIRS. To localize the infarct area and morphological information, the 2D scanning system with DOI technique was developed. The optical properties of the tissue i.e. absorption and scattering coefficient, representing the photon transmission model in the tissue, were applied to observe the structural changes of the brain tissue. In the pilot study, we have applied FDNIRS to monitor the changes of blood oxygen level controlled by neurovascular coupling in ischemic brain. By comparing the FDNIRS measurement result of the sham group with the middle carotid arterial occlusion (MCAO) group stroke rat on post operation day 3. We found that there is a high correlation between the reduced scattering coefficient and the infarct area tissue. The system has been applied to observe the *in-vivo* experiment of ischemic animal stroke model. Further study can develop optical techniques for monitoring and diagnosing the progress of stroke and the novel therapy methods.**

Keywords—**stroke, infarction, near infrared spectroscopy, diffuse optical imaging, hemodynamic.**

I. INTRODUCTION

Human brain plays an important role as a control center of human body. The damage of brain tissue not only caused disability, moreover, it may take our lives. The brain tissue damage could result from many causes. In this paper, we concerned about the reason which is brain vascular related syndrome — stroke. Generally, we categorized stroke into two types, one is hemorrhagic stroke, and the other is ischemic stroke. Ischemic stroke is the major type of this syndrome, which induced by brain infarction. The insufficient blood flow caused by infarction will damage the brain tissue due to lack of nutrient and oxygen. It has been reported that there were approximately 90% of strokes was identified with this type (Sacco et al., 1998).The damaged brain tissue caused by ischemia could separate into two region, one is infarct core and the other is penumbra region. The infarct core is the earliest illness position during the acute ischemic stroke syndrome, which is irrecoverable area regardless of reperfusion. On the other hand, the area surrounds infarct core, which is called penumbra region, defined as a hypoperfused tissue whose blood flow is too low to keep the electric activity except the ion channel mechanism. The infarct core continuously propagating the harmful metabolic process to the nearby region such as excitotoxicty, spreading depolarization, inflammatory response(Ramos-Cabrer et al., 2011). Whether or not the brain tissue is intact depends on one of these responses which is called spreading depolarization (Kawauchi et al., 2014; Strong et al., 2007). This key phenomenon will extend the infarction area of ischemic stroke from the infarct core as time goes on. Massive ionic migration in the cellular membrane often accompanies with spreading depolarization. This mechanism influences the morphological properties of the tissue which may also directly affects the light transmission in the tissue, especially the scattering event.

Near infrared spectroscopy (NIRS) has been used as a clinical diagnostic technique since 1977 by Dr. Jobsis who had tried to use NIR light to get the oxidation and circulation in the superficial tissue (Jobsis, 1977). The NIRS is a technique which applied the infrared light wavelength in the range from 600 nm to 1000 nm, and then detect the light that transmitted through the sample or tissue. The light propagating in the media was described by radioactive transport equation (RTE). It is hard to derive an analytical solution form RTE so the simplified form base on some assumption was developed and applied to NIRS study. This form was called diffuse approximation (DA), with appropriate boundary condition, we could get the analytical model of light intensity and optical properties. By analyzing the change of light signal, NIRS could provide a quantitative parameter to monitor real time hemodynamic response. The main optical properties of tissue – absorption coefficient (μ_a) and reduced scattering coefficient (μ_s'), which are

J. Goh and C.T. Lim (eds.), *7th World Congress on Bioengineering 2015*,
IFMBE Proceedings 52, DOI: 10.1007/978-3-319-19452-3_19

wavelength dependent, are the key indices that reflect the physiological information of the tissue or sample(Zhao et al., 2005). The absorption rapidly decrease the intensity of the light, meanwhile the multiple scattering also happened and randomize the propagating direction of the photon. This two factor let us hard to detect the residue light which escape the tissue. In biomedical materials or tissues, the major substances which absorb the light are water, the others are some lipid(Hillman, 2007). However, these substances s have relative lower absorption coefficient in the near infrared region which we called biomedical window(Nissilä et al., 2005). By applying proper optical model, for example, modified Beer-Lambert's law, we could easily compute the change of hemoglobin concentration. In order to monitor and acquire the spatial and temporal information while the ischemic stroke processing, we develop a non-invasive, real-time, fast scanning system which based on NIRS technique.

II. Materials and Methods

A. 2D Galvos Mirror Scanning System

To achieve fast and two dimensional scanning, we used the Galvos Mirror TSH-8203 (Sunny Technology, Inc.) which controlled by LabVIEW program through NI-DAQ 6363 (National Instrument, Inc.). The scanning type could choose two different mode, one is pixel by pixel, the other is line by line. The pixel size was 1mm depend on the diameter of the focused laser light beam (690 nm). The scanning angle and range could be set by the maximum input voltage. Figure.1 shows the scanning track of this two different mode. The fast moving speed of pixel by pixel is 1 ms, and the scanning speed of line by line mode could reach 1 ms each line.

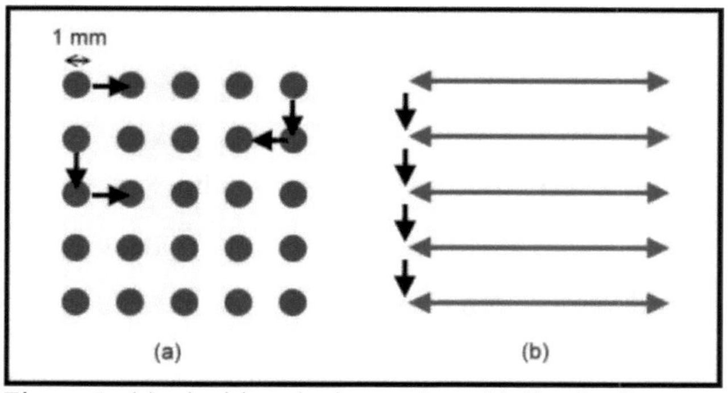

Figure 1. (a) pixel by pixel scanning, (b) line by line scanning.

B. NIR signal measurement

The back scattering light was collected by a home-made multi-mode (MM) optical fiber probe with diameter 200 μm. The one end of the probe was coupled with FC connector and then fixed on the avalanche photon diode (APD) C54060-01 (Hamamatsu, Inc.), and the other end was directly contact the surface of phantom to avoid the surface reflection noise. The APD output a DC voltage signal which was acquired by NI-DAQ 6363 with LabVIEW user interface. The scanning and acquisition system was showed in Figure 2.

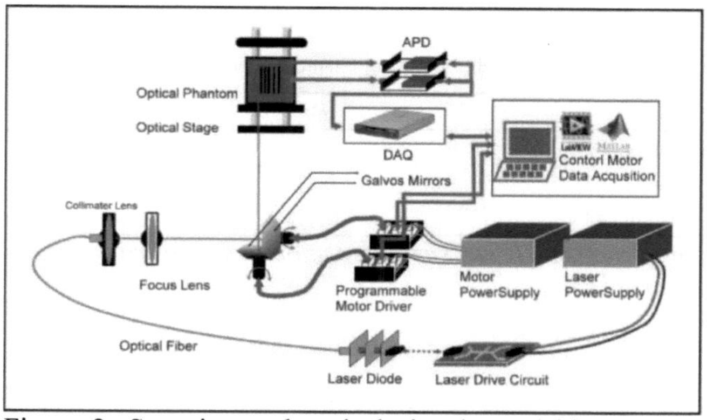

Figure 2. Scanning and optical signal acuqisition system diagram.

C. Optical Phantoms & Simulation Pattern

The phantoms was consist of PDMS subtrate (Sigma, Inc.) with mixture of TiO_2 solution and diluted black ink. This two matters form a basic scattering and absorption coeffiecient, respectively. The scattering and absorption coeffcient was measured ISS imagent frequency domain NIRS (ISS, Inc.). The absoulte oxygen saturation of the simulation phantoms were also measured. The test pattern was used to simulate the change of optical properties of the tissue. In this study, we use bar code print on a transparent plastic paper as a test pattern to observe the NIR reflected signal change by measuring the light intensity. The bar code pattern will paste on and burry in the phantoms to test the depth limitation.

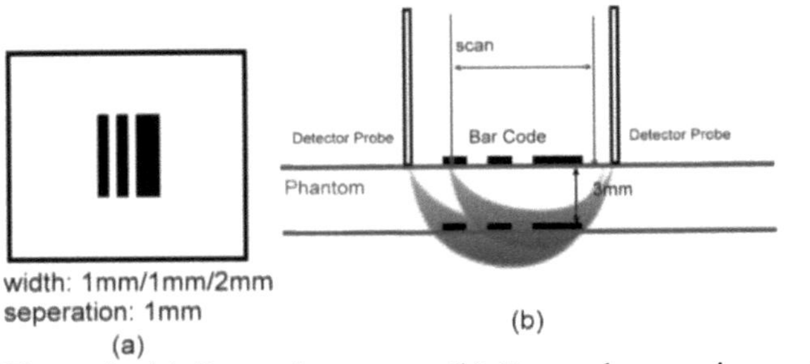

Figure 3. (a) Bar code pattern, (b) Bar code scanning and two different positions.

Table 1 Optical Phantom Content

No.	Ink	TiO2
Phantom #1	0.2%	0.2%
Phantom #2	0.2%	0.4%
Phantom #3	0.1%	0.4%

The phantom #1 was simulate the optical properties of normal rat brain.

III. RESULTS

A. μ_a , $\mu_s{}'$ *of optical phantoms*

We use commercial instrument to measure the phantom optical properties, we collect the 690 nm to correspond with the scanning light source. In here we compute the mean value of this two coefficient with different time length. Figure 4. shows the short time period variation of absorption and scattering coefficient. From Table 1 and Table 2, we know that while the absorbent(Ink) concentration is fixed, the scattering difference slightly influence on the mean value of μ_a, on the other hand, we could see obvious change and instability from both μ_a and $\mu_s{}'$ in the fixed high scattering concentration phantom.

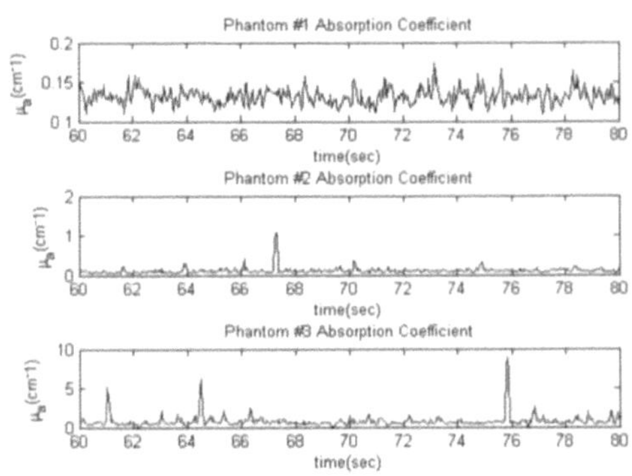

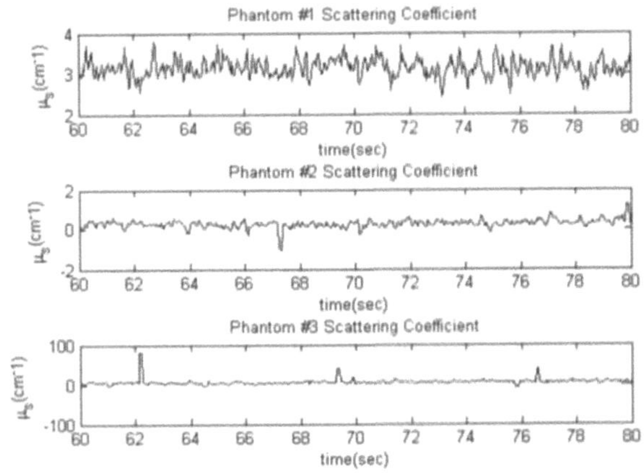

Figure 4. Measurement of absorption and scattering coefficients of different phantoms.

Table 2 Phantom Absorption and Scattering Mean Values

No.	μ_a (cm^{-1}) 90sec/20sec	$\mu_s'(cm^{-1})$ 90sec/20sec
Phantom #1	0.1324/0.1320	3.1891/3.1467
Phantom #2	0.1196/0.1125	0.2638/0.2701
Phantom #3	1.6510/0.7601	3.8878/5.3222

B. Scanning Pattern

We could observe the main trend of the light intensity signal attenuation through Fig. 5. The bar code pattern shows good SNR in superficial scanning, the deep pattern (3mm) has poor SNR but the intensity change correspond to the bar code pattern could still be identified.

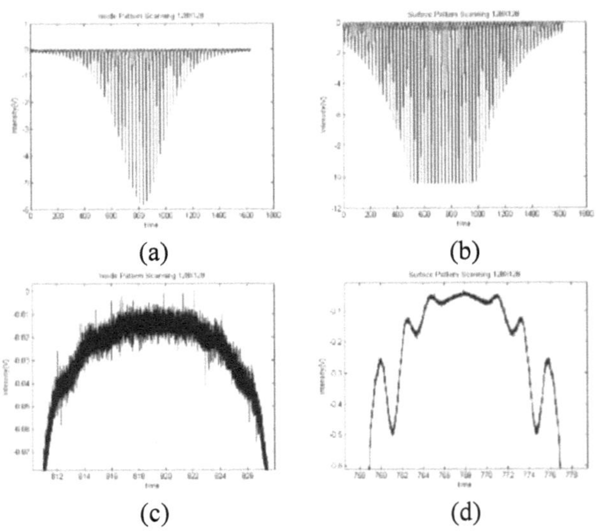

Figure 5. (a)(b) Scanning inside and superficial pattern, (c)(d) one period of scanning inside and superficial pattern.

IV. CONCLUSIONS

The pattern scanning presents good sensitivity between light intensity and optical property. The scanning depth could reach to 3 mm. The μ_a, μ_s' of optical phantoms were not stable due to the non-uniform mixture and solubility this may result in uncertainty while calculating oxygen saturation and hemoglobin concentration change.

ACKNOWLEDGMENT

Format the Acknowledgment and References headlines without numbering.

REFERENCES

[1] Sacco, R. L., Wolf, P. A., & Gorelick, P. B. (1998). Risk factors and their management for stroke prevention: outlook for 1999 and beyond. Neurology, 53(7 Suppl 4), S15-24.

[2] Ramos-Cabrer, P., Campos, F., Sobrino, T., & Castillo, J. (2011). Targeting the ischemic penumbra. *Stroke, 42*(1 Suppl), S7-11. doi: 10.1161/STROKEAHA.110.596684

[3] Kawauchi, S., Nishidate, I., Nawashiro, H., & Sato, S. (2014). *Near-infrared diffuse reflectance imaging of infarct core and peri-infarct depolarization in a rat middle cerebral artery occlusion model.* SPIE BiOS.

[4] Jobsis, F. F. (1977). Noninvasive, Infrared Monitoring of Cerebral and Myocardial Oxygen Sufficiency and Circulatory Parameters. Science, 198(4323), 1264-1267.

[5] Zhao, J., Ding, H. S., Hou, X. L., Zhou, C. L., & Chance, B. (2005). In vivo determination of the optical properties of infant brain using frequency-domain near-infrared spectroscopy. J Biomed Opt, 10(2), 024028. doi: 10.1117/1.1891345

[6] Hillman, E. M. (2007). Optical brain imaging in vivo: techniques and applications from animal to man. J Biomed Opt, 12(5), 051402. doi: 10.1117/1.2789693

[7] Nissilä, I., Noponen, T., Heino, J., Kajava, T., & Katila, T. (2005). Diffuse Optical Imaging. Advances in electromagnetic fields in living systems: Springer US.

Use macro [author address] to enter the address of the corresponding author:

Author: Jia-Jin Jason Chen
Institute: National Cheng Kung University
Street: No.1, University Road, Tainan City 701, Taiwan (R.O.C.)
City: Tainan
Country: Taiwan
Email: jasonbiolab@gmail.com

Development of Low Cost Set Up Microscopic Skin Imaging Using Structured Light Projection

Naila Zahra, Suprijanto, and Endang Juliastuti

Engineering Physics, Institut Teknologi Bandung, Bandung, Indonesia

Abstract— **Microscopic measurement of skin topography is essential both in health and cosmetics industry, for example in scar healing monitoring and efficacy assessment of beauty products. Limitation of direct visual assessment using human's eyes leads to development of many optical devices to help assessing skin condition. One of the common devices used in skin assessment is dermatoscopy, which basically is a digital microscope used to magnifies skin profile by utilizing light interaction with skin surface. However, this method has certain limitation. It cannot yield the exact depth or height information, thus it is not sufficient enough for specified needs. Another method to acquire depth information of skin topography is by measurement of light transmission through transparent skin replica made from a silicon gel. Therefore, this method requires contact with respondent's skin during replication.**

Through this paper, development of a method involving projection of structured light is presented as non-contact microscopic skin imaging. The system, built of commercially available devices, consists of a projector and a camera configured in certain angles. The processing consists of phase extraction using phase shifting interferometry (PSI) and phase unwrapping process to obtain wrapped phase and absolute phase value respectively. Experimental work was done on different areas of the skin surface each represents different skin roughness to test the feasibility of the system. Quantitative analysis was done by texture analysis based on entropy calculation to compare intensity and phase images.

Keywords—**Microscopic Skin Imaging, Structured light projection, Phase Shifting, Phase Unwrapping.**

I. Introduction

Human skin naturally undergoes changes due to aging and environmental conditions. The changes can be appeared as depigmentation, roughening, wrinkles, and so on. This phenomenon affects human needs of skin aging prevention and retardation. One of the efforts made is by using a variety of skin care products that offers anti-aging formula. With the increased demand for such beauty products, cosmetics industry is growing rapidly. In order to gain recognition and acknowledgement of its product, cosmetic industries need to test their product before it comes out in the market, which is known as efficacy assessment.

Over the last few decades, technologies to quantify the health condition of skin have been developed. One of it is to use silicone replica-based method. It is done by measuring surface profile of an artificial skin topography made of a transparent silicone material. The measurement can be done either by mechanical scanning or by measuring the distribution of light transmission. Disadvantages of this method are the possibility of distortion due to the flexibility of the skin surface during the replicating process, time consuming, and unavoidable contact with the skin of patients [1][2]. Another method is to use a digital dermatoscopy, which in principle is a digital handheld microscope utilizing optical magnification and skin-light interaction [2]. Reflection is one of skin-light interactions that have been used for skin documentation process [3]. However, this method is lack of spatial resolution, in the sense that this method cannot provide information of the depth of skin surface

One method that can also be utilized as skin surface topography measurement is structured light projection [1][4]. This method has been proved to offer accurate measurement and has been applied in various fields of measurement [5]. Unfortunately, the established structured light projection-based devices are rather expensive. Thus, assembling the device from commercially available tools can be a good solution to get the same performing technology [6].

This paper offer a microscopic measurement of the skin surface that can provide good depth information by utilizing commercially available devices to assemble a low cost measurement based on structured light projection technique. The method used in this research was phase measurement using structured light projection technique. Phase extraction was conducted using a four-steps phase shifting profilometry. Followed by the phase unwrapping to obtain the absolute phase value. Performance measurement is done in the frequency domain by calculating the entropy values, which is value often used as a quantification tools of texture analysys of surfaces with different roughness level. The entropy of phase image will show how the data represent skin sur-

J. Goh and C.T. Lim (eds.), *7th World Congress on Bioengineering 2015*,
IFMBE Proceedings 52, DOI: 10.1007/978-3-319-19452-3_20

face compared to intensity measurement. Through this paper the feasibility of a low cost set up measurement system for application of microscopic skin imaging using structured light projection will be further discussed.

II. MATERIALS AND METHODS

A. Principles of Structured Light Projection Technique

Structured light projection based measurement utilized triangulation configuration between the light source, image acquisition device, and the object plan. First, the pattern generated using computer programming is projected on the object surface by the projector. In this research, a set of sinusoidal fringe pattern is used as the light source. The pattern would be distorted due to depth variation of the object. The depth information itself would be modulated into phase information. A digital camera, positioned on a certain angles from projector, captures each distorted patterns. By having several images of distorted fringe pattern, the depth information can be obtained by phase shifting algorithm. Figure 1 shows the configuration between camera, projector, and the object plan for structured light projection technique.

The image captured by camera can be modeled as:

$$I(x,y)= a(x,y)+b(x,y)\cos[\phi(x,y)+\alpha] \quad (1)$$

I (x, y) is the intensity image captured by camera with a(x,y) has the background light intensity information. The second component contains the shape or depth information modulated in ϕ (x,y), α is the given phase initial value, b(x,y) contains intensity value [4][6].

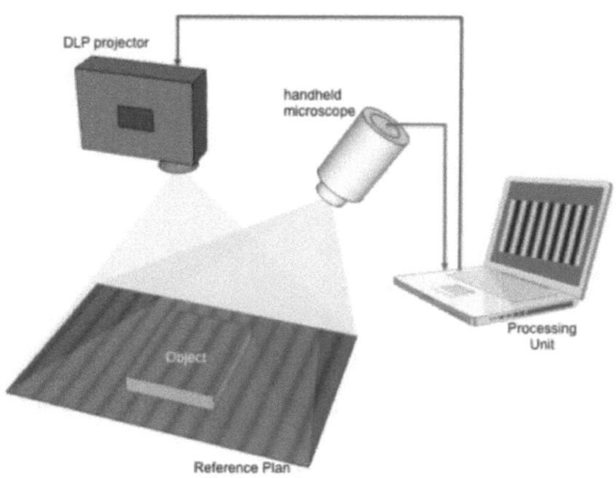

Fig.1 Configuration of structured light projection device

B. Phase Shifting Interferometry (PSI)

There are various ways of phase extraction that can be used in structured light-based measurement. Two of many methods are the phase shifting interferometry (PSI) and Fourier transform profilometry (FTP). Phase shifting interferometry is known to have better accuracy and smaller computational load when compared to FTP method. In this study, the four steps phase shifting interferometry was used. Thus, the distorted fringe patterns are:

$$I_1(x,y)= a(x,y)+b(x,y)\cos[\phi(x,y)] \quad (2)$$

$$I_2(x,y)= a(x,y)+b(x,y)\cos\left[\phi(x,y)+\frac{\pi}{2}\right] \quad (3)$$

$$I_3(x,y)= a(x,y)+b(x,y)\cos[\phi(x,y)+\pi] \quad (4)$$

$$I_4(x,y)= a(x,y)+b(x,y)\cos\left[\phi(x,y)+\frac{3\pi}{2}\right] \quad (5)$$

In four steps phase shifting interferometry, the phase information is obtained by the following equation:

$$\phi(x,y)=\tan^{-1}\frac{\sum_{i=1}^{N} I_i\sin(\alpha_1)}{\sum_{i=1}^{N} I_i\cos(\alpha_1)} \quad (6)$$

With tangential operation involved in the process of phase extraction, the phase detected will be wrapped and the value is limited from $-\pi$ to π and so, a phase unwrapping process is needed to obtain the real phase value.

C. Phase Unwrapping

Because the phase value obtained from the extraction process is limited in range [$-\pi$, π], phase unwrapping is necessary to get a continuous phase value. Basically, the relationship between wrapped and unwrapped phase is defined as:

$$\phi(x,y)=\phi_w(x,y)\pm 2\pi m(x,y) \quad (7)$$

ϕ is the unwrapped phase, which can be obtained from ϕ_w added or subtracted by multiplication of 2π, and m is the integer-valued multiplication number. However, phase unwrapping will be more complicated when the wrapped phase contains discontinuities due to high-frequency noise or false phase jumps [7].

III. EXPERIMENT AND RESULT

Experiment had been conducted to test the feasibility of the structured light projection for skin topography measurement. Figure 2 shows the workflow of the experiment. First, the fringe pattern used for the experiment was sinusoidal fringe generated by computer program. The pattern

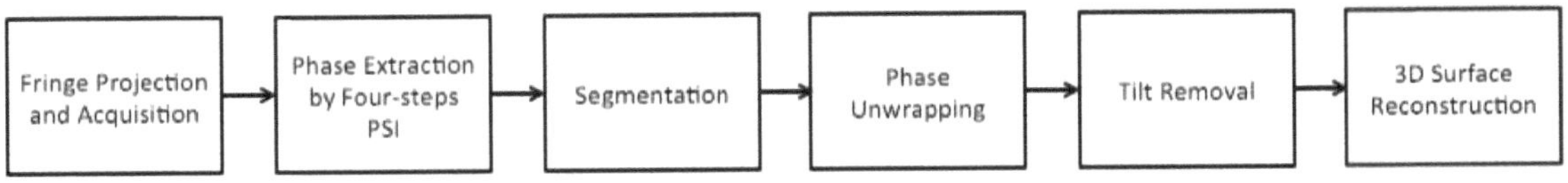

Fig. 2 Experiment workflow

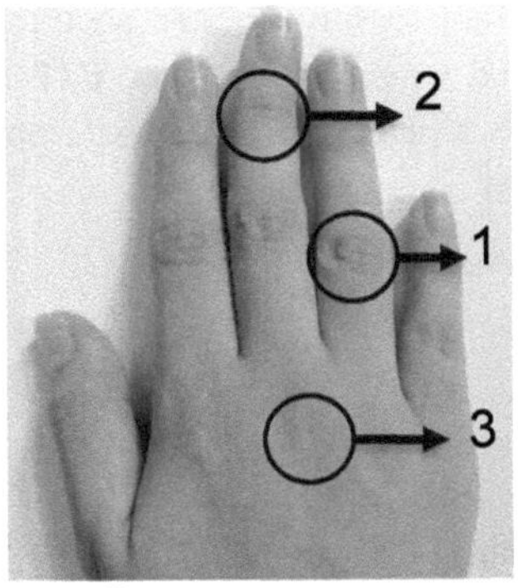

Fig. 3 Observed test area

was projected on the test area using a commercially available pico-projector Optoma PK320 with Digital Light Processing (DLP) technology. Image acquisition was done with Dinolite AD-413ZT, a handheld microscope commonly used as dermatoscopy device. The image then processed using a computer. It starts with four-steps PSI to extract phase information from the intensity images followed by segmentation to choose the observed area only, so that the unimportant area can be excluded from the next step and help reducing the computational load. The next step is phase unwrapping to obtain the continuous phase. Tilt removal is conducted as a correction to get absolute phase value, since the continuous phase contains a slope due to phase unwrapping process. Finally, the data is reconstructed into 3D surface profile. Figure 3 shows the observed areas in this research. There were three different test areas having visually different depth profile. The test area 1, 2, and 3 each represent deep, medium, and shallow profile respectively.

Figure 4 (a) shows the wrapped phase and (b) shows the unwrapped phase of area 1. Figure 5 shows the 3D reconstruction of each test areas from phase data of the experiment. As can be seen from Figure 5, the reconstruction image has a very near representation of the topography of the test area.

One vertical line was plotted to see how the depth varies in phase value. Figure 6 represents data of column 200th from test area 1 and 3. As can be seen, area 1 has wider phase value deviation while area 3 has smaller deviation ranging only from 0 to a slightly more than 0.6. It was also shown from the variance of each test area, which is 0.065, 0.022, and 0.019 for test area 1 to 3 respectively. From the variance and the graph below, it can be derived that the depth information of test areas are proportional to phase value as a result of fringe projection based measurement.

Moreover, like another method of skin documentations, in this research texture analysis was also done based on the entropy value of image. High value of entropy is often defined as high roughness level and low entropy value represent smooth surface [2].

(a) (b)

Fig. 2 (a) wrapped phase, (b) unwrapped phase

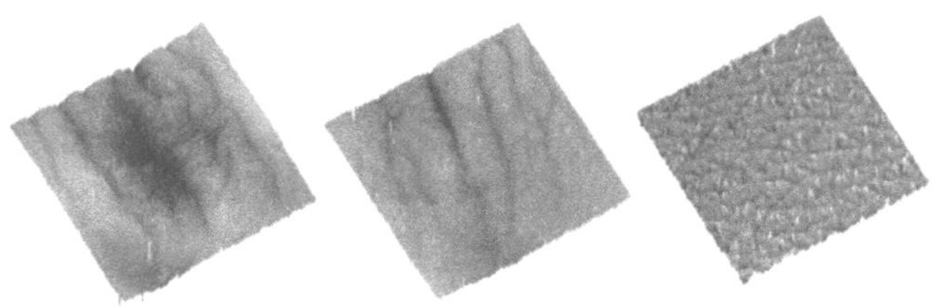

Fig. 5 3D surface reconstruction of test area 1, 2, and 3

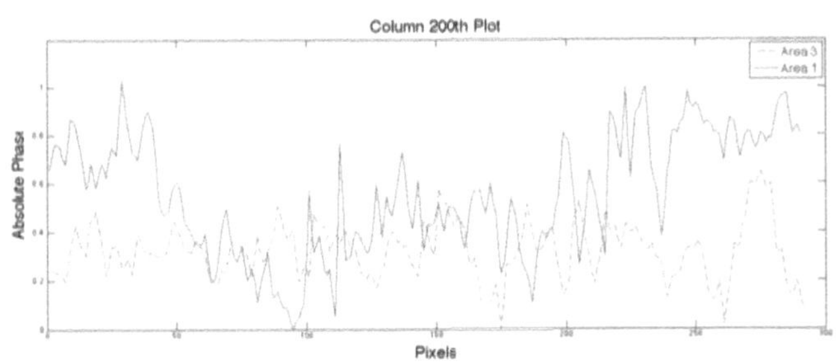

Fig. 6 Plot of column 200th of test area 1 and 3

Table 1 shows the result of entropy calculation for each test area. It also shows the entropy calculation for the intensity image, which is the same as obtained image of dermatoscopy system.

Table 1 Entropy values for phase and intensity image

Test Area	Entropy of Phase Image	Entropy of Intensity Image
1	0.59	0.34
2	0.74	0.21
3	1.89	0.58

From the entropy value of phase image, it shows that the test area 1 has the least value, followed by area 2 and area 3. Meanwhile, the entropy value of intensity image shows that area 3 has the biggest entropy value, but it was followed by area 1. Area 2 had the least entropy value, which in other words it was the softest skin out of three areas.

The different trend in entropy value of phase image and intensity image shows that there was different information perceived by phase-based measurement method. From the entropy calculation it can be derived that phase-based measurement obtains a better details than intensity-based measurement. Therefore, the entropy of phase image shows that roughness level of area 2 is higher than area 1 but lower than area 3.

IV. Conclusions

This paper demonstrates the feasibility study and development of low cost microscopic skin imaging using structured light projection technique to obtain similar performance to the established device. The feasibility study was done by observing three areas having different visual depth information. The study shows that the result of depth measurement system is proportional to phase value. Moreover, both phase and intensity based measurement were evaluated by the calculation of entropy value commonly used as texture analysis. It can be derived that phase based measurement using structured light projection can capture better details compared to intensity-based measurement. Thus, from the result of the feasibility study it can be concluded that the low cost set up utilizing commercially available device has a great potential as microscopic measurement of skin surface.

Acknowledgment

This research paper is made possible through the help and support of Indonesia Endowment Fund for Education (LPDP) – Ministry of Finance.

Conflict of Interest

The authors declare that they have no conflict of interest.

References

1. Tchvialeva L, Zeng H, Markhvida I et al. (2010) Skin Roughness Assessment. New Developments in Biomedical Engineering. Intech Open, Canada
2. Suprijanto, Nadhira V, Lestari DA, Juliastuti E, Darijanto ST. (2011) Digital Dermatoscopy Method for Human Skin Roughness Analysis. ITB J.ICT, 5(1):57–71
3. Kolias N, Skin Documentation with Multimodal Imaging or Integrated Imaging Approaches. Bioengineering of the Skin: Skin Imaging and Analysis. Informa Healthcare, USA. 221–246
4. Jaspers S, Fringe Projection for In Vivo Topometry. Bioengineering of the Skin: Skin Imaging and Analysis. Informa Healthcare, USA. 137–148
5. Gorthi SS, Rastogi, P. (2010) Fringe Projection Techniques: Whiter are We?. Optics and Lasers in Engineering, 48(2):133-140
6. Koo JM, Cho TH. (2011) Accurate 3-D Measurement System Using the Pico Projector-based Phase Shifting Techniques. IAPR MVA Proc. June:13–15
7. Malacara D, Servín M, Malazara Z. (2005) Interferogram Analysis for Optical Testing. Taylor & Francis, Boca Raton, FL. 499–506

Author 1: Naila Zahra
Institute: Engineering Physics, Institut Teknologi Bandung
Street: Jalan Ganesha 10
City: Bandung
Country: Indonesia
Email: nailazahra.nz@gmail.com

Author II: Suprijanto
Email: supri@tf.itb.ac.id

Author III: Endang Juliastuti
Email: yuliast@tf.itb.ac.id

Towards a Micro Pneumatic Actuator with Large Bending Deformation for Medical Interventions

Xinquan Liang[1,2], Chengkuo Lee[2] and Hongliang Ren[1]

[1] Biomedical Engineering, National University of Singapore, Singapore
[2] Electrical & Computer Engineering, National University of Singapore, Singapore

Abstract— **The micro actuator has been studied for its application in micro operations such as manipulating cellular aggregate in human body, clutching the tissues or drug delivery. The traditional actuation methods include thermo-mechanical actuation, electromagnetic actuation, electrostatic actuation and pneumatic actuation. Among these actuation methods, pneumatic actuation has the advantage of not generating heat and current during actuation. In this study, a new micro pneumatic actuator consisting of two biocompatible materials is designed, fabricated and tested. The actuator has one bending degree of freedom and the largest bending deformation is about 115°.**

Keywords— **Micro Actuator, Pneumatic Balloon Actuator, MEMS.**

I. Introduction

Modern bio-medical applications and surgeries require flexible and miniaturized actuators to realize certain functions, such as manipulating cellular aggregate in human body, clutching the tissues, or drug delivery [1][2]. Therefore, many kinds of actuators with different actuation principles were designed and studied, including thermo actuation [3][4], electromagnetic actuation [5][6], electrostatic actuation [7][8] and pneumatic actuation[9][10]. However, all these actuation methods excluding pneumatic actuation would generate current or heat during the actuation [8], which is not desirable for biomedical applications and surgeries. Pneumatic actuation has advantages of not generating current and heat, and the materials are usually soft and flexible which do not scratch human tissue [11]. However, the pneumatic actuation also has disadvantages such as difficult to control and small exerting forces. Therefore, pneumatic actuators are more promising in biomedical application and surgeries for minimally invasive operations [1][11].

In this study, a new micro pneumatic actuator is designed, fabricated and tested. The actuator consists of two biocompatible materials, PDMS and Ecoflex. By taking advantages of the different properties of the materials, a large bending deformation is achieved. This paper will focus on the design and fabrication of this micro pneumatic actuator.

II. Design

A. Material selection

Current studies on micro pneumatic actuators mainly use Polydimethylsiloxane (PDMS) as materials because of its strong properties and biocompatibility [12]. The Young's Modulus of PDMS is about 1.32-2.97 MPa, depending on its mixture ratio, curing temperature and curing duration [13]. Thus PDMS is a very suitable material for micro fabrication. However, the maximum strain of PDMS is below 200% [13]. There's some deformation limitations by using PDMS alone, which is not very favorable in actual biomedical applications. On the other side, another biocompatible material, Silicone Ecoflex, has low Young's Modulus, while the elasticity is very good (strain up to 900% [14]). Thus the combination of these two materials is able to provide large bending deformation due to the mechanical property differences.

B. Actuator design

The micro pneumatic actuator consists of two attached layers, one PDMS layer and one Ecoflex layer. The PDMS layer is demolded from a SU-8 mold with patterns, so this layer contains pneumatic spaces to form pneumatic balloon actuator. The patterns consist of a connection channel and

J. Goh and C.T. Lim (eds.), *7th World Congress on Bioengineering 2015*,
IFMBE Proceedings 52, DOI: 10.1007/978-3-319-19452-3_21

four balloon actuators linked by an air channel as illustrated in figure 1. When pneumatic pressure is applied in the micro actuator via the connection channel, the air go through the air channel and fill up the balloon actuators. On the other hand, the Ecoflex layer is flat, and it is able to deform much more than the PDMS layer because the Young's Modulus of Ecoflex is much smaller than PDMS. Therefore the actuator will bend under the different deformations of the two layers. The actuator is designed to have a total length of 14 mm and width about 3 mm. The actual bending length is about 10mm, excluding the connection channel. The thickness of the actuator is depended by the thickness of the two layers. The dimension of the actuator is designed for easily operation process. The actuator is adhered on a PDMS lump for air tube connection.

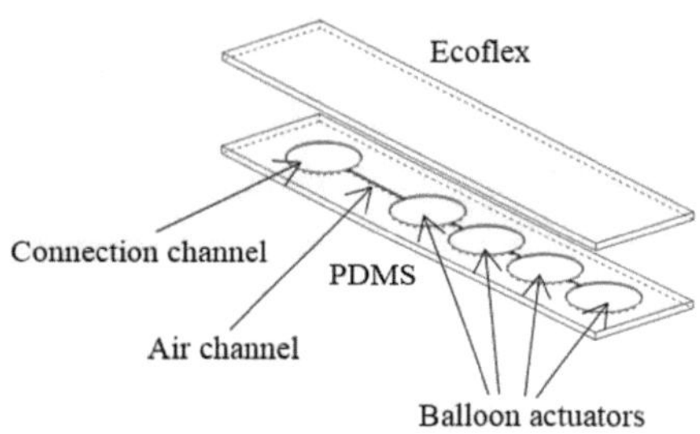

Fig. 1 Schematic of the micro pneumatic actuator

III. Fabrication

A. Mold fabrication

The mold is fabricated using SU-8 based on Microelectromechanical System (MEMS) technology. First, a glass substrate is prepared and washed using de-ionized water before it is put into the plasma cleaner for further surface cleaning. Then the substrate is dripped by several SU-8 droplets and undergoes spin coating process at a spinning rate of 2500 rpm. A very thin SU-8 film is formed as the base of the mold. The substrate is baked at a temperature of 90°C for about 5 minutes for hardening. Then UV light is exposed to the whole substrate to stabilize the chemical property of SU-8. Therefore the base of the substrate is prepared for further mold fabrication process.

The substrate is then dripped by more SU-8 droplets and goes through second spin coating process with spinning rate of 1700 rpm to form a film of thickness about 100μm. Then it is baked at 95°C for 15 minutes. In addition, a mask is designed with certain patterns to form balloon actuators and air channel. The substrate then undergoes photolithography process with the UV light passing through the mask in order to form patterns on the SU-8. After that, the substrate goes through post exposure bake process for 10 minutes at 95°C. After the substrate cools down, it is immersed in SU-8 developer for 10 minutes to remove the unexposed materials and form the balloon actuators and air channel followed by rinsing process with isopropyl alcohol (IPA) and de-ionized water. At last hard bake process is done to maintain the properties of SU-8 mold in actual use. Finally the SU-8 mold is ready to be used for next step.

B. Material preparation

The PDMS prepolymer (Sylgard 184, Dow Corning, Inc) is mixed by a 10:1 mixing ratio. Then it is put into mixer to fully mix the PDMS and curing agent. After that it is put into a vacuum chamber for 20 minutes to degas. Then the PDMS is dripped on the glass substrate with SU-8 mold and spun coated to form a thin layer. The layer thickness is pre tested and observed under Scanning Electron Microscope (SEM) at an angle of 45 degrees for different spinning speed, as illustrated in figure 2. The spinning speed is chosen to be 400 rpm to form a layer of about 200μm. At last the substrate with PDMS is put into oven for 2 hours at 70°C for curing process.

The mixing ratio of silicone Ecoflex is 1:1. Then it is put into the mixer followed by vacuum box. To make a thin film of Ecoflex, a glass slice is used as the substrate. For easy peel-off process, a very thin layer of surfactant is spun on the substrate. After that the Ecoflex is spun onto the substrate to form a thin layer and sent for curing process.

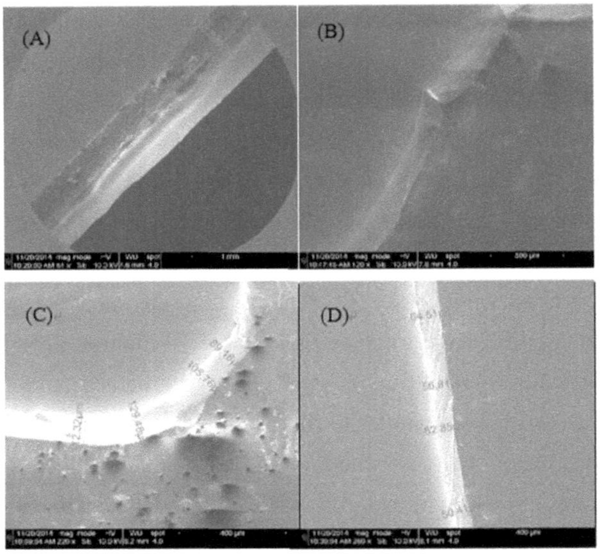

Fig. 2 PDMS thickness under SEM at 45 degree at different spinning speed (A) 200rpm (B) 400rpm (C) 600rpm (D) 800rpm

C. Bonding process

The two layers are cut off from the substrates and bonded together. Several bonding methods are tested to achieve the optimal bonding strength. First, a normal oxygen plasma etching process is conducted to both the PDMS layer and Ecoflex layer and then they are attached together. The result suggests that the two layers cannot be bonded together. Then another bonding test is conducted using the half-cured Ecoflex layer adhering the PDMS layer followed by heating process in oven. The result shows that the two layers are bonded together, but not very tightly. Besides, the two layers have internal stress, so the actuator is not flat. Finally, the bonding test is conducted with the half-cured Ecoflex bonded to PDMS that went through oxygen plasma etching process. The bonded layers are placed under room temperature instead of oven to prevent the stress in layers.

Therefore the Ecoflex layer is put in oven of 90°C and half cured for 3 minutes. It becomes half liquid half solid after heating. At the same time the PDMS layer is put into plasma cleaner for 10 minutes. The oxygen flow rate is set to 20 sccm and the power is 50 watt. When both layers are prepared ready, the PDMS layer is cut off from the substrate and carefully attached on the half-cured Ecoflex layer, avoiding bubbles between two layers as well as internal stress of the two layers. The bonded actuator is placed in room temperature 23°C for 24 hours to fully cure the Ecoflex as well as tightly bonding the two layers.

After the actuator is fabricated, it is paste on a lump PDMS to connect to an air tube. The air tube is connected to a syringe which can provide pneumatic pressure.

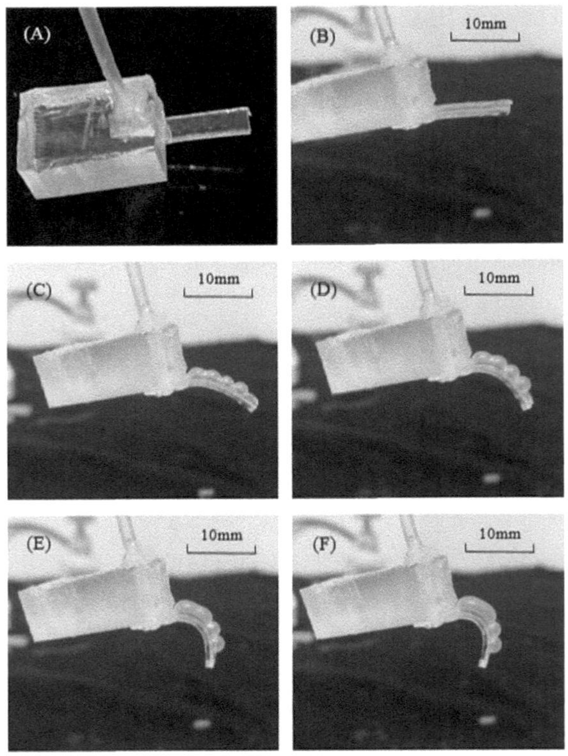

Fig. 3 Captured image of the micro pneumatic actuator (A) Top view (B)-(F) Actuation process

IV. RESULTS

The dimension of the actuator is about $\mathbf{14 \times 3 \times 0.5}$ mm. The pneumatic pressure is applied through the syringe manually. The actuation process is shown in figure 3. The maximum bending angle is about 115°. When air pressure is large, the bonding issues appear at the middle area of the actuator. The actuator remains soft during actuating, indicating that the actuator is safe for biomedical applications.

V. CONCLUSION AND FUTURE WORK

In this study, a new kind of micro actuator is designed and fabricated. The actuator is fabricated using two materials, PDMS and silicone Ecoflex, so a large bending deformation about 115° can be achieved due to the different mechanical properties of the materials. The actuator has a potential to be used in biomedical applications and surgeries for tissue manipulation

Future works will focus on optimizing fabrication process especially bonding process, comparison between different patterns, and miniaturizing the actuator.

ACKNOLEDGEMENT

This work is in-part supported by Singapore Academic Research Fund (AcRF), under Grants R397000166112.

The authors would like to thank Mr. Sun Yi and Mr. Wang Hao at National University of Singapore for their helpful support and guidance during the experiments.

CONFLICT OF INTEREST

The authors declare that they have no conflict of interest.

REFERENCE

1. Teramachi, Y., Shimomura, S., Tonomura, W., Tajima, S., Tabata, Y., & Konishi, S. (2013, June). Cellular aggregate catcher using fluidic manipulation in high compatibility with widespread micro-well-plate. In Solid-State Sensors, Actuators and Microsystems (TRANSDUCERS & EUROSENSORS XXVII), 2013 Transducers & Eurosensors XXVII: The 17th International Conference on(pp. 2209-2212). IEEE.
2. Ok, J., Lu, Y. W., & Kim, C. J. (2006). Pneumatically driven microcage for microbe manipulation in a biological liquid environment. Microelectromechanical Systems, Journal of, 15(6), 1499-1505.
3. Chronis, N., & Lee, L. P. (2005). Electrothermally activated SU-8 microgripper for single cell manipulation in solution. Microelectromechanical Systems, Journal of, 14(4), 857-863.
4. Meszmer, P., Hiller, K., Hartmann, S., Shaporin, A., May, D., Rodriguez, R. D., ... & Wunderle, B. (2014). Numerical characterization and experimental verification of an in-plane MEMS-actuator with thin-film aluminum heater. Microsystem Technologies, 20(6), 1041-1050.
5. Basantkumar, R. R., Stadler, B. H., Robbins, W. P., & Summers, E. M. (2006). Integration of thin-film galfenol with MEMS cantilevers for magnetic actuation.Magnetics, IEEE Transactions on, 42(10), 3102-3104.
6. Wright, J. A., Tai, Y. C., & Chang, S. C. (1997, June). A large-force, fully-integrated MEMS magnetic actuator. In Solid State Sensors and Actuators, 1997. TRANSDUCERS'97 Chicago.,1997 International Conference on (Vol. 2, pp. 793-796). IEEE.
7. Mu, X., Zhou, G., Yu, H., Tsai, J. L., Neo, D. W., Kumar, A. S., & Chau, F. S. (2013). MEMS Electrostatic Double T-Shaped Spring Mechanism for Circumferential Scanning. Microelectromechanical Systems, Journal of, 22(5), 1147-1157.
8. Li, J., Fang, J., Du, M., & Dong, H. (2013). Analysis and fabrication of a novel MEMS pendulum angular accelerometer with electrostatic actuator feedback.Microsystem technologies, 19(1), 9-16.
9. Jeong, O. C., & Konishi, S. (2006). All PDMS pneumatic microfinger with bidirectional motion and its application. Microelectromechanical Systems, Journal of, 15(4), 896-903.
10. Gorissen, B., Chishiro, T., Shimomura, S., Reynaerts, D., De Volder, M., & Konishi, S. (2014). Flexible pneumatic twisting actuators and their application to tilting micromirrors. Sensors and Actuators A: Physical, 216, 426-431.
11. Shimomura, S., Teramachi, Y., Muramatsu, Y., Tajima, S., Tabata, Y., & Konishi, S. (2014, January). Pinching and releasing of cellular aggregate by microfingers using PDMS pneumatic balloon actuators. In Micro Electro Mechanical Systems (MEMS), 2014 IEEE 27th International Conference on(pp. 925-926). IEEE.
12. Tsai, N. C., & Sue, C. Y. (2007). Review of MEMS-based drug delivery and dosing systems. Sensors and Actuators A: Physical, 134(2), 555-564.
13. Johnston, I. D., McCluskey, D. K., Tan, C. K. L., & Tracey, M. C. (2014). Mechanical characterization of bulk Sylgard 184 for microfluidics and microengineering. Journal of Micromechanics and Microengineering, 24(3), 035017.
14. Yamamoto, M., Witt, U., Skupin, G., Beimborn, D., & Müller, R. J. (2005). Biodegradable Aliphatic - Aromatic Polyesters: "Ecoflex®". Biopolymers Online.

A Study of Frictional and Biomechanical Forces in a Trapezoidal Bracket versus Rectangular Bracket and Periostin Activity

Teresa Cobo[1], Alberto A. Suárez[2], Juan Cobo[1], Santiago Cal[3], and Álvaro J. Obaya[4]

[1] Department of Surgery and Medical-Surgical Specialties, University of Oviedo and Instituto Asturiano de Odontología. Asturias. España
[2] Department of Mechanical Engineering. Escuela Politécnica Superior de Ingeniería. University of Oviedo. Asturias. España
[3] Department of Biochemistry and Molecular Biology. University of Oviedo. Asturias. España
[4] Department of Fuctional Biology. University of Oviedo. Asturias. España

***Abstract* – This paper presents a study of the frictional and biomechanical forces in a trapezoidal versus rectangular bracket. The main aim is to validate the novel and innovative bracket design (utility model U-2013 30 854 applied for on: 8th July 2013. The scope of protection of this utility model covers Spain, France and Portugal).**

Keywords - **orthodontics, trapezoidal bracket, biomechanical, periostin.**

I. INTRODUCTION

Orthodontic treatment is characterized by the application of forces to teeth with the intention of moving them to a predetermined desired position. To efficiently establish tooth movement, the forces applied to the teeth mist be within a certain range. Forces which are too high can result in rapid, painful tooth movement or ankyloses, while forces are too low can result in slow or non-existent tooth movement.

Our hypothesis is the following: given that the archwire is not only supported by the bottom of the slot, but also by the walls, could a change in the morphology of the slot and/or archwire achieve an improvement in the behavioural relationship between the two in different scenarios? At first the bracket had a standard, rectangular face, and now, why not a trapezoidal face?

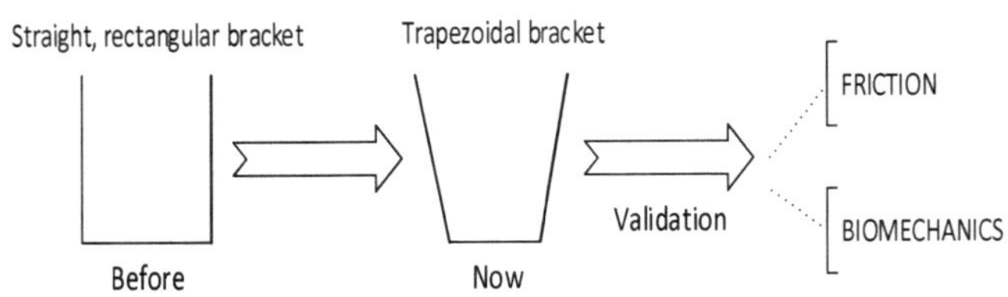

Fig. 1 Evolution of the bracket

II. METHOD

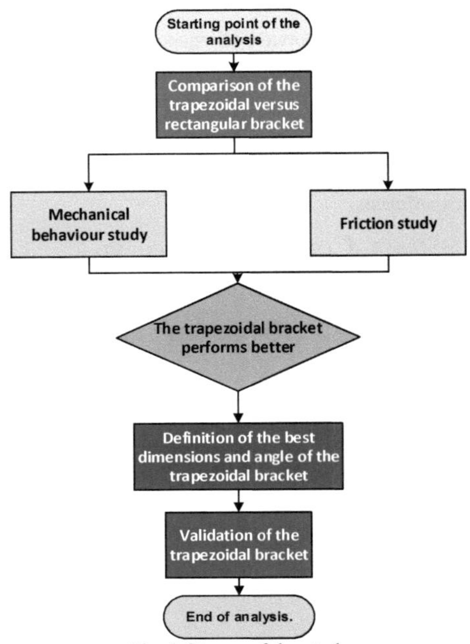

Fig. 2 Stages of the study

A. STUDY OF THE MECHANICAL BEHAVIOUR OF A TRAPEZOIDAL VERSUS RECTANGULAR BRACKET

A numerical simulation was performed using the 3D finite element method of three models of a dental bracket with different geometries for the cross-section of the archwire/slot: one "rectangular" in shape; one with trapezoidal geometry, $\alpha = 5^o$ in the area of interest; and a third with trapezoidal geometry, $\alpha = 8^o$.

Fig. 3 Trapezoidal bracket

J. Goh and C.T. Lim (eds.), *7th World Congress on Bioengineering 2015*, IFMBE Proceedings 52, DOI: 10.1007/978-3-319-19452-3_22

B. COMPARATIVE STUDY OF ARCHWIRE FRICTION ON DIFFERENT BRACKET GEOMETRIES. 1st CASE

Three case studies were conducted of the bracket/archwire assembly, varying its geometry.

-Rectangular bracket with rectangular archwire.

-Trapezoidal bracket with rectangular archwire.

-Trapezoidal bracket with rectangular archwire supported on one side.

To obtain comparable results, all cases were simulated under the same conditions of friction, estimating a single coefficient of dynamic friction. Similarly, the forces applied to the archwire were always equal and constant for all studies, the estimated slip velocity between the bracket and archwire taking an equal and constant value in all cases.

C. COMPARATIVE STUDY OF ARCHWIRE FRICTION ON DIFFERENT BRACKET GEOMETRIES – VARYING THE ANGLE OF THE TRAPEZOIDAL SLOT BETWEEN 1° AND 10°. 2nd CASE

Considering a common state of loading, a statics study of the archwire-bracket system was conducted on 4 possible hypotheses of impending movement.

Ten bracket geometries were studied in each of the four analysed cases, with the angle of the trapezoidal slot taking the values 1°, 2°, 3°... up to 10°.

After defining the relative position between archwire and bracket, different scenarios may be postulated based on hypotheses regarding the possible impending movement the archwire may suffer within the slot.

Specific reactions or frictional forces are generated for each case in each of these scenarios.

Four different cases were postulated, analysing in each one the 10 bracket geometries under study (trapezoidal slot angle varying between 1° and 10°). These cases are described below.

1. The archwire is supported on one wall of the bracket slot, considering no state of impending movement. Hence, no frictional forces appear along the direction of the contact faces.
2. The archwire is supported on one wall of the bracket slot, considering a state of impending movement to exist on this surface. Hence, a frictional force appears along the direction of the slot wall.
3. The archwire is supported on the wall of the bracket slot, considering a state of impending movement to exist on the base of the slot (the archwire tends to be displaced in an anti-clockwise direction. In this case, a frictional force appears along the direction of the base.
4. The archwire is supported on the wall of the bracket slot, considering a state of impending movement to exist simultaneously on the base and the slot wall (the archwire tends to turn in a clockwise direction. In this case, frictional forces appear along the direction of both the base and the contact slot wall.

III. RESULTS

A. RESULTS OF THE STUDY OF THE MECHANICAL BEHAVIOUR OF A TRAPEZOIDAL VERSUS RECTANGULAR BRACKET

RESULTS		
	Max. Stress (Pa)	**Max. Displacement (mm)**
Straight Bracket	11864806	1,55076 E-07
Trapez. Bracket 5°	11872979	1,59855 E-07
Trapez. Bracket 8°	11884190	1,63252 E-07

Fig. 4 Results of the study

It may be concluded that the geometry of the archwire in the states of loading torque and compressive force:

1. Has a local effect only in terms of the distribution of forces around the slot.
2. The distribution of forces in the base of the bracket is independent of the local geometry of the archwire and the slot.
3. The actions or forces transmitted to the tooth are independent of the local geometry of the archwire and the slot.

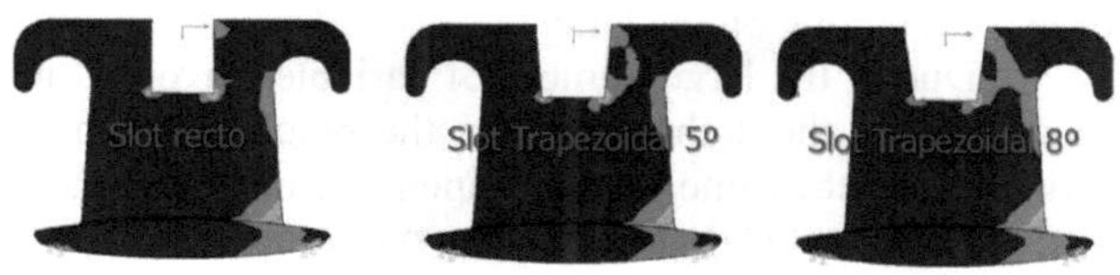

Fig. 5 Comparison of the 3 cases

B. RESULTS OF THE COMPARATIVE STUDY OF ARCHWIRE FRICTION ON DIFFERENT BRACKET GEOMETRIES. 1st CASE

RESULTS	
	Frictional force [N]
Rectangular bracket with rectangular archwire	4.9504
Trapezoidal bracket with trapezoidal archwire	4.9068
Trapezoidal bracket with rectangular archwire supported on slot wall	3.5005

Fig. 6 Results of the comparative study

As can be seen from the above table, the results of the frictional forces vary considerably depending on the loading case. Although the value of the applied forces is the same in all cases, the geometry of the bodies on which they are applied varies, resulting in a variation in the normal components associated with these external forces.

From these results, it can be concluded that:

Brackets with rectangular slots generate greater frictional forces than those with a trapezoidal geometry.

For small angles of the trapezoidal slot (between 0.1°-1°), the influence on the resulting frictional force is very small, as this depends directly on its cosine component.

In the case of rotation of the archwire relative to the bracket, we assume that contact occurs on a single surface. The component of the reaction force, perpendicular to the surface, is clearly smaller than the sum of the normal components applied to both faces. The frictional force associated with the former normal component will therefore also be less than in the case of contact with two surfaces. Comparing the "trapezoidal bracket/trapezoidal archwire" with the "trapezoidal bracket/rectangular archwire on the slot wall", a decrease of 29% is observed when only one contact surface is involved instead of two.

C. RESULTS OF THE COMPARATIVE STUDY OF ARCHWIRE FRICTION ON DIFFERENT BRACKET GEOMETRIES – VARYING THE TRAPEZOIDAL SLOT ANGLE BETWEEN 1° AND 10°. 2nd CASE

For all hypotheses of movement studied here, it can be concluded that the frictional force always decreases with increasing angle of the trapezoidal slot.

In term of the different postulated cases, the lowest frictional force is found in the archwire that slides along the slot wall until it is supported on its base, without the possibility of it sliding on the latter.

Due to the large number of variables involved in the movement of the archwire inside the bracket, it cannot be affirmed that the same case of impending movement always occurs. That is, the 4 cases of impending movement are not exclusive and, at any one time, one and/or another may occur.

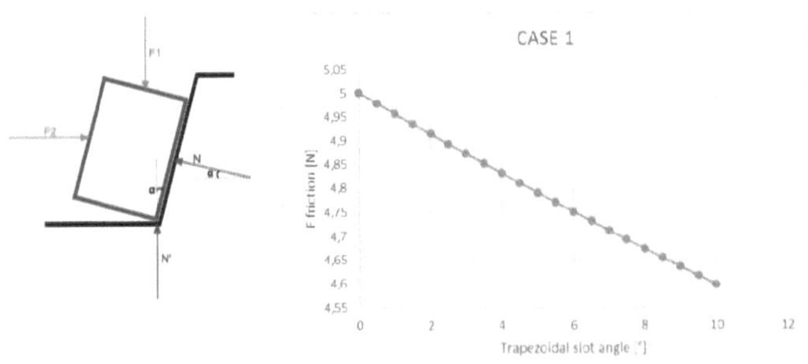

Fig. 7 Results of Case 1

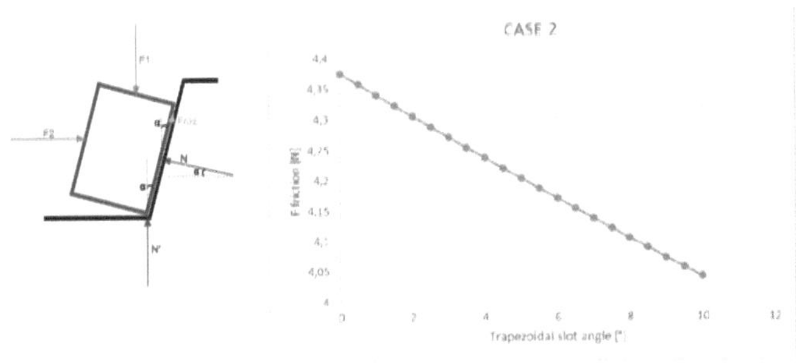

Fig. 8 Results of Case 2

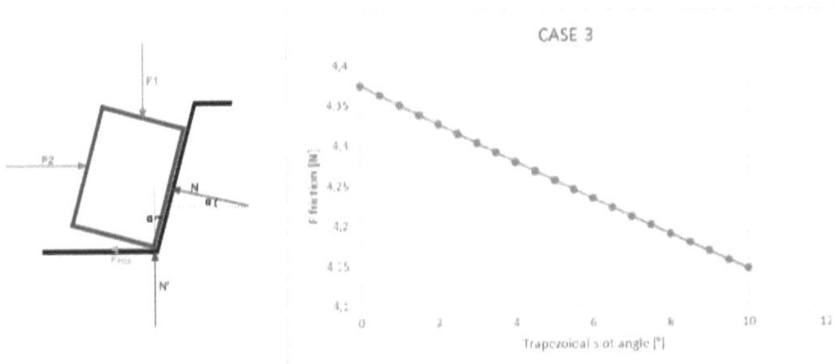

Fig. 9 Results of Case 3

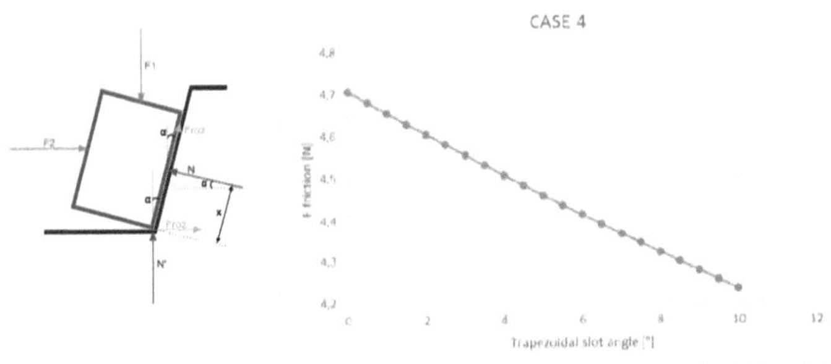

Fig. 10 Results of Case 4

Carrying out a percentage comparison, the most notable decrease in frictional force is produced in the case of the archwire that slides on the slot wall until it is supported on its base, without the possibility of sliding on the latter. This decrease was calculated with respect to the value of the theoretical frictional force (5N), which would be the value in the case of a rectangular bracket and rectangular archwire.

If a torque is produced in the archwire, what is most likely to occur (failing a study to verify this) is that the increasing angle would constitute an adverse factor hindering the rotation of the tooth.

If the direction or sign of the assumed forces are varied (once again failing a study to verify this), increasing the angle of the bracket slot could be counterproductive.

IV. Conclusions

The distribution of forces transmitted to the tooth is more uniform in the trapezoidal than in the rectangular bracket.

Brackets with rectangular slots generate higher frictional forces than those with a trapezoidal geometry.

The frictional force always decreases with increasing angle of the trapezoidal slot.

The lowest frictional force is produced when the archwire slides along the slot wall until it is supported on its base, without the possibility of sliding on the latter.

The study of the frictional forces between the bracket and archwire show that a greater bracket angle (regardless of the assumed hypothesis) produces a decrease in the frictional force of advancement in both the rectangular and trapezoidal archwire.

The maximum recommended angle to manufacture brackets is 8°.

All the preceding results lead us to conclude that the best combination is the trapezoidal bracket with a trapezoidal archwire.

Furthermore, our research provides new evidence on the impact of periostin in the physiology of tooth movement resulting from the application of orthodontic forces using an archwire bracket device with rectangular and trapezoidal slot walls.

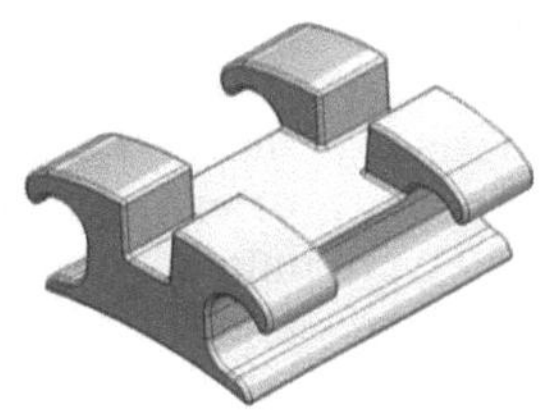

Fig. 11 Design of the innovative trapezoidal bracket

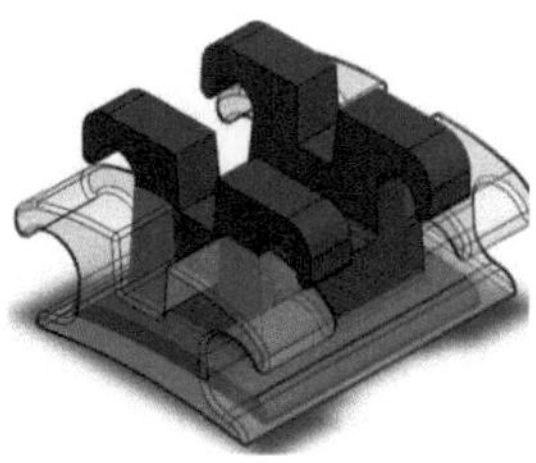

Fig. 12 Comparison of the change in dimensions of the trapezoidal versus conventional bracket

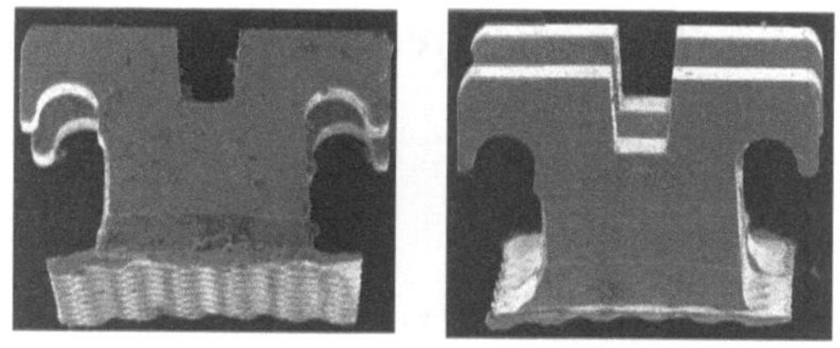

Fig.13 Surface topography of the bracket using a scanning electron microscope.

Acknowledgment

This research was supported by Instituto Asturiano de Odontología, Principado de Asturias, Spain.

Conflict of Interest

The authors declare that they have no conflict of interest.

References

1. Justin K Wong, Dan L Romanyk, Roger W Toogood, Giseon Heo, Jason P Carey and Paul W Major. (2014). The effect of perturbations on resistance to sliding in second-order moments comparing two different bracket types. J Dent Biomech. 5, 1-8.
2. Maria Regina Guerra MONTEIRO, Licinio Esmeraldo da SILVA, Carlos Nelson ELIAS, Oswaldo de Vasconcellos VILELLA. (2014). Frictional resistance of self-ligating versus conventional brackets in different bracketarchwire-angle combinations. J Appl Oral Sci. 22(3), 228-34.
3. Yumi Yanasea, Hideki Loib, Masato Nishiokaa, Ichiro Takahashi. (2014). Effects of sliding velocity on friction. An in vitro study at extremely low sliding velocity approximating orthodontic tooth movement. Angle Orthodontist. 84 (3), 451-458.
4. Raıssa Costa Araujoa; Lıvia Monteiro Bicharab; Adriana Monteiro de Araujob; David Normando. (2014). Debris and friction of self-ligating and conventional orthodontic brackets after clinical use Angle Orthodontist. DOI: 10.2319/012914-80.1.
5. Jun-ya Tominaga, Hiroya Ozaki, Pao-Chang Chiang, Mayumi Sumi, Motohiro Tanaka, Yoshiyuki Koga, Christoph Bourauel, Noriaki Yoshida. (2014). Effect of bracket slot and archwire dimensions on anterior tooth movement during space closure in sliding mechanics: A 3-dimensional finite element study. American Journal of Orthodontics and Dentofacial Orthopedics. vol.146, no.2, Pág.166-174.

Author: Teresa Cobo Díaz
Institute: Department of Surgery and Medical-Surgical Specialties. University of Oviedo, Spain
Email: teresa@iaodontologia.es

Author: Alberto A. Suárez
Institute: Department of Mechanical Engineering. Escuela Politécnica de Ingeniería de Gijón, Spain
Email: suarez@uniovi.es

Author: Juan M. Cobo Plana
Institute: Department of Surgery and Medical-Surgical Specialties. University of Oviedo, Spain
Email: jcobo@uniovi.es

Application of Cortical Optogenetic Stimulation of Animal Model for Modulating Motor Plasticity

Cho-Han Hsieh, Chun-Wei Wu, and Jia-Jin Jason Chen

Department of Biomedical Engineering, National Cheng-Kung University, Tainan, Taiwan

Abstract—**Cortical theta burst stimulation (TBS) can modulate motor plasticity via long-term potentiation/depression (LTP/LTD)-like mechanisms and enhance motor performance, which make TBS a potential non-invasive therapy for motor deficit diseases such as Parkinson's disease (PD). In previous rodent study, we had demonstrated that cortical electrical stimulation (CES)-TBS protocols were capable to change motor-evoked potentials (MEPs), and enhance motor performance in chronic PD rats. However, the cellular mechanism of TBS is still unknown. CES excited all types of neurons surrounding the electrode in cortex, it is difficult to differentiate the effect of TBS on specific neural circuit that responses to MEP modulation. LTP/LTD occurred dominantly at glutamatergic synapse, therefore the cell type-specific stimulation will be applied using optogenetic approach to reveal the functional roles of glutamatergic neuron in motor plasticity.**

CaMKIIalpha promoter driven channelrhodopsin-2 (CaMKIIalpha-ChR2) was expressed in glutamatergic pyramidal neuron in primary motor cortex (M1). Optogenetic stimulation was achieved using blue laser guided by optical fiber. Local field potentials (LFPs) and MEPs were collected during and after optogenetic TBS treatment. The results showed that MEPs amplitude were increased after optogenetic TBS treatment, which indicated that motor plasticity could be modulated by applying TBS on glutamatergic neurons in M1. However, there was no significant change in cortical excitability revealed by LFPs. In summary, these results suggested that LTP/LTD-like effects induced by cortical TBS treatment might be located at glutamatergic projections downstream of M1.

We had demonstrated the modulation of motor plasticity using cell type-specific TBS scheme in M1. This finding may contribute to develop high efficient therapies for neural disorders via targeting specific neural circuit.

Keywords—**theta burst stimulation, Parkinson's disease, optogenetics, neuroplasticity.**

I. INTRODUCTION

Cortical stimulation is developed to change cortical activity in clinical treatment. Differ to traditional stimulation protocol such as low and high frequency stimulation, theta burst stimulation (TBS) is a stimulation paradigm which has been verified in repetitive transcranial magnetic stimulation (rTMS) study. It could lead to long-lasting effect of cortical excitability by short period of stimulation. The excitability of motor cortex can be measured by the motor evoked potentials (MEPs) which can be elicited by a single pulse magnetic stimulation [1]. In clinical study in healthy subject, continuous TBS (cTBS) inhibits the MEP amplitude reflecting long-term depression (LTD)-like plasticity and the intermittent TBS (iTBS) increases the MEP amplitude indicating long-term potentiation (LTP)-like effect [2]. This plasticity effect seem to have therapeutic potentials in neurological disease such as stroke, Parkinson's disease and traumatic brain injury [3]. However, the mechanism of TBS are not understood well. In vivo animal studies on TBS might help to understand the mechanism that how TBS works. In the rTMS studies, the main discrepancy between animal and human is the ratio of the size of magnetic coil to head. The relative large size of the coil may cause low spatial resolution in the brain of rodent. For understanding the cellular mechanism of TBS, a fast and precision stimulation was developed. Optogenetic is a novel technique that transfer a light sensitive ion channel on the neural cell membrane called channelrhodopsin-2 (ChR2) which can be driven by light illuminating. Approach of optogenetics could achieve the specific neural cell type stimulation.

Previous study has shown that high frequency optical stimulation on motor cortex can decrease the times of amphetamine-induced ipsilateral rotation and increase the path length in open field test. However, low frequency stimulation did not show significant effect. The aims of this study is to establish an animal model of optogenetics for applying optogenetic-TBS to evaluate the effects in cortical plasticity via MEPs in the normal rats. We target the glutamatergic neuron which were concerned about the synaptic plasticity. The outcomes from normal rat would be further used to evaluate the effect of TBS-elicited plasticity in parkinsonian rat model. We want to compare the efficiency between the specific TBS and previous study.

J. Goh and C.T. Lim (eds.), *7th World Congress on Bioengineering 2015*,
IFMBE Proceedings 52, DOI: 10.1007/978-3-319-19452-3_23

II. Material and Methods

A. Viral production

Second generation packaging system is used for virus transfection, psPAX2 is packaging plasmid and pMD2.G is envelope plasmid. The insert gene used are pLenti-EF1a-GFP, pLenti-CaMKIIa-ChR2(H134R) and pLenti-EF1a-ChR2(H134R). The plasmids were acquired from a non-profit organization: addgene (www.addgene.com). For producing lentiviral particles, the 293FT cells were used for transfection using calcium chloride ($CaCl_2$) method. 10μl DNA contain 5 μg insert gene, 1.75μg pMD2.G and 3.25μg psPAX2 mix with 215 μl ddH_2O. And mixd with 250 μl calcium chloride. And the mixed solution was added in 500μl HeBS buffer (pH7.05) with vortex. Stand for 30 mins, the 500 μl mix solution will add to 6 cm dish contain 293FT cell. The medium was collected after 16 to 24 hours, and concentrate the viral solution by ultracentrifuge.

B. Viral transduction

For viral transduction, an anesthetized rat was first fixed on the stereotactic apparatus. The rat's head skin was opened to expose the bregma. A hole was drilled in the left skull region to the primary motor cortex of the right forelimb (AP: +1.5 mm, ML:-3.0 mm) according to functional brain mapping in rats. 2μl viral solution was injected in two different depths, 2.0 mm and 2.5 mm. Optotrode including a stainless canula contained 200 μm diameter optical fiber and electrode were implanted. A stainless screw electrode was also implanted as a reference electrode in the hindlimb (Figure.1A) in front of brain. After implantation, the optotrode and electrode were coverd with dental cement (Figure.1B).

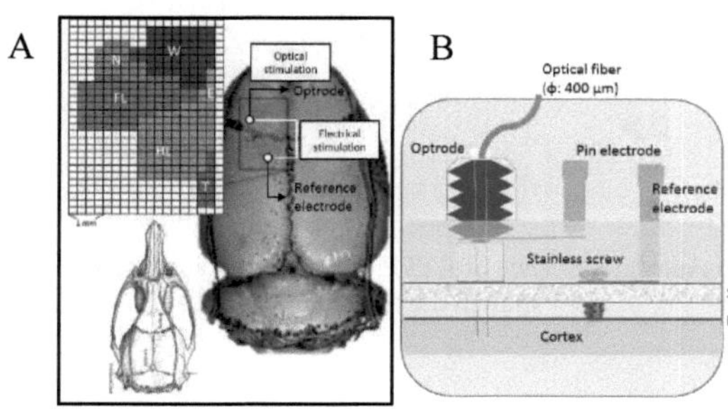

Figure. 1 Schematic paradigm of optrode implantation. (A) Location of optrode and electrode. (B) Completed implantation.

C. Optogenetics stimulation and signal recording

The function of ChR2 for optogenetics stimulation was verified at 3 weeks after virus injection. The rat were anesthetized and fixed on the stereotactic apparatus and the event-related signals were recorded during optical stimulation (Figure 3). The laser generator was triggered by TTL control of LabVIEW program. The wavelength of the laser is 473 nm that can active the ChR2. The optical stimulation are set at 0.1 Hz in frequency and 1 ms in pulse width. During 5 mins stimulation, the signals were amplified and bandpass filtered at 10 to 600 Hz for local field potential (LFP) recording. Signal was processed by MATLAB to average all. The local field potential was shown in Figure.2.

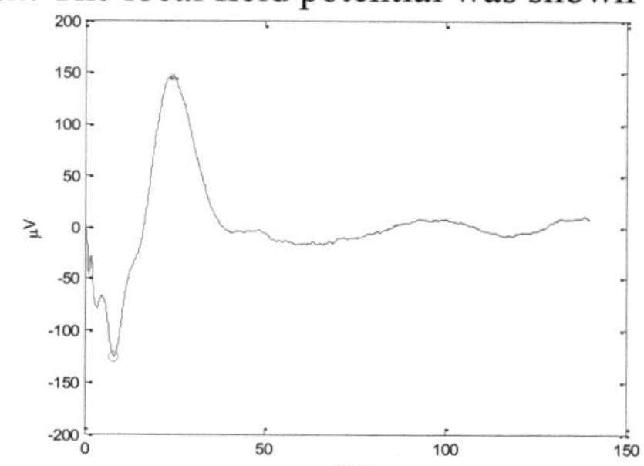

Figure. 2 Optical stimulation elicited local field potentials.

D. Cortical electrical stimulation elicit MEPs

A monopolar uninsulated stainless steel needle electodes (Axon System Inc., Hauppauge, NY) was inserted in to the brachhioradialis muscle for electromyographic (EMG) signals recording [6]. The motor evoked potentials (MEPs) were elicited by single pulse electrical stimulation which were reflected the cortical excitability. The EMG signals were amplified (1000X), filtered by 60 Hz notch and 10Hz-1kHz bandpass filter with 10kHz sampling (MP36, BIOPAC System). The intensity of electrical stimulation was determined by the resting motor threshold (RMT). RMT is the minimal intensity of electrical stimulation for eliciting MEPs greater than 20 μV in 5 out of 10 consecutive trials. For MEPs recording, the intensity was given in 120% RMT with 0.1 Hz.

E. Theta burst stimulation protocol

The theta burst stimulation (TBS) consisted of 3 pulses at 50 Hz and repeated at 200 ms interval which was described by Huang et al. (2005) [2]. The intermittent TBS (iTBS) protocol was a 2 sec train repeated every 10 sec for 190 sec (total 600 pulses). In the continuous TBS (cTBS) protocol was a continuous train for 40s (total 600 pulses) (Figure.3.).

The intensity of electrical stimulation was given in 80% RMT and optical stimulation was maxial of laser generator (~2 mW). The TBS protocols were performed in anesthetized rats and assessed the MEPs/LFPs amplitude before TBS 10 mins (baseline) and after 30 mins.

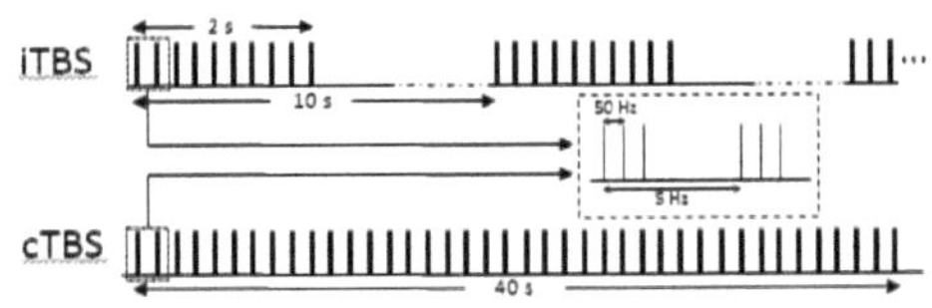

Figure. 3 Schematic paradigm of TBS protocol

F. Histology assessment

The rats were deeply anaesthetized and perfused with saline and 4% paraformaldehyde, and then take the brain for frozen section. The brain is fixed by 4% paraformaldehyde overnight and dehydrated by 30% sucrose 2 days. The thickness of the frozen section was 8 μm. For estimating the expression range, we direct observed the autofluorescence.

III. RESULTS AND DISCUSSION

A. Electrophysiological recordings

After virus transduction 3 weeks, optical stimulation would be performed for LFPs recording. The LFPs amplitude was increase through the increase light power. We compared the MEPs amplitude before and after Opto-iTBS and CES-iTBS (Figure.4). However, the results were similar between optical and electrical TBS treatment. The effect of LFPs were compared further step. The LFPs were recording and compared before and after CES-iTBS. And there were no different after CES-iTBS treatment in LFPs. (Figure.5).

B. Histology assessment

The autofluorescence of EYFP in the brain was shown in Figure.6. The expression range was about 1 mm^3 under the lesion site which was implant electrode.

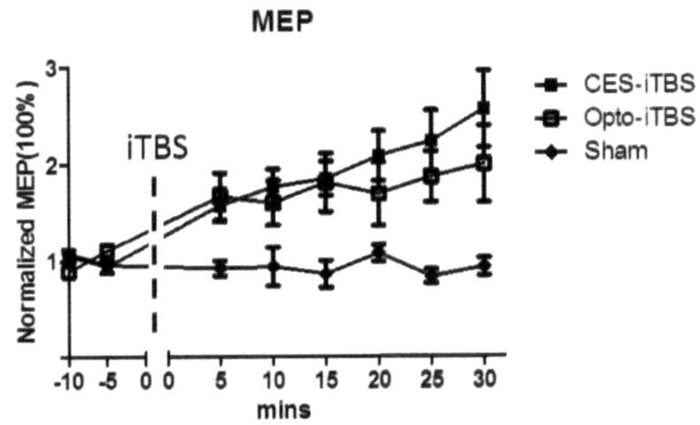

Figure. 4 Plasticity effect with CES- and Opto-iTBS treatment. (CES-iTBS(n=11), Opto-iTBS(n=1 with 4 repetitive), Sham (n=2))

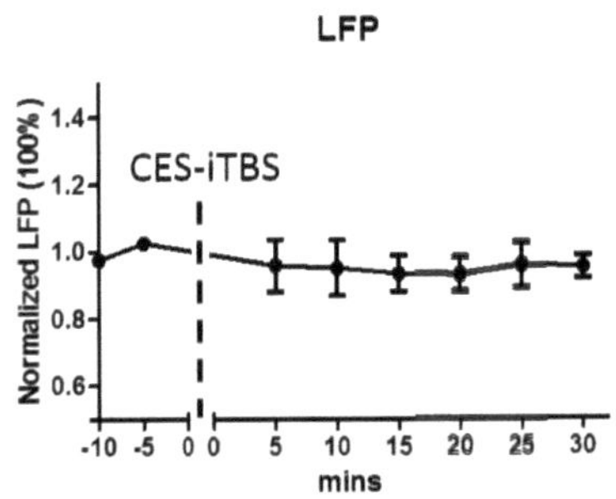

Figure. 5 The change of LFPs after CES-iTBS treatment.

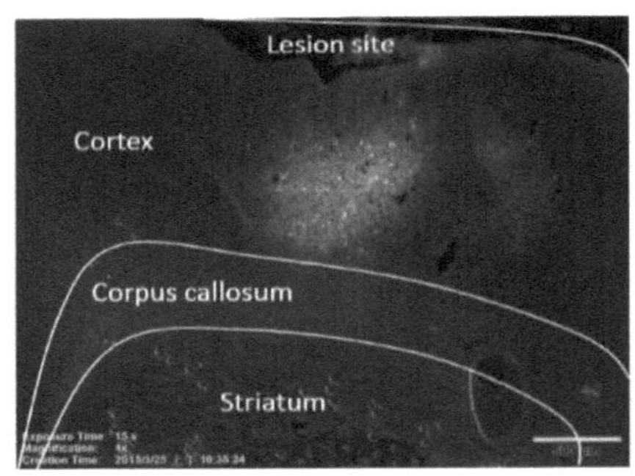

Figure. 6 Autofluorescence of EYFP. The expression range was about 1 mm^3 in the motor cortex

IV. CONCLUSIONS

The MEPs amplitude were increased after opto-TBS treatment like CES-TBS increase the MEPs amplitude. The optical stimulation was specific in the glutamatergic neuron. However, the cortical stimulation may concerned about the glutamatergic neuron. But the CES-iTBS was not affect the LFPs amplitude. Maybe the plasticity effect was occur in the downstream of the glutamatergic neuron in the basal ganglia. According to previous study, the depletion of dopamine in dopaminergic neuron, the plasticity effect would be abolished [6]. The further steps would perform the optogenetic stimulation in the treatment of parkinsonian rat.

ACKNOWLEDGMENT

The authors would like to thank the Ministry of Science and Technology of R.O.C. for financially supporting this research under contract No.NSC 101-2320-E-006 -018 - MY3.

CONFLICT OF INTEREST

The authors declare that they have no conflict of interest.

REFERENCES

1. Pascual-Leone A, Tormos JM, keenan J, Tarazona F, Cañete C, Catalá MD. (1998) "Study and modulation of human cortical excitability with transcranial magnetic stimulation." J Clin Neurophysiol. Vol 15:333–343.
2. Huang YZ, Edwards MJ, Rounis E, Bhatia KP, Rothwell JC. (2005) "Theta burst stimulation of the human motor cortex." Neuron. Vol 45:201–206.
3. Elahi B, Behrad Elahi MD and Chen R (2008) "Effect of transcranial magnetic stimulation on Parkinson motor function Future of modeling and simulation." Vol 24:357–363
4. Gradinaru V, Mogri M, Thompson K R. Henderson J M, Deisseroth K. (2009) "Optical Deconstruction of Parkinsonian Neural Circuitry." Science. Vol 324:354–359
5. Boyden E S., Zhang F, Bamberg E, Nagel G and Deisseroth K. (2005) "Millsecond-timescale, genetically targeted optical control of neural activity. Nature neuronscience." Vol 8:1263–1268
6. Hsieh TH, Huang YZ, Rotenberg A, Pascual-Leone A, Chiang YH, Wang JY and Chen JJ. (2014) "Functional Dopaminergic Neurons in Substantia Nigra are Required for Transcranial Magnetic Stimulation-Induced Motor Plasticity." Cereb cortex.

Author: Cho-Han Hsieh
Institute: Biomedical Engineering, National Cheng-Kung Unversity
Street: No.1, University Road
City: Tainan City 701
Country: Taiwan (R.O.C.)
Email: hank.jasonbiolab@gmail.com

A Coupled Mathematical Model of Glioblastoma Growth, Pre-existing Vessel Co-option, Angiogenesis and Blood Perfusion

Y. Cai[1,2,*], Z.J. Zhou[1,2], J. Wu[3], Z.Y. Li[1,2], and Q. Long[4]

[1] State Key Laboratory of Bioelectronics, Southeast University, Nanjing, China
[2] School of Biological Science and Medical Engineering, Southeast University, Nanjing, China
[3] School of Naval Architecture, Ocean and Civil Engineering, Shanghai Jiaotong University, Shanghai, China
[4] Brunel Institute for Bioengineering, School of Engineering and Design, Brunel University, Uxbridge, Middlesex, UK

Abstract—**Mathematical modelling has been used for the study of the interaction between tumour growth and local microenvironment for many years. We started our work in the field by studying the interaction between local haemodynamics and angiogenesis in a 2D model and a 3D model. The inclusion of tumour cell evolution into the original haemodynamics study forms our second generation of model in which the tumour growth, vessel dynamics and blood perfusion were included. The present study is a further development aiming to simulate tissue specific tumour growth phenomena in early development of glioblastoma. We propose a coupled mathematical modelling system to investigate glioblastoma growth in response to dynamic chemical and haemodynamic microenvironments caused by pre-existing vessel co-option, remodelling, collapse and angiogenesis. A typical tree-like architecture network with different orders for vessel diameter is designed to model pre-existing vasculature in host tissue. The chemical substances including oxygen, vascular endothelial growth factor, extra-cellular matrix and matrix degradation enzymes are calculated based on the haemodynamic environment which is obtained by coupled modelling of intravascular blood flow with interstitial fluid flow. The haemodynamic changes, including vessel diameter and permeability, are introduced to reflect a series of pathological characteristics of abnormal tumour vessels such as vessel dilation, leakage, angiogenesis, regression and collapse. Migrating cells are developed as a new phenotype to describe the migration behaviour of malignant tumour cells. The simulation focuses on the avascular phase of tumour development and stops at an early phase of angiogenesis. The model is able to demonstrate the main features of glioblastoma growth in this phase such as the formation of pseudopalisades, cell migration along the host vessels, the pre-existing vasculature co-option, angiogenesis and remodelling. The model also enables us to examine the influence of initial conditions and local environment to the early phase of glioblastoma growth.**

Keywords—**modelling of glioblastoma growth; pre-existing vessel co-option and remodelling; tumour microvasculature and blood perfusion; coupled mathematical model.**

I. INTRODUCTION

Gliomas are the most common central nervous system tumours and carry high rates of morbidity and mortality. As the most malignant and also the most frequent gliomas, Grade IV tumours including glioblastoma (GBM) and gliosarcoma, can develop from a lower grade tumour, metastasize from other tumours or directly from glioblastoma cells and have a mortality rate close to 100%.

An important pathological feature that distinguishes GBM from lower-grade brain tumours are the necrotic foci which are typically surrounded by hypercellular zones referred to as pseudopalisades[1]. Pseudopalisades are pathophysiologically linked with adjacent microvascular hyperplasia, which is another important distinction between GBM and lower-grade gliomas. As with many other solid tumours, hypoxic pseudopalisading cells can express high levels of angiogenic regulators including vascular endothelial growth factor (VEGF). The new vasculature in response to the upregulated angiogenic factors provides essential nutrients for rapid neoplastic expansion. As well as this generally known mechanism of neovasculature, the "co-option" of pre-existing vessel networks plays a significant role in glioma progression. Holash et al[2] studied early angiogenic events using the rat C6 glioma model. The results showed that even the smallest C6 gliomas at just 1 week after implantation were well vascularized by co-option of pre-existing blood vessels. Further experiments revealed that when a small number of tumour cells were implanted into healthy tissue, they managed to co-opt and migrate along host vessels, as well as produce many chemical substances, such as VEGF, Ang-1, Ang-2, to change the microenvironment around the host vessels. This can induce immature changes in the host tissue vasculature, including vessel dilation, increased capillary permeability and tortuosity. With tumour growth, cancer cells migrate along the blood vessels, compressing and destabilizing them, which leads to vessel regression and reduced blood perfusion.

The present study aims to develop a mathematical model which is capable of simulating the dynamic processes of tumour cell proliferation, migration, co-option of pre-existing vessels and angiogenesis, coupled with blood perfusion at the early stage of GBM growth. The model adopted the following assumptions based on the corresponding experimental and clinical observations: (a) the migration

* Corresponding author.

J. Goh and C.T. Lim (eds.), *7th World Congress on Bioengineering 2015*,
IFMBE Proceedings 52, DOI: 10.1007/978-3-319-19452-3_24

speed of small groups of cancer cells along the host vessel longitudinal direction is faster than in a radial direction from the vessel, (b) vessel maturation is estimated by the vessel dilation and the increased permeability of the vessel wall, (c) vessel diameter is determined by a relationship between the intravascular pressure, interstitial pressure and the collapse pressure, (d) the initial collapse pressure is proportional to the diameter of the pre-existing vessel network, and changes with the level of vessel maturation. One intention of this paper is to reproduce the observed pathophysiological phenomena such as (i) glioma cell migration along the host vessel, (ii) formation of pseudopalisades, (iii) pre-existing vessel co-option, remodelling and collapse.

II. Method

A. Pre-existing vessel network

For the morphological analysis we consider vessel segments within a cube simulation domain Ω of 1mm3. A basic grid of 100x100x100 is generated uniformly in the cube with a centre to centre length of 10μm between the neighbouring nodes [Figure 1]. The pre-existing vasculature we designed in the basic model has a typical tree-like architecture network in accord with the features of human cerebral microcirculation in Cassot et al's experimental work[3]. For the topological analysis, we classify vessel branches according to the Strahler system[3], a well-established method for describing stream order [Figure 1]. In our model, there are three Strahler orders to show a brief tree architecture of an arteriolar branching pattern. The main stem of trees, with a Strahler order 3, grow approximately in vertical direction from plane x=100 to x=0, and have the biggest value of vessel diameter.

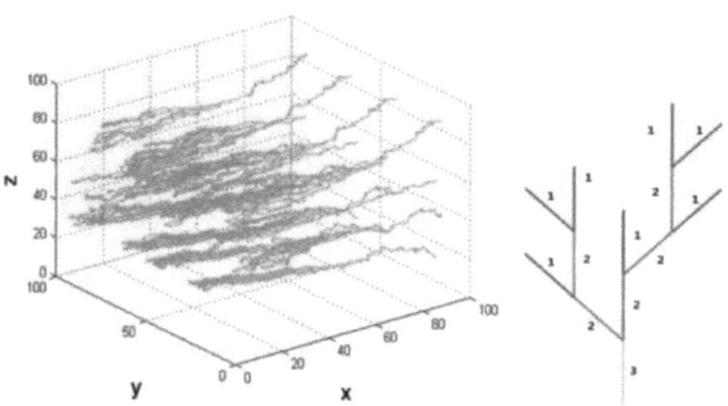

Fig. 1 The typical tree-like architecture network (left) and the Strahler system (right) of pre-existing vessels in the basic model.

B. Haemodynamic calculation

The haemodynamic model in this study is based on our previous work on the coupled modelling of intravascular blood flow with interstitial fluid flow[4]. Briefly, the basic equation for the intravascular blood flow is the flux concentration and incompressible flow at each node. Flow resistance is assumed to follow Poiseuille's law in each vessel segment. The interstitial fluid flow is controlled by Darcy's law. The intravascular and interstitial flow is coupled by the transvascular flow, which is described by Starling's law. Blood viscosity is a function of vessel diameter, local haematocrit, and plasma viscosity. In addition, vessel compliance and wall shear stress are correlated to vessel remodelling and vessel collapse.

C. Chemicals concentration calculation

The glioma cell and endothelial cell behaviours are coupled by the changes of the chemicals in the extra-cellular matrix (ECM), such as oxygen, VEGF and MDEs. The transport of these chemicals (oxygen, VEGF and MDEs) are modelled by quasi-steady reaction-diffusion equations. The ECM is treated as a continuous substance and can be degraded by MDEs, while the MDEs are governed by diffusion, produced by TCs and ECs, and the decay of itself. VEGF is assumed to diffuse, decay and be consumed by angiogenic sprouts. The production of VEGF is assumed to be proportional to TCs and ECM, representing the secretion of VEGF by TCs and the up-regulated level of VEGF in the ECM.

To obtain a more realistic oxygen concentration field, the advection and diffusion of oxygen in the vessel network are introduced, based on the work of Fang et al[5]. The computational space is separated into three domains to characterize three distinct physiological processes, which are (a) the oxygen advection equation inside the vessel, (b) the oxygen flux across the vessel wall and (c) the free oxygen diffusion in the tissue.

The initial condition of ECM density is set to be 1 and other chemicals' concentration (oxygen, VEGF and MDEs) are 0. No-flux boundary conditions are used in the simulation field. Since chemicals are transported much faster than the characteristic time for cell proliferation and migration, the chemicals' concentrations are solved to steady state at each time step of the simulation with an inner iteration step of 5s.

D. Tumour cell phenotype

We assumed four different phenotypes of glioma cells: the proliferating cells (PC), the quiescent cells (QC), the necrotic cells (NC) [6] and the migrating cells (MC). Initially, we put 20 proliferating cells in the central area. Two thresholds of oxygen concentration for cell proliferation (θ_{prol}) and cell survival (θ_{surv}) are introduced to describe the effects of oxygen field on the tumour cell actions.

As mentioned in the introduction, glioma cells can migrate along the pre-existing vessels in the early stage of microtumour growth. We assumed a specific phenotype, called the migrating cell (MC), to reflect this experimental observation. When local oxygen level is higher than θ_{surv} but lower than θ_{prol} and a space is available, a proliferating cell has a probability (50%) to become a migrating cell, and will migrate to a neighbouring space which has the highest oxygen concentration in the neighbouring elements. It was also assumed that the migrating cells adjacent to the pre-existing vessel wall have higher probability of moving in the longitudinal direction (vessel axial direction) than the radial direction. The migration speeds of the two directions are the same, i.e., 10μm per time step. After a migrating cell completes its movement, the space it originally occupied will be released for other cells. The relationships of the four phenotypes of glioma cells with the local microenvironment are shown in a diagram. Each phenotype of tumour cell has a different coefficient of oxygen consumption rate and the production rate of VEGF and MDEs[6].

E. Vessel co-option, remodelling, collapse and angiogenesis

We consider vessel dilation as the first sign of a pre-existing vessel becoming an immature vessel. A vessel segment has a VEGF concentration larger than a threshold θ_{VEGF} will increase its radius R with the rate of 0.4μm/h which will stop when the vessel radius reaches the maximum value of Rmax=10 μm. At the same time, the permeability of the vessel wall L_p is increasing in a dilation vessel.

The pressure value that will cause a vessel to collapse is defined as P_c which represents the ability of a vessel segment remaining structurally intact under the trans-wall pressure difference. In this study, initial P_c values were predefined for each vessel according to the Strahler order of the vessel segment in the pre-existing vessel network. The vessel with the larger diameter has a higher initial P_c. We assume that P_c decreases with increasing permeability of vessel wall L_p in the immature vessel. For a pre-existing vessel, once vessel dilation occurs, the vessel segment is treated as an immature vessel with increased L_p and decreased P_c. In the simulation, vessel wall compliance is defined by the radius changing under the influence of intravascular and interstitial pressures and collapse pressure. when the vessel segment becomes immature, L_p will increase which causes lower P_c, and consequently Pi will increase, both of the changes can cause vessel compressing. A compressed vessel, on the other hand will induce a higher flow resistance, lower flow which will then decrease the wall shear stress (WSS) level for the vessel. Vessel collapse will occur by either WSS criteria (as described below) or a significant reduced R. As defined in our previous work [6] and others' that vessel will collapse due to a long period of low WSS status in which the apoptosis of EC dominate the collapse process. WSS is used to estimate this kind of vessel regression.

III. Results

Figure 2 shows 3D global pictures of the GBM distribution and vessel network at different time phases during the growth period. The red tubes are the capillaries, while the blue region shows the invasion area of GBM to the surrouding tissue. It clearly indicates that tumour develops around well perfused regions at the early phases of growth. Althrough not presented, there is hardly any tumour cell growth in the region with no pre-existing vessel even at T=150.

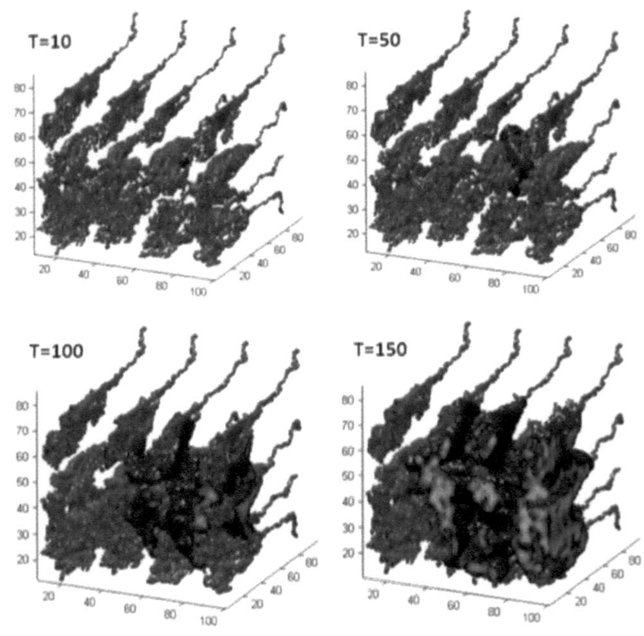

Fig. 2 3D global pictures of GBM distribution and vessel network at different time phases during the growth period.

The growth history curves in Figure 3A show the development history and characteristics of tumour cells and vessel segment more directly. At the early growth stage (T<60), there are limited neo-vasculature vessels which suggests that the pre-existing vessel network supplied sufficient nutrients to satisfy the requirement of an early GBM cluster.

However, the total number of vessel segments decreased at around T=40 due to the co-option and remodelling of pre-existing vessels (arrows in Figure 3A). To have a better comparison, the starting point of angiogenesis for a simulation is defined when the neo-vessels account for 5% of total vessel segments (the broken line shown in Figure 3A). After T=100, the angiogenesis phase occurs due to the increasing hypoxic area. The neo-capillaries have a rapid growth forming well perfusion region for GBM cells to enter. The tumour is then in an accelerating development phase. When analyzing the TC growth curve in detail (Figure 3B), it can be seen that the number of quiescent cells is normally small, about 10% of total number of TCs, while the growth curves for proliferating cells and necrotic cells are parallel with a difference of about 10k. Also in the early phase (as shown in the close up view from T=40 to T=80 in Figure 3B), the number of quiescent cells is slightly higher than the number of necrotic cells at T=40, which is due to the sufficient oxygen supply but limited available space in the microenvironment. However, a higher rate of TC growth due to the neo-vasculature made the number of necrotic cells overtake the quiescent cells at about T=60.

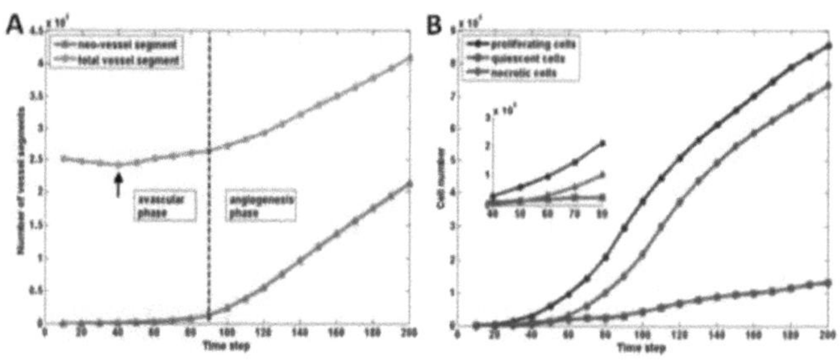

Fig. 3 A. The growth curves of angiogenic vessels and total tumour vessels in the baseline model. B. The growth curves of different phenotypes of tumour cells in the baseline model. A close up view at the early phase from T=40 to T=80 is shown in the inserted panel.

IV. Conclusions

In this work, we have proposed a dynamic mathematical modelling system to investigate the early growth process of glioblastoma by coupling the chemical and haemodynamic microenvironment caused by pre-existing vessel co-option, remodelling, collapse and angiogenesis. A 3D tree-like architecture network with different orders for vessel diameter is generated as pre-existing vasculature in host tissue. The model confirms the two different mechanisms of pseudopalisade formation by study of the dynamic relationships between the pseudopalisades, migrating cell distribution and the co-option and remodelling of pre-existing vessels. The model is not only able to provide the global results given above, but also to investigate the local immature vessel remodelling such as co-option, dilation, leaky, angiogenesis, regression and collapse and its influence on the local micro-environment and GBM growth.

Acknowledgment

The authors like to thank Mr. Justin Halls for his kind help on manuscript preparation. This research is supported by the National Basic Research Program of China (973 Program) (No. 2013CB733800), the National Nature Science Foundation of China (No. 11302050, No. 11272091), the Nature Science Foundation of Jiangsu Province (No. BK20130593), the Fundamental Research Funds for the Central Universities (YAN CAI, SEU), and the Priority Academic Program Development of Jiangsu Higher Education Institutions (YAN CAI, SEU).

Conflict of Interest

The authors declare that they have no conflict of interest.

References

1. Brat DJ, Van Meir EG (2004) Vaso-occlusive and prothrombotic mechanisms associated with tumor hypoxia, necrosis, and accelerated growth in glioblastoma. Laboratory Investigation 84: 397-405.
2. Holash J, Maisonpierre PC, Compton D et al. (1999) Vessel co-option, regression, and growth in tumors mediated by angiopoietins and VEGF. Science 284: 1994–1998.
3. Cassot F, Lauwers F, Fouard C et al. (2006) A novel three-dimensional computer-assisted method for a quantitative study of microvascular networks of the human cerebral cortex. Microcirculation 13: 1-18.
4. Wu J, Long Q, Xu SX et al. (2009) Study of tumor blood perfusion and its variation due to vascular normalization by anti-angiogenic therapy based on 3D angiogenic microvasculature. J. Biomech. 42: 712–721.
5. Fang Q, Sakadžić S, Ruvinskaya L et al. (2008) Oxygen advection and diffusion in a three dimensional vascular anatomical network. Opt. Express. 16: 17530-17541.
6. Cai Y, Xu S, Wu J et al. (2011) Coupled modelling of tumour angiogenesis, tumour growth and blood perfusion. J. Theor. Biol. 276: 90-101.

Author: Yan Cai
Institute: School of Biological Science and Medical Engineering
Street: Sipailou 2
City: Nanjing
Country: China
Email: yancai@seu.edu.cn

The Study of Contact Pressure of a Customized Knee Spacer Using Finite Element Analysis

Alex C.T. Choh[1,2], Jaryl C.K. Ng[1,2], Andy K.S. Yew[3], Chia Shi-Lu[3], and Desmond Y.R. Chong[1,2]

[1] Engineering Design and Innovation Centre, National University of Singapore, Singapore
[2] Department of Biomedical Engineering, National University of Singapore, Singapore
[3] Department of Orthopaedic Surgery, Singapore General Hospital, Singapore

Abstract — **With one-third of patients having osteoarthritis predominantly in one compartment of the knee, unicompartmental knee spacers have been introduced as a less invasive alternative to total knee replacement. However, patients with the knee spacer implanted were seen to have persistent pain, resulting in high revision rates. A static computational finite element model of the knee at full extension and 90° flexion, mimicking the peak tibio-femoral force at heel strike during gait and peak joint forces during an ascending motion from a squatting position was used to study the tibio-femoral contact pressure of different designs and proper-ties of the knee spacer. Three designs (contoured - conventional, flat and C-shaped) of the knee spacers were created and simulated with three different materials (cobalt-chromium, UHMWPE and polyurethane). The average contact pressures acting at the medial compartment of the femoral condyle were extracted for analysis.**

Spacers made out of soft material exerted lesser contact pressure onto the femur cartilage. In the full extension model, using a soft material of polyurethane, the C-shaped spacer exhibited the lowest femoral cartilage contact pressure with a 31% and 18% decrease from the flat and contoured spacers respectively. In the 90° knee flexion model, the contoured spacer had the lowest contact pressure among the other spacers. Contoured spacers made of polyurethane exhibited a 13% and 20% decrease from the flat and C-shaped spacers respectively.

Therefore, it is suggested that the knee spacers should be made out of a softer material that can withstand the joint forces in various motions and have a geometry that would increase the contact area of the between the joints.

Keywords— **Knee spacer, Contact pressure, Biomechanics, Finite element analysis**

I. INTRODUCTION

With the increasing demand by the younger generation (age below 50) suffering from osteoarthritis (OA) [1], the loss in mobility and discomfort faced with the gold standard treatment of the implantation of total knee replacements (TKR), unicompartmental knee replacements (UKR) and high tibial osteotomies (HTO) is unfavorable as this patient group is generally more active and functionally more demanding after the surgery. To account for the initial onset of OA in younger patients, the interpositional knee spacer has recently been introduced as a minimally invasive alternative to the total knee replacement, with the hopes that this implant would provide a more natural experience to patients as a minimal portion of bone is resected this procedure. Furthermore, the implantation of knee spacers are easier to be done and are also easier to be removed from the system as compared to standard current practices should there be further degeneration in the knee and a need for implant revision arises. These knee spacers preserve the stock of the bone and work on the concept of early intervention of OA, aiming to replace the function of the cartilage and meniscus that has been lost due to mild OA in the worn out side (usually the medial side) of the knee, keeping the ligaments intact. These spacers resemble the shape and form of the tibia plateau and provide a smooth and polished surface for the articulation of the femoral condyle. The spacers are left free floating utilizing a unique shape that helps hold the implant in place on the tibia plateau.

However, there has been mixed reactions stating the unsuitability of the knee spacer due to its high revision rate. Clarius et al.'s study [2] on the radiographs of 18 patients before and after the implantation of the UniSpacer showed that during a five-year follow-up period, 21% of spacers were converted to a UKR or TKR due to persistent pain. The mean time for revision was 23.8 ± 18 months. The mechanical failure mode reported was mainly anterior dislocation (or spitting out) of the knee spacer. This could be due to unclear biomechanical factors in which the anterior shear force is acting on the spacer by the femoral condyle. In another study by Bailie et al. [3], it was found that with an average follow up of 17 months, 17 out of the 18 patients who received the UniSpacer implants had persistent pain within the first 3 to 6 months post-operation. 12 patients required co-intervention and there were 8 implant failures. It is noted that most of the literature studies reported the survivorship of the knee spacer, with little or none investigating the exact mechanism of how the knee spacer would subluxate out of the knee capsule.

It is hypothesized that the material property and geometry of the knee spacer would have an impact on the survivorship of the implant. With little cartilage removal during a

J. Goh and C.T. Lim (eds.), *7th World Congress on Bioengineering 2015*, IFMBE Proceedings 52, DOI: 10.1007/978-3-319-19452-3_25

knee spacer surgery, hard contact between the knee spacer and the cartilages of the bone would occur, thereby increasing the contact pressure and frictional force. A mismatch in material properties may cause further wear (or arthritis) in the cartilage, causing pain and discomfort to the patients. Furthermore, proper design of the shape and geometry and introducing appropriate undersurface coating of the knee spacer could help in better anchoring of the implant to the knee joint, hence reducing the likelihood of implant subluxation. Therefore, this study aims to investigate the influence of new design factors of material properties and shape and geometry on the contact pressures in the knee joint.

II. Materials and Methods

A. Acquisition of geometrical models

3D models of a healthy knee belonging to a 78 year old male and a knee with varying interpositional knee spacers in the medial compartment of the tibio-femoral joint at full extension was created using Mimics 17.0 (Materialise, Belgium), a medical image processing software which allows for the visualization of 3D models using medical images. The hard tissue components (femur and tibia) of the knee at full extension were modelled from CT images while the soft tissues (ligaments, meniscus and articulating cartilages) were modelled from the MRI images obtained. The CT and MRI images were superimposed onto one another in order to obtain the full geometry of the bones and soft tissues in the knee joint and to ensure proper positioning of the bones relative to the soft tissues. In the case of the knee models with the implant, the medial meniscus and medial portion of the tibia cartilage were replaced with the knee spacer. A small portion of bone was resected in the medial side of the tibia to accommodate the placement of the spacer. All components of the model were then assembled and meshed in 3-Matic 9.0 (Materialise, Belgium (Figure 1).

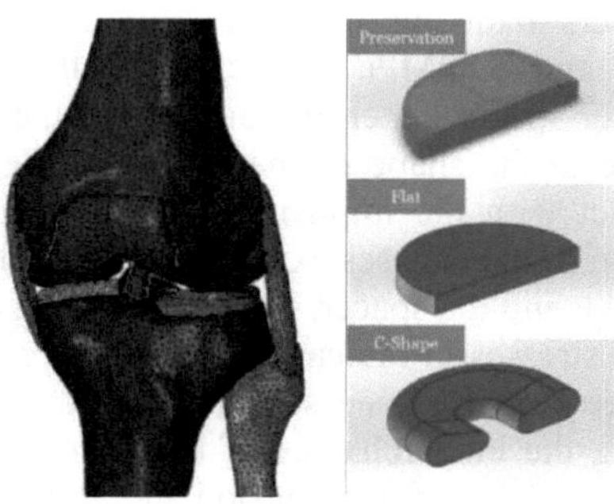

Figure 1 The intact FEA knee model and the 3 knee spacers modelled

Isotropic material assignment of the bones was done in Mimics 17.0 based on the Hounsfield Units obtained from the CT scans. Equations obtained from a study by Rho et al. [4] were utilized to relate the Hounsfield Units of each pixel in the CT scan to a density value. Each bone models was segregated into 30 sets based on their density. Sets that had a density of above 1.03g/cm^3 were classified as cortical bone and sets that were below 1.03g/cm^3 were classified as cancellous bone. From the density obtained, the Young's Modulus of Elasticity of each set was calculated based on previous literature review (Table 1) [4]. Equations differed between cortical and cancellous bone. However, in the case of the soft tissues, a homogenous isotropic material property was given to cartilage, ligaments and meniscus [5-7].

Table 1 Material properties of the components in the finite element model

Item	Density (kg/m^3)	Young's Modulus (MPa)	Poisson's Ratio
Femur (Cortical)	p = 139 + 1.205CT#	E = -6.143 + 0.014p	0.3
Tibia (Cortical)	P = 114 + 0.916CT#	E = -3.842 + 0.013p	0.3
Femur (Cancellous)	p = 139 + 1.205CT#	E = 367 + 6.28CT#	0.3
Tibia (Cancellous)	P = 114 + 0.916CT#	E = 296 + 5.20CT#	0.3
Co-Cr	8500	258000	0.3
UHMPE	924	500	0.46
PU	1200	148	0.3

3 different shapes of the interpositional knee spacer were modelled in SolidWorks. One spacer design mimicked the top contour of a pre-existing unicompartmental knee arthroplasty, the Preservation® Uni-compartmental Knee System (DePuy Synthes, United States), while the other two designs were a simple flat design and a C-shaped design. Coupled with the different shapes, the materials of the implants and the surface coating were also varied. Three different materials (cobalt-chromium, Co-Cr, ultra-high molecular weight polyurethane, UHMWPE, and polyurethane, PU) of the knee spacer were simulated based on materials currently used in knee implants. The material properties and densities used for each material are also listed in Table 1.

B. Finite element analysis

The solid models obtained, consisting of tetrahedral elements were then imported into Abaqus CAE (Dassault Systèmes, France) where contact, loading and boundary conditions were created. C3D4 elements were chosen to reduce computational time. The model consisted of 604,000 elements with maximum edge lengths of 3mm. Contact interactions between surfaces consist of two general components involving finite sliding of pairs of curved, deformable surfaces; the normal contact behavior and the tangential contact behavior. All throughout the model, 3 main interaction types were established; cartilage to cartilage interactions, cartilage to implant interactions and bone to implant interactions. Cartilage to cartilage interactions includes the contact between the femur cartilage and tibia cartilages and meniscus and they have a friction coefficient of 0.02. Contact between implant to bone were assigned a friction coefficient of 0.6. Contact between spacers with cobalt-chromium, UHMWPE and polyurethane properties and the femur cartilage were assigned friction coefficients of 0.55, 0.3 and 0.15 respectively [8-10].

Two loading and boundary conditions were assigned to mimic the i) peak joint forces in a gait cycle for the full extension model and ii) the knee joint at 90° flexion during an ascending motion from a squatting position. A loading of 3 × body weight (BW) was applied to the proximal femur shortly after heel-strike at about 15% of the gait cycle for the full extension models [9]. For the 90° flexion model, an inferior and posterior force of 65% and 50% BW respectively were applied to the femur along the line of tibio-femoral contact [11]. For all models, the tibia was fully constrained in all degrees of freedom and the soft tissues were tied to their respective attachment points. However, for the 90° flexion model, the ligaments were modelled as 2D linear springs [12, 13] as the position and the structures of the ligaments in a 90° knee flexion setup were unknown. The knee spacers were left unconstrained and free to translate during the simulations. The contact pressure and contact area of the femur cartilage was recorded and compared.

III. Results

Tables 2 – 4 contain the results obtained from the computational simulations of the knee joint with and without the implantation of the spacers.

Table 2 Average and peak contact pressures experienced by the intact knee model

Average medial femur cartilage contact pressure (MPa)	Average lateral femur cartilage contact pressure (MPa)	Peak medial femur cartilage contact pressure (MPa)	Peak lateral femur cartilage contact pressure (MPa)
1.94	2.50	5.37	6.54

IV. Discussion

This preliminary study aimed to look into the biomechanics of varying spacers in the knee joint. To do so, the average contact pressures and contact areas of the medial and lateral compartments of the femur cartilage after each simulation were observed. Validation of the model was performed by comparing the peak contact pressures in the intact knee model at full extension to that of literature. Marzo et al. reported peak contact pressures of 3.84 MPa ± 1.24 MPa and 5.08 MPa ± 0.76 MPa in the medial and lateral compartments of the knee respectively [14]. Comparing these values to that obtained in the current model (5.37 MPa and 6.54 MPa in the medial and lateral compartments respectively), the magnitudes and values did not differ much, with the discrepancies being explained by the slight differences in loading conditions and different bone properties.

Based on the full extension simulations, the introduction of an implant that had stiffer material properties than that of the articulating cartilages did prove to have a big impact in the biomechanics of the knee, with average contact pressures in the medial femoral condyle having an 88% – 419% increase as compared to the intact knee model. Interestingly, only a small variation in the lateral compartment (7% – 10% difference) was observed. Furthermore, the contact area in the medial compartment of the knee was seen to reduce when a spacer was introduced into the model as compared to the intact knee model (71% – 83% decrease). The reduction in contact area in the models with an implant introduced would result in a large increase in contact pressure as the load distribution is focused onto a smaller area, as compared to the intact models with the meniscus and cartilages dampening and distributing the force applied.

It was also noticed that in both the full extension and 90° models, as the elastic modulus of the spacer decreased, the average contact pressure on the medial femur cartilage decreased, and at the same time, the contact area between the femur cartilage and implant increased. In the full extension model, the knee model with the polyurethane spacer was highly favorable as the contact pressure was the lowest (4.49MPa averaged across all geometries) and closest to mimicking the contact pressure experienced by the femur cartilage in the intact knee (1.94MPa) as compared to the other spacers made out of cobalt chromium or polyethylene (7.39MPa and 6.89MPa respectively). It is hypothesized that a softer spacer material would allow the femur to conform more to the shape of the spacer as the femur comes into contact with the implant. This allows for an optimal distribution of the stresses experienced by both the articulating cartilage and the spacer, thereby reducing the magnitude of contact stresses in the cartilage and spacer. Besides that, the geometry of the knee spacer also played a part in the influencing the biomechanics of the knee. Results differed across the models in full extension and in 90° flexion. However, contrasting effects of the geometry of the spacer was observed between the contoured and C-shaped spacer in the full extension model and the 90° flexion had the lowest con-

Table 3 Average contact pressures (MPa) acting between the medial femoral condyles and the knee spacer at full extension and 90° flexion

	Full Extension			90° flexion		
	Preservation	Flat	C-Shape	Preservation	Flat	C-Shape
Co-Cr	6.15	10.08	5.91	2.65	3.71	3.29
PE	7.81	7.59	5.28	2.70	3.21	2.95
PU	4.49	5.31	3.66	2.43	2.80	3.05

Table 4 Total contact area (mm^2) between the medial femoral condyles and the knee spacer at full extension and 90° flexion

	Full Extension			90° flexion		
	Preservation	Flat	C-Shape	Preservation	Flat	C-Shape
Co-Cr	66.51	45.09	56.06	47.50	33.46	40.92
PE	75.24	61.86	65.67	47.62	33.53	42.25
PU	75.35	69.80	72.64	52.68	43.61	43.98

tact pressure among the other geometries of spacers. Using a soft material of polyurethane, the C-shaped spacer exhibited the lowest femoral cartilage contact pressure with a 31% and 18% decrease from the flat and contoured spacers respectively. In the 90° knee flexion model, the contoured spacer had the lowest contact pressure among the other spacers. Again, using polyurethane as a base comparison, the contoured spacer exhibited a 13% and 20% decrease from the flat and C-shaped spacers respectively. The difference in results between the full extension model and the 90° models again can be attributed to the difference in contact area between the femur cartilage and the implant. It was noticed that all models that had lesser contact pressures acting on the femoral cartilage had an increase in contact area between the cartilage and implant. Therefore, it is suggested that the knee spacers should be made out of a softer material that can withstand the joint forces in various motions and have a geometry that would increase the contact area of the between the joints.

However, these suggestions are generally based on a simplified knee model where the material properties of the soft tissues were simplified, without considering its viscoelasticity behavior. Viscoelasticity of the ligaments, meniscus and articulating cartilages could be looked into in the subsequent studies. In addition, the current study only encompasses a static model at full extension and at 90° flexion. While this model may be sufficient as a preliminary model as the forces simulated are set to represent the peak joint forces acting on the knee joint during gait and during a specific point during knee flexion, it does not take into account the various changes in loadings across different flexion angles and the movement of the femur relative to the tibia when the knee undergoes a dynamic movement.

References

1. Singapore, M.o.H., *National Health Surveillance Survey 2007*, E.D.C. Division, Editor. 2007. p. 140.
2. Clarius, M., et al., *The UniSpacer: correcting varus malalignment in medial gonarthrosis.* Int Orthop, 2010. **34**(8): p. 1175-9.
3. Bailie, A.G., et al., *The Unispacer knee implant: early clinical results.* J Bone Joint Surg Br, 2008. **90**(4): p. 446-50.
4. Rho, J., M. Hobatho, and R. Ashman, *Relations of mechanical properties to density and CT numbers in human bone.* Medical engineering & physics, 1995. **17**(5): p. 347-355.
5. Wangerin, S., *Development and validation of a human knee joint finite element model for tissue stress and strain predictions during exercise.* 2013, California Polytechnic State University, San Luis Obispo.
6. Zielinska, B. and T.L.H. Donahue, *3D finite element model of meniscectomy: changes in joint contact behavior.* Journal of biomechanical engineering, 2006. **128**(1): p. 115-123.
7. Donahue, T.L.H., et al., *A finite element model of the human knee joint for the study of tibio-femoral contact.* Journal of biomechanical engineering, 2002. **124**(3): p. 273-280.
8. Murakami, T., et al., *Tribological behaviour of artificial cartilage in thin film lubrication.* Tribology Series, 2000. **38**: p. 317-327.
9. Morrison, J., *The mechanics of the knee joint in relation to normal walking.* Journal of biomechanics, 1970. **3**(1): p. 51-61.
10. Rancourt, D., et al., *Friction properties of the interface between porous-surfaced metals and tibial cancellous bone.* Journal of biomedical materials research, 1990. **24**(11): p. 1503-1519.
11. Nagura, T., et al., *Mechanical loads at the knee joint during deep flexion.* J Orthop Res, 2002. **20**(4): p. 881-6.
12. Fishkin, Z., et al., *Changes in human knee ligament stiffness secondary to osteoarthritis.* Journal of orthopaedic research, 2002. **20**(2): p. 204-207.
13. Race, A. and A.A. Amis, *The mechanical properties of the two bundles of the human posterior cruciate ligament.* Journal of biomechanics, 1994. **27**(1): p. 13-24.
14. Marzo, J.M. and J. Gurske-DePerio, *Effects of medial meniscus posterior horn avulsion and repair on tibiofemoral contact area and peak contact pressure with clinical implications.* Am J Sports Med, 2009. **37**(1): p. 124-9.

Patellofixator Rig – Design Specifications of a Device Assisting Medial Patello Femoral Ligament [MPFL] Reconstruction

Sarthak Patnaik[1] and Sudesh Sivarasu[2]

[1] Learning Trauma Med, Santander, Spain.
sarthakpatnaik@hotmail.com

[2] MRC/MIRU, Division of Biomedical Engineering, Department of Human Biology, Faculty of Health Sciences, University of Cape Town, South Africa
sudesh.sivarasu@uct.ac.za

Abstract— **Patella dislocation is one of the most common pathology of the knee joint. Incidence of patellar dislocation is 5.8 to 77.8 in 100000. Medial patella femoral ligament plays a pivot role in patellofemoral instability. All acute dislocations causes some degree of macroscopic MPFL damage and studies shown lateral dislocation is impossible without damage to MPFL. 44% non-conservative treatment patient go on to suffer redislocation, along with painful symptoms of chronic instability. MPFL reconstruction surgeries prevent further dislocation of patella by adopting one of the four surgical techniques.There been no special intra operative method to determine the anatomical location though an intra operative xray image can help in identifying the exact MPFL insertion site but there are high chances of radiation exposure to the patient, surgeon, assistant and other technical staffs.We have designed a zig which can fit to the normal patella of any size and can be adjusted to determine the MPFL attachment site in Patella.**

The test rig (Patellofixator), which is divided into an upper and lower section, has been designed to be placed for guiding tools for performing the surgical procedure. Its pin slots are located to the MPFL attachment to the patella. Some surgeon prefers the position of the MPFL patella footprint at the Centre. Going through the various papers regarding the anatomy and correct position of the patella foot print our ZIG will help the surgeon to maintain their anatomical location. Considering the chances of the patella fracture as complication we are determined to give the surgeons a zig which will ease their anchor or tunnel placements in the patella.

This small and economical device will assist surgeons to adapt to varying patellar size of the patients and effectively reconstructing the medial patella femoral ligament.

Keywords— **Patella Surgery, Fixation, Medical Device, Design**

I. Introduction

Patella dislocation is one of the most common pathology of the knee joint. Incidence of patellar dislocation is 5.8 to 77.8 in 100000[3]. Medial patella femoral ligament plays a pivot role in patella-femoral instability. All acute dislocations causes some degree of macroscopic MPFL damage [1,2] and studies shown lateral dislocation is impossible without damage to MPFL. 44% non-conservative treatment patients go on to suffer redislocation [4], along with painful symptoms of chronic instability. The MPFL usually ruptures at its femoral origin upon lateral dislocation of patella typically under rotational force with knee extended. Upto 94% of patients suffer from MPFL rupture after acute patella dislocation. Over 130 distinct techniques for alleviating the patella femoral instability and pain of all origins have been described since 1915 and studies have demonstrated encouraging results with early follow up.

Anatomy of Patella: The MPFL is defined as a thick band like condensation of tissues extending from patella to medial femur. As per the Anatomical study [5]. The Average width of patella insertion of MPFL – 17mm. (14 – 20). Avg width of femoral origin on medial epicondyle – 15.4 mm (11- 20). Avg Patella height – 43.8 mm (40-47).Vertical distance from superior pole of patella to superior edge of MPFL at its patella insertion – 6.1 mm (mostly supero-medial). Percentage of total patella width of MPFL at patella – 38.8%. Distance from superior pole of patella to the inferior edge of MPFL at its patella insertion Avg 23.1 mm. (15 -29mm).

Isometry of MPFL : Points of reference on the patellar attachment did not differ significantly from each other whereas the isometry point in femur is the superior aspect of femoral attachment of MPFL [5]. For example Steinsen et al found out that the portion of MPFL extending from its inferior aspect of its patella attachment to superior aspect of its femoral attachment (C-D) was most isometric . During knee flexion 0-90o the total change in length was 1.1 mm between these 2 points. Second isometric point, middle of patella insertion to superior edge of femoral origin. from 0-90o flexion the total change in length was 1.8m.

II. Materials and Methods

MPFL reconstruction surgeries prevent further dislocation of patella by adopting one or the other following surgical techniques. 1. MPFL reconstruction with a divergent patellar trans 2- tunnel technique.; 2. Y graft – (fixation of

J. Goh and C.T. Lim (eds.), *7th World Congress on Bioengineering 2015*,
IFMBE Proceedings 52, DOI: 10.1007/978-3-319-19452-3_26

femur first) ; C graft – fixation of patella first. (2 incision, first 3cm longitudinal incision was along the proximal medial border of the patella and other 1cm above the MPFL insertion.) ; 3. V shaped tunnel – drill 1.5 cm at the supero medial half of patella with sufficient distance between tunnels to avoid fracture. ; 4. Docking technique for MPFL reconstruction

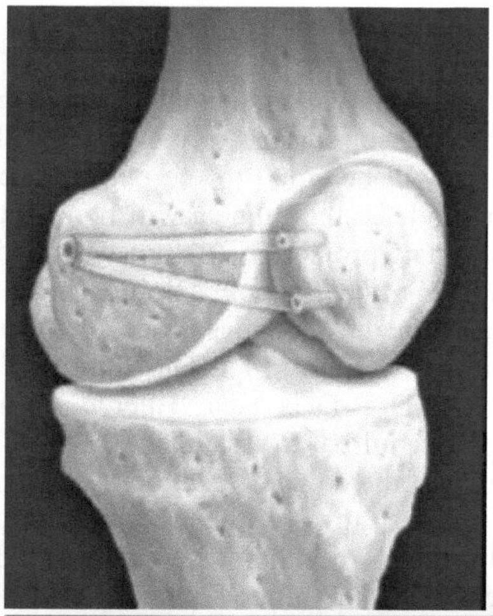

(a)

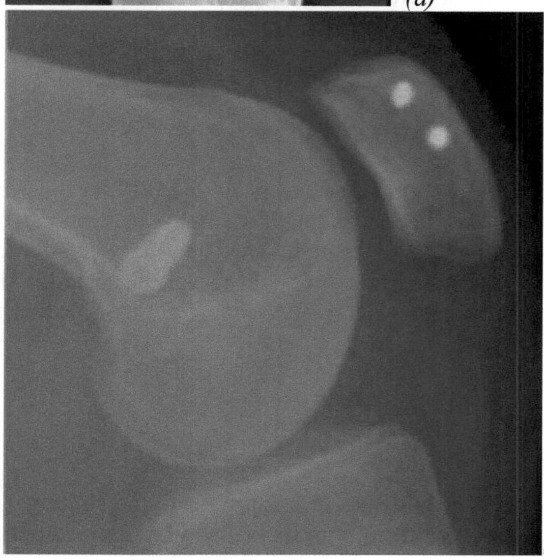

(b)

Fig. 1 (a) &(b) MPFL Reconstruction Surgery

In all the above methods, patella was approached blindly with no special rig for identifying the exact anatomical landmark of MPFL on the patella. There has been special scale (Arthrex) to determine the position of attachment of MPFL on the femur under radiographic guidance, though MPFL can be localized without radiological interventions by determining the point between the adductor tubercle and medial epicondyle of the femur. There been no special intra operative method to determine the anatomical location though an intra-operative x-ray image can help in identifying the exact MPFL insertion site but there are high chances of radiation exposure to the patient, surgeon, assistant and other technical staffs.We have designed a rig which can fit to the normal patella of any size and can be adjusted to determine the MPFL attachment site in Patella. A CAD model of the test rig is shown in Fig. 2. The test rig, which is divided into an upper and lower section, has been designed to be placed for guiding tools for performing the surgical procedure. Its pin slots are located to the MPFL attachment to the patella. Some surgeon prefers the position of the MPFL patella footprint at the Centre. Going through the various papers regarding the anatomy and correct position of the patella foot print our RIG will help the surgeon to maintain their anatomical location. Considering the chances of the patella fracture as complication we are determined to give the surgeons a rig which will ease their anchor or tunnel placements in the patella.

III. RESULTS & DISCUSSIONS

Patella is a sesamoid bone which is still strong. To do a MPFL repair, we just need a small incision on the skin on medial surface of patella. The patella generally moves when we try to hold it, so there can be slight loss in orientation. The device which will be built has to sit on the top of the knee, above the skin such that it forms a tight grip around the patella with special knobs which will be holding the patella with few thin grasper like thing (may be needle diameter as that doesn't make big impression on the skin). The rig is such built that it can be adjustable to the length and breadth of the patella. After the patella rig sits on the patella,we have to make sure that its upper margin is at the level of superior pole of patella and similarly its lower margin at the level of inferior pole of patella similarly the medial and the lateral edges.

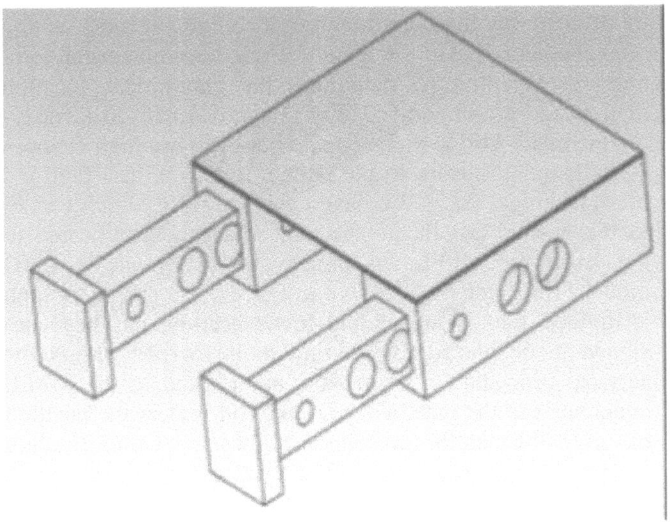

(a)

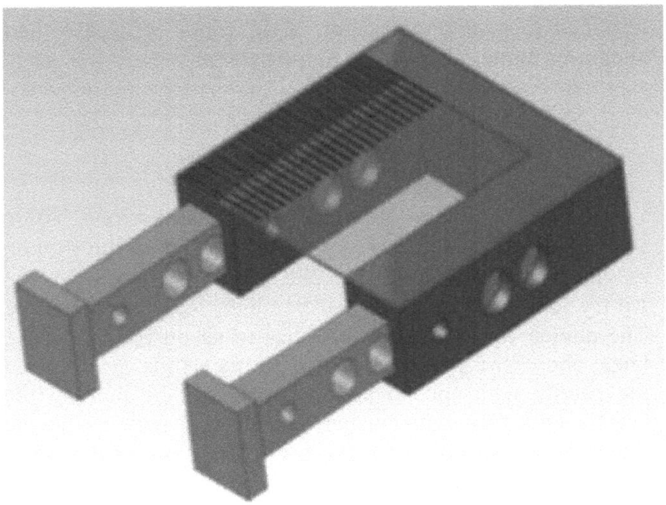

(b)
Fig.2 (a) & (b) MPFL PatelloFixator Design

Once the rig sits, the only way we can see the anterior surface of patella through the transparent radiolucent material on the top of the rig which lies adjacent to the anterior surface of the patella. This surface do bears a scale on top of it whose markings will be radio opaque and then determining the exact central position of the specific patella the drill channels will be adjusted. Since the upper surface is translucent , when we pass the drill bit and take an image intra op we can determine the position of the drill bit in Image intensifier, confirming the position of the tunnels (if a surgeon require it or an amateur surgeon) The most important thing there might be the necessary of two scales on the anterior surface, one scale which determine the total length of the patella the other will be vernier scale separately or imprinted on the first which will help in correct establishment of the tunnels. Though the size of the patella been mentioned initially. The distance between the tunnels is nearly 1 cm. The device need not be required to bend while putting the anchors or suture in knee flexion

It will look with a slight convex on the anterior side and minimal concave at the anterior surface of the patella. The total thickness of the rig will be equal to the patella thickness if not possible then atleast the tunnels will be slightly anteriorly placed. The small projection (x) present on the posterior surface acts like a small hook helping to get it attached to the patella. On the medial surface of the rig there will be 4 holes (A,B – 7,8mm) (C,D- 7,8mm). The distance between A-C AND B-D is 1cm. 1thereby maintained the distance gap required to establish the patella tunnel.(Fig 2).The frame of the device has to be rigid frame to sustain the drill bit and also to give it a structure, but the body has to be radiolucent so that the scale can be readable.it will help in determining the tunnels intra op with the help of imaging intensifier.

IV. CONCLUSIONS

Despite the preliminary success with MPFL reconstruction No rig has been designed specially to recreate the isometry of the native ligament. This small and economical device will assist surgeons to adapt to varying patellar size of the patients and effectively reconstructing the medial patella femoral ligament.

ACKNOWLEDGMENT

The authors would like to acknowledge the support and assistance received from both their institutes for conducting this study.

CONFLICT OF INTEREST

"The authors declare that they have no conflict of interest".

REFERENCES

1. Christopher S Ahmed, Gabriel D. The docking technique of medial patellofemoral Ligament Reconstruction: Surgical Technique and Clinical Outcome. American Journal of Sports Medicine 2009: vol-37,No 10:2021-27.
2. Sally PI, Poggi J, Speer KP, Acute dislocation of the Patella: A correlative Pathoanatomic study . Am J Sports medicine.1996;24:52-60.
3. Sillanpaa P,Mattila VM,Livonen T. Incidence and risk factors of acute traumatic primary patella dislocation. Med Sci Sports Exerc.2008;40(4):606-611.
4. Trikha SP,Acton D, O'Reilly Acute lateral dislocation of the patella: correlation of ultrasound scanning with operative findings. Injury 2003;34:568-571.
5. Robert N Steensen, Ryan M. Dopirak, William G. The Anatomy and Isometry of the Medial Patello femoral Ligament. Am J Sports Medicine. 2004,Vol 32,No 6,1509-1513.
6. Sudesh Sivarasu, Sarthak Patnaik - Novel Stress Radiography Device for Measuring Knee Laxity at Various Flexion Angles – Laxmeter, Journal of Medical Devices: ASME transactions, 2014, Vol.8(2) :020938-020938-2.MED-14-1024
7. Sudesh Sivarasu, Sarthak Patnaik - Design and prototype development of a novel low cost stress radiographic device - Laxmeter (An Indo- Africa Invention) Appropriate Healthcare Technologies for Developing Countries – AHT2014. The 7th International Conference – World Health and Wellbeing, 17 – 18 Sep, 2014. London, UK Paper no. 0036 DOI: 10.1049/cp.2014.0776

Author: Dr. Sarthak Patnaik
Institute: Learning Trauma Med,
City: Santander
Country: Spain
Email: sarthakpatnaik@hotmail.com

Development of Wireless Sensing and Optical Stimulation Module for Optogenetic Animal Study

Meng-Chun Liu[1], Yu-Ting Li[2], and Jia-Jin Jason Chen[1]

[1] Institute of Biomedical Engineering, National Cheng-Kung University, Tainan, Taiwan
[2] Instrument Technology Research Center, National Applied Research Laboratories, Hsinchu, Taiwan

Abstract— **Optical stimulation of target neurons, known as optogenetics, is an effective method to selectively control the genetically-modified neurons using different wavelength of light. The optical stimulation can provide spatial specificity for target cells and eliminate electrical artifacts. In this study, we develop a wireless sensing and optical stimulation module which can wirelessly control the parameters of optical stimulation for exciting dopaminergic neurons and provide fast-scan cyclic voltammetry (FSCV) for detecting the phasic dopamine (DA) release in the dorsal striatum of rat. The optical stimulator system can be remotely controlled using LabVIEW computer interface for delivering blue light (473nm) in the brain via optical fiber to evoke DA releasing. FSCV provides higher time resolution for measuring transient dopamine changes and causes less tissue damage using carbon fiber microelectrodes (7μm) as electrode for brain insertion. The measured local field potential (LFP) and dopamine signals are unidirectionally transmitted from a commercially available wireless module to the host unit. Our study first validated that the carbon fiber microelectrode coated with Nafion has increased the electrode sensitivity and improved the selectivity of DA detections. We also found that the responsive currents are increased linearly in proportion to DA concentrations (0.01-0.64 μM) with a correlation coefficient of 0.999. In an *in vivo* test, the wireless system delivered the blue light for exciting LFP which is transmitted to the host via a wireless transmitter. The occurrence of measured LFP was in synchronizing with the optical stimulation which was averaged to show the LFP template. The developed wireless system is proven to be a useful experimental tool for the continuous monitoring of LFP and DA. A miniature FSCV module might be feasible using ASIC approach in the future. The proposed wireless system will be extended as an experimental platform for Parkinson disease experiment in an ongoing project.**

Keywords— **Fast-scan cyclic voltammetry, Carbon fiber microelectrode, Dopamine, Local filed potential, Wireless.**

I. Introduction

Parkinson's disease (PD) is caused by the loss of neurotransmitter dopamine (DA) that is related to motion control. DA has been one of the important neurotransmitters in the biological system, linking to psychosis and locomotion modulation in the biological system [1, 2]. So there are many studies have been investigated the correlations between the motor functions and dopamine, including medication, DBS, rTMS, tDCS and some new approaches. However, the limitations of all clinical treatments are poor temporal resolution and synaptic plasticity to clear the mechanisms between the DA and the motor functions, and the optical stimulation and FSCV can provide them in the same time. *Optogenetic stimulation* technique can provide spatial specificity for target cells which express light-sensitive ion channel, channelrhodopsin-2(ChR2), and it would be activated to induce depolarization by the millisecond pulses of blue light at 473nm wavelength and the irradiance is between 0.1 and 1 mW/ mm^2 [3]. And *fast-scan cyclic voltammetry* (FSCV) is the common methodology for real time detection of electrochemically active compounds with great sensitivity [4]. The one purpose of this study is in an attempt to coated Nafion onto the surface of microelectrode to enhance better sensitivity and selectivity for DA detections [5].

Apart from the surface modification of microelectrodes, this study would integrate a wireless FSCV system and optical stimulation module which can be applied to transmit the measured dopamine signal and the electrophysiological signal which evoked by optical stimulation with TBS protocols as output patterns in the rat from the wireless module to the host unit.

II. Materials and Methods

A. Fabrication and Modification of dopamine sensing microelectrodes

The two-electrode, comprised of the working electrode (WE) and reference electrode, are used for fast-scan cyclic voltammetry DA sensing system.

Working electrodes: A cylindrical carbon fiber microelectrode was fabricated by pulling a glass capillary around a 7-um-diameter single carbon fiber (product no. C005722, Goodfellow Corporation, Berwyn, USA). The microelectrode was then back-filled with Graphpoxy-PX (Dylon Industries, Cleveland, Ohio, USA) to connect the carbon fiber to the copper wire, after curing overnight, cut off carbon fiber around 250 um which was exposed from the tip of glass capillary by Focused Ion Beam (FIB) (FEI, Nova-200, NanoLab Compatible), the carbon fiber microelectrodes are illustrated in Figure 1. The microelectrodes were prepared for

J. Goh and C.T. Lim (eds.), *7th World Congress on Bioengineering 2015*, IFMBE Proceedings 52, DOI: 10.1007/978-3-319-19452-3_27

use by soaking in 2-propanol purified by pretreatment with 0.1 g/mL Norit A activated carbon fiber for 20min [6]. And in order to enhance the selectivity and sensitivity of electrodes, we applied to immersed the tip of electrodes in 5% Nafion solution (Sigma, 274704) for 15 seconds, then dried it at the 85°C oven for 5 minutes, and repeated three times to sure the thickness of Nafion be unanimous [7]. Then, we applied to test the traditional characterizations of electrodes, and compare it before and after coating the Nafion films on the surface of electrodes.

Reference electrodes: Ag/AgCl electrode is a common material for reference electrode in related study. The reference electrode (RE) is fabricated with silver wire which coated on AgCl (Ag/AgCl). The Ag/AgCl was coated by applying 1.5 V for 1 min in 3 M KCl. Finally, the carbon fiber electrode and reference electrode were allowed to glue with a connector.

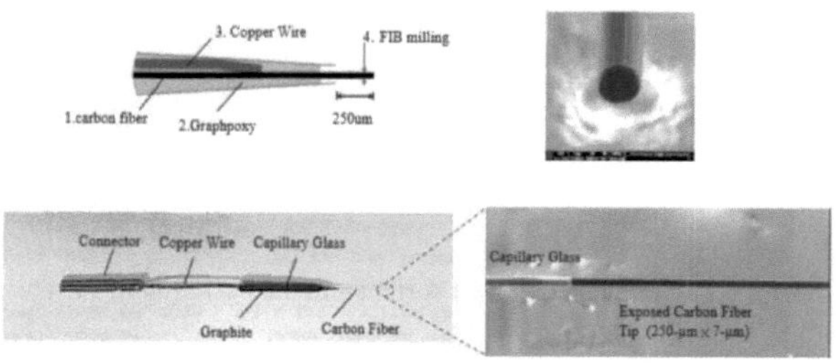

Fig. 1 Structure of carbon fiber microelectrodes.

B. Optical stimulation system

The wireless optical stimulation system is composed with two units: transmitter and receiver. There are two kinds of theta burst stimulation protocol are transmitted by 27M Hz radio frequency transmitter, and the two patterns are the continuous TBS (cTBS) and the intermittent TBS (iTBS) produced by micro controller, the first one sustained 40 second train of TBS for a total number of 600 pulses. In contrast, the iTBS, 2 second train of TBS repeated at 0.1 Hz and total number of 600 pulses. The receiver unit is minimized to be a backpack for the rat, which involve the 27M Hz radio frequency receiver, amplifier and Schmitt circuits to sure the integrity of the patterns from radio frequency module, and filter circuit to avoid the noise from environment. Finally, applying a current regulator to 30 mA for high intensity LED chips (EZ500 Gen 2, Cree). Then using LED-coupled fiber to deliver the light into the brain via the cannula of optotrode. And the optotrode is composed with a stainless canula that contain 200 μm optical fiber and electrode. Thus, the stainless electrode can detect the electrophysiological signal like local field potentials (LFP). And the whole block diagram is shown in Figure 2.

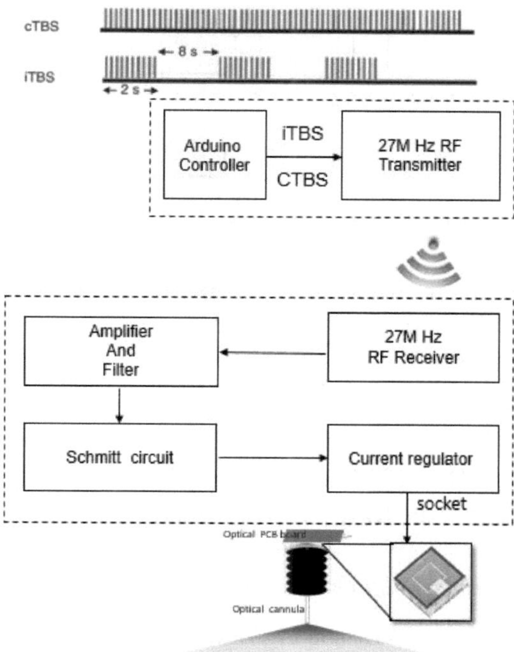

Fig. 2 Block diagrams of wireless optical stimulation system and the two patterns are produced by microcontroller.

C. Wireless dopamine recording system

Block diagrams of wireless voltammetry system is shown in Figure 3. In the FSCV system, the triangular potential is generated via a digital-to-analog converter (DAC) (AD8801) controlled by a microcontroller unit (MCU) (PIC18F452). The baseline of the triangular potential is shifted using a subtractor circuit and then the amplitude of the triangular potential is amplified and low-pass-filtered at a 3-kHz cut-off frequency. This system provide 10-Hz triangular potential (-0.3 to 1.4V versus Ag/AgCl electrode, 340 V/s) between the carbon fiber microelectrode and the reference electrode that produces an adequate time resolution of 0.1 sec for dopamine measurement. A head-mounted pre-amplifier is implemented to convert the voltammetric current to a voltage. And the volammetric signal of electrochemical compounds and the local field potentials which evoked by optical stimulation are transmitted to the host unit via the wireless module (Triangle Biosystems Inc., Durham, North Carolina, USA). Then the signal is received by the wireless receiver and sampled at 80 kHz via the ADC (DAQ pad-6221, National Instruments). The voltammetric signal, triangular waveform, dopamine response signal and LFPs are displayed in real-time on a graphical user interface and recorded in the computer by LabVIEW. Finally, those signal are analyzed with MatLab.

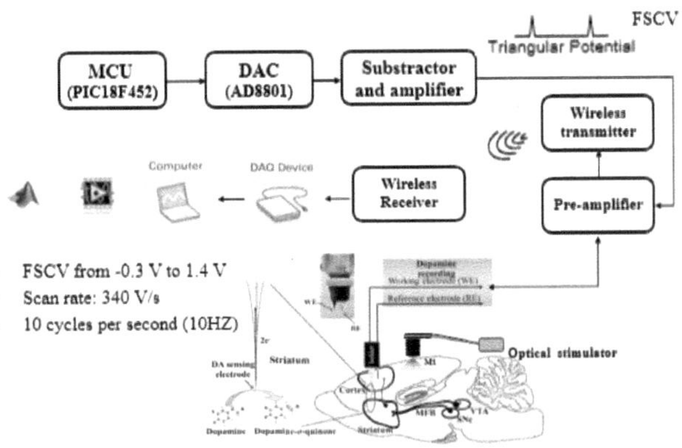

Fig. 3 Block diagrams of the wireless FSCV system and illustrations of implanted carbon fiber, reference electrode for in vivo study.

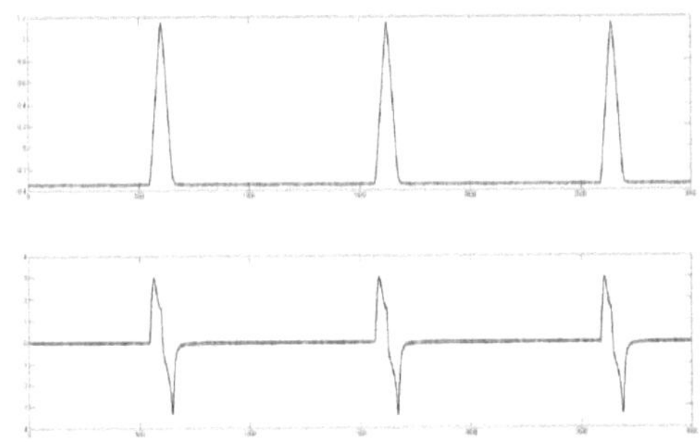

Fig. 5 (a) Redox potential for DA rise from -0.4 V to 1.3 V generated by microprocessor, and recorded current, (b) Detect redox current changes follow with redox potential.

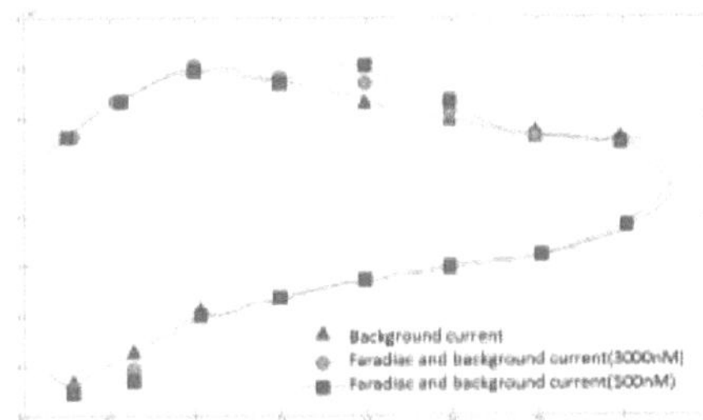

Fig. 6 The faradaic oxidized current and the background current.

III. RESUT AND DISCUSSION

A. Characterizations of carbon fiber microelectrodes

Carbon fiber (WE) and Ag/AgCl (RE) electrodes were glued into a connector as shown in Figure 4 (A). To obtain the lower noise-ground, carbon fiber was electrically treated, which was immersed in Tris buffer (pH 7.4) and carried out with CV from 0 V to 1.0 V (*vs.* Ag/AgCl) until the CV curves were stable. An obvious peak in Figure 4 (B) shows that the sensing electrode, manipulated with CV, was verified the oxidized potential for DA at around +0.5 V (*vs.* Ag/AgCl).

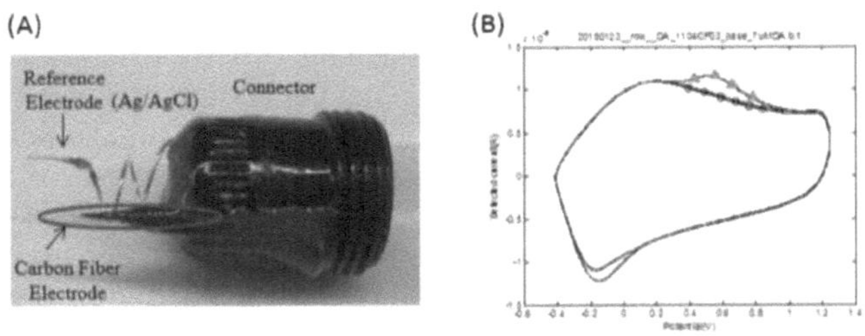

Fig. 4 **(A)** Representative photograph of carbon fiber electrode and Ag/AgCl electrode arrange into connector.
(B) CV spectrum in the absence (○) and presence (△) of 7 μM DA. Tris buffer (pH = 7.4); scan rate = 10 V/s

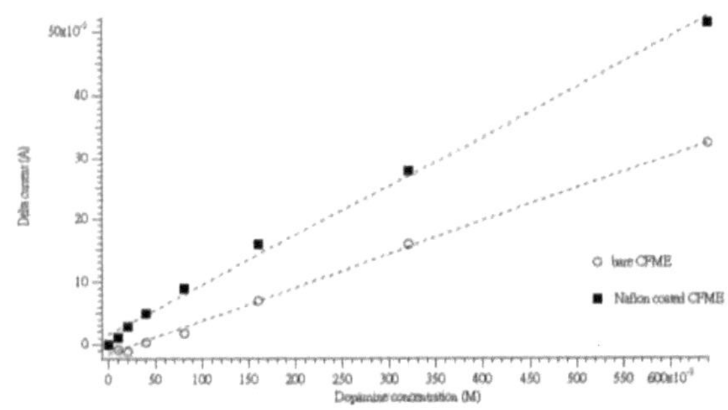

Fig. 7 The calibration curves for DA detections in order of 0.01μM, 0.02μM, 0.04μM, 0.08μM, 0.16μM, 0.32μM, 0.64μM DA additions. And compares the linear sensitivities of bare electrodes (○) and Nafion coated electrodes (■)

In-vitro experiments, FSCV detected oxidized current of background and faradaic current. The faradaic current is known to be proportional to DA concentration. Figure.5 and 6 show the micro DA sensor system with FSCV recorded in-vitro experiment. And in Figure 7 shown that the coated Nafion electrode's sensitivity is significantly increase compare to bare carbon fiber microelectrodes.

B. Wireless sensing and optical stimulation system

The wireless sensing and optical stimulation module is a miniaturized circuits for backpack of rat. To avoid electrical interference from the cable, a miniature voltammetric amplifier with dimensions of 1.5 cm (L) × 1 cm (W) × 0.5 cm (H) was implemented. A wireless RF transmitter module has been modified to adapt the DA sensing and optical stimulation modules, the module had dimensions of approximately 4.7 cm (L) × 3.3 cm (W) × 4 cm (H), the module is illustrated in Figure 8, which makes it suitable for experiments on freely moving rats. And the illumination power of optotrode is 5.09 mW/mm^2, power by 3.7-V Li-ion batteries. The top view and side view of optotrode as shown in Fig 9.

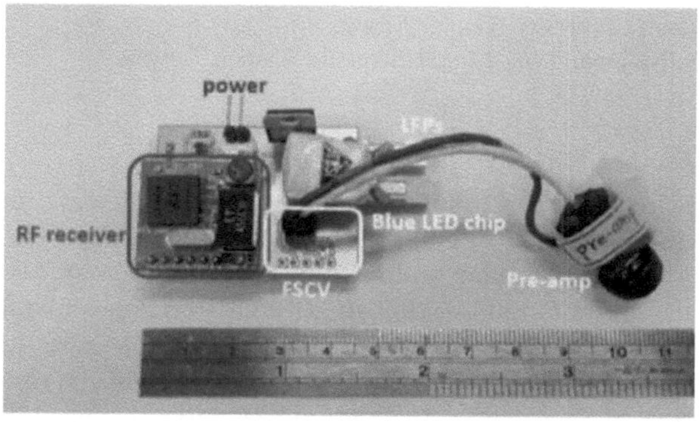

Fig. 8 Photograph of miniaturized circuits of wireless sensing and optical stimulation module.

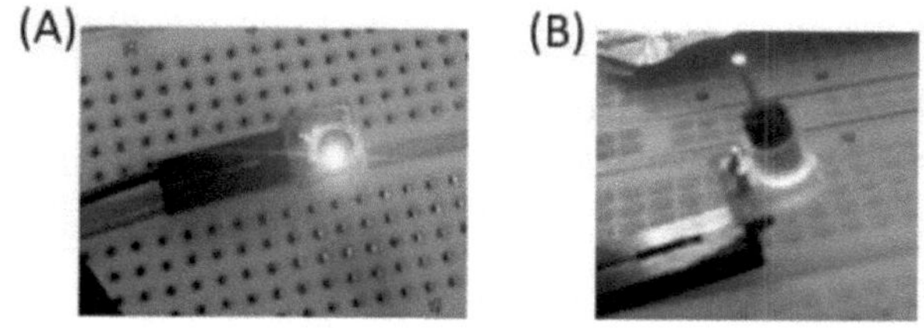

Fig. 9 The top view (A) and the side view (B) of optotrode.

IV. Conclusions

Apply the technique of focus ion beam milling the electrode to fix the redox area, and modify the Nafion on the surface of carbon fiber microelectrodes thereby to accomplish the sensitive and specific DA detections resulting from Nafion coated microelectrode not only reduces AA interference, but enhance the sensitivity of DA. The developed backpack circuits which integrate the wireless sensing and optical stimulation module, and supply enough illumination to activate optogenetic animal model, which express light-sensitive ion channel, channelrhodopsin-2(ChR2) in the brain. Our future work is applying this wireless module to evaluate the relationship between behavior of freely moving PD rats' and DA changes under optical stimulation.

Acknowledgment

The authors would like to thank the National Science Council (NSC) of R.O.C. for financially supporting this research under contract NSC 101-2221-E-006 -006 -MY3.

Conflict of Interest

The authors declare that they have no conflict of interest.

References

1. J. R. Cooper, F. E. Bloom, and R. H. Roth, (2003) The biochemical basis of neuropharmacology, 8th ed. New York: Oxford University Press.
2. Michael AC, Borland LM et al (2007). Electrochemical methods for neuroscience. Boca Raton: CRC Press; 2007. pp. 407–450.
3. Bernstein, J. G. and E. S. Boyden (2011). "Optogenetic tools for analyzing the neural circuits of behavior." Trends in cognitive sciences 15(12): 592-600.
4. P. Peter, J. A. Stamford, "Real-time monitoring of endogenous noradrenaline release in rat brain slices using fast cyclic voltammetry: 1. Characterisation of evoked noradrenaline efflux and uptake from nerve terminals in the bed nucleus of stria terminalis, pars ventralis," Brain Research, vol. 587, pp. 137-146, 1992
5. Capella, P., et al. (1990). "Nafion-coated carbon fiber electrodes for neurochemical studies in brain tissue." Electroanalysis 2(3): 175-182.
6. Bath B.D., Michael D.J., Joseph J.D et al (2000) Subsecond adsorption and desorption of dopamine at carbon-fiber microelectrodes.2000, pp 5994–6002.
7. erhardt, G. A. and A. F. Hoffman (2001). "Effects of recording media composition on the responses of Nafion-coated carbon fiber microelectrodes measured using high-speed chronoamperometry." Journal of neuroscience methods 109(1): 13-21.

Author: Meng-Chun Liu
Institute: Biomedical Engineering, National Cheng-Kung University
Street: No.1, University Road
City: Tainan City 701
Country: Taiwan (R.O.C.)
Email: emma.jasonbiolab@gmail.com

A Design of an Ambulatory Training Device with External Cues for Stroke Patients

C.-F. Chen[1], C.-H. Yu[2], and C.-H. Yang[2]

[1] Department of Product Design, Ming Chuan University, Taipei, Taiwan
[2] Department of Physical Therapy and Assistive Technology, National Yang-Ming University, Taipei, Taiwan

Abstract—**Stroke is one of the most common diseases among older population. During the rehabilitation process, it is important for stroke survivors to retrieve independent mobility. Gait training in clinical practice largely relied on therapists' clinical experiences. Proper cues are usually offered to stroke patients to facilitate better gait patterns. However, in the absence of therapists, stroke patients may develop abnormal walking patterns due to lack of cues, which would affect their future rehabilitation largely. So the purpose of this study was to develop a novel assistive device which can provide external cues during gait training to stroke patients. We hope this device could help stroke patients to enhance gait training efficiency. In this study, a novel assistive device with external cues using sensors and embedded system was established. Three stroke patients were recruited and their gait performances and gait symmetry as outcome measures were collected by Vicon motion system.**

The results showed that when use the assistive device, subjects' stride length increased 17.12% and 20.16% that used 1 and 2 lacer target respectively, step length of affected side increased 9.53% and 20.6%, step length of sound side increased 34.15% and 36.2%, single foot swing time of affected side increased 15.17% and 27.58%, and single foot swing time of sound side increased 21.29% and 10.18%. The step length symmetry also increase 10% and 26%. All of the above-mentioned parameters revealed clinical importance.

The device we developed has been basically proved to be helpful in stroke gait performance. It may be potentially useful in general clinical rehabilitation for normalizing walking patterns and performances to stroke patients.

Keywords—**cerebral vascular accident (CVA), stroke, hemiplegia, visual cue, assistive device, gait.**

I. Introduction

Basically, stroke may affect patients widely with psychological, cognitive, emotional change, while it can also cause damage to the sensory and motor function, particularly in muscle control and muscle strength which may cause a direct impact on the ability to walk. Mostly, stroke patient's life cannot be independent as they need to rely on others to take care for a long-term period.

The recovery of walking ability for stroke patients in clinical training program, including actions to induce tension treatment, muscle endurance training, static and dynamic balance training, proprioception training, walking training, training and bearing lower limb motor coordination training [1]. After the Brunnstrom Stage I as the weak period for the stroke patient, physical therapists will begin actively training of walking ability. Until the restoration of basic motor skills of stroke patients, depending on the level of muscular endurance and balance of stroke patients, the therapist will suggest different assistive devices such as walkers, quadricane, regular cane and others to assist the patient to walk in order to provide different levels of support. When using these crutches for training, therapist will teach the stroke patient way of walking as stride patterns can be as step to and step through which can further be classified into methods of two-point gait and three-point gait.

When the patient walks, therapist will give clear guidance and use the foot to remind the patient's pace and step distance generally while observe whether the patient walks with unstable bearing load and difficult situations such as motion control. The clinical therapist will assess a patient's ability to walk if there is a progress, and will also be able to observe the patient's step distance, pace, bearing proportion feet, gait symmetry and other indicators [2]. These stroke rehabilitation and training methods, mostly rely on the therapist's clinical experience. If the therapist is lack of experience or cannot make the continuous training because of the patient's often transferring, rehabilitation effect may not be as good as expected. By the way, rehabilitation exercise rooms in hospitals are often crowded and manpower of the therapist is limited. If the patient is unable to take the time for intensive training, the effect and the quality of rehabilitation therapy is greatly reduced.

In order to provide the stroke patient when performing daily activities, such as walking, with no therapists accompany the side, this study has developed an assistive device as a gait training aid for providing stepped position as therapists provide reminders and bearing lower limb function which could increase training efficiency. So that patients can continue to train at home, ward and other environments, to accelerate its rehabilitation progress and efficiency. Moreover, through the actual stroke patients to verify the assistive device compared to relative traditional

J. Goh and C.T. Lim (eds.), *7th World Congress on Bioengineering 2015*,
IFMBE Proceedings 52, DOI: 10.1007/978-3-319-19452-3_28

training methods in the gait performance parameters such as gait symmetry training effect.

II. LITERATURE REVIEW

Stroke patients' walk contains some issues, including slow walking speed, fewer cadence, shorter stride length and change of gait kinematics and kinetics [3]. In recent years, scholars using dynamic analysis system, stroke patients in addition to the average gait speed slower, while step length and cadence are lower than in healthy person. The ratio of stance phase and swing phase as well as affected side standing on the percentage of time and the proportion of the gait cycle are shorter than those on the sound side. In addition, stroke patients whose gait pattern asymmetry problems have been widely discussed in recent years, which even be considered as an important indicator to assess whether the patient can return to an independent living community [4]. In 2010, Patterson had compared a variety of gait symmetry index and clinical parameters and found that parameters of stride length, swing and stance phase of gait symmetry for the patients are more sensitive. Symmetry ratio is the most suitable index among them which will be an important reference for choosing parameters in this study [5].

Biofeedback has been applied to gait training for stroke patient for a long time, which uses visual or auditory feedback for the patient as a clear indicator. The most common feedback signal of EMG (electromyographic biofeedback) is applied to training patients' muscle strength of lower limb and gait performance.

Based on the literature review, we believe that an assistive device having a guidance and easy to operate can be a choice for increasing the effect of gait training coupled with improved visual and auditory crutches guideline. The user may not have to take too much effort to learn and can use in the patient's home rehabilitation plan, and enhance the effectiveness of patients' gait rehabilitation.

III. DEVELOPMENT OF AN AMBULATORY TRAINING DEVICE

A. Introduction of system concept

After literature review, this study conducts an experiment using the concept of visual and auditory tips for clinical physical therapists to help patients perform gait training with a sensor, laser projection, and an embedded system fixed on commonly used in standard cane and verify its feasibility through clinical trials.

In this study, we use three-point gait training mode. Stroke patients conduct gait training with new developed cane which contains external visual cues. When the new developed cane touches the ground, laser projection device will be triggered and then launch laser beams on the ground to provide a clear target for the patient's stride. An early functional prototype is shown n Fig. 1.

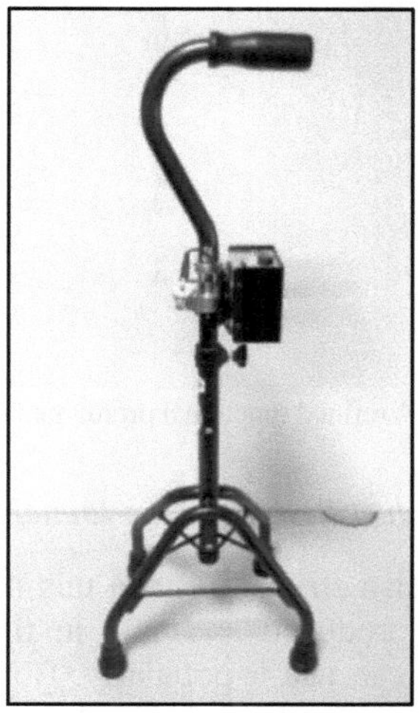

Fig. 1 An early functional prototype of a cane with external cues for stroke patient's gaiting training

B. Hardware development for new developed device

According to Beauchamp's study to patients with gait asymmetry, the use of standard canes can significantly increase the performance of their gait and gait symmetry [6]. Thus a standard cane is used as the system base for carrying other units.

For achieving a better gait symmetry effects, this system applies two laser modules for providing both feet stepping cues. In addition, If we take into account the laser module installed directly on canes, then the position of the laser projector will fluctuate because of changes in the angle of canes. To solve this problem, we installed a base at the bottom of a cane. In addition to placing the battery and embedding external systems, the base also increase the stability of the cane. Most importantly the shaft design between the base and the cane will be able to overcome the problems walking movements of the laser projection position.

Some parameters are considered, including the best height adjustment of the cane is set as greater trochanter, the cane for both right-handed and left-handed use, laser projection position adjusted by therapists manually according to clinical standard. One of authors as a physical therapist conducted the experimental process and asked the subjects to follow his treatment. A refined functional prototype used for the experiment is shown in Fig. 2.

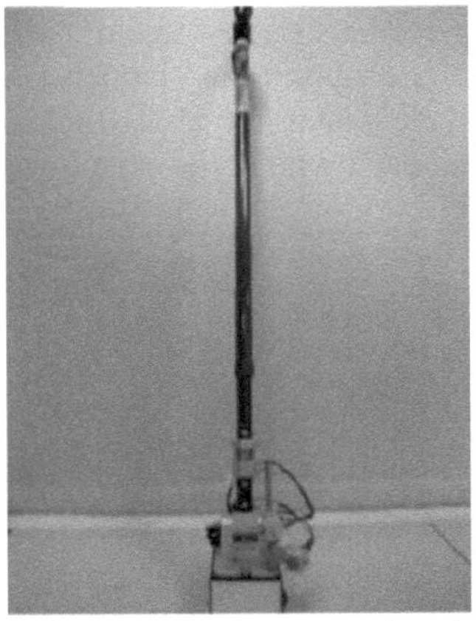

Fig. 2 A refined functional prototype for gait training

C. Further product design for the launching to market

This study also strives to have this new developed assistive device produced and sale in the market so that stroke patients can use it popularly. Both of the usability and form design are focused. CAD drawings and animation are both produced for verification. A multidisciplinary team was formed including two product designers. Two re-designed prototypes are produced as shown in Fig. 3. Through the product design process, this ambulatory training device is more close to market and users' needs.

Through this product design process, this newly designed technical assistive device has proven that different professionals can work together, such as industrial design, bioengineering and clinical therapist. Industrial designers can be more creative on usability and form ideas while therapist is more medical effectiveness. This is a cooperative mode for developing assistive device.

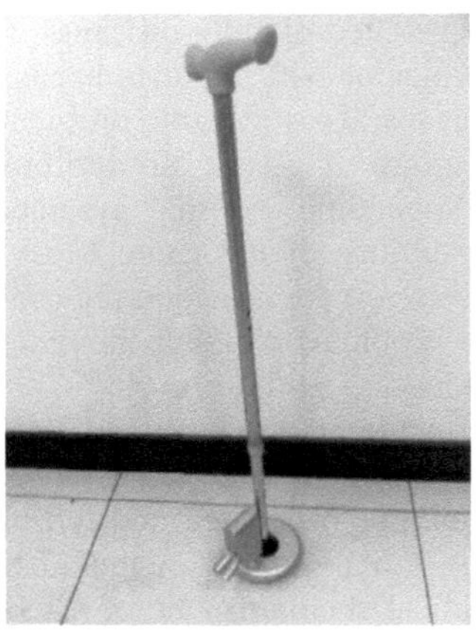

Fig. 3 A functional prototype focusing on product form design and usability

IV. Conclusions

Based on the experimental result, clinical therapist oral guidance with external cues can remind stepping positions effectively and increase stride length. Overall, this assistive device can play as a walking aid and rehabilitation training aid. After a long period of intensive use, stroke patients will have a positive stride ability.

According to the result of questionnaire survey after experimental tests, subjects think this assistive device will have a good effect on stroke patients on gait training. Subjects keep positive comments that the two-point stepping prompt stride clears guideline than a single-point stride hint.

Acknowledgment

Authors would like to express sincere thanks to Ministry of Science and Technology, Taiwan, for partially funded under the research grant of MOST 103-2420-H-130-004-MY2.

Conflict of Interest

The authors declare that they have no conflict of interest.

References

1. Duncan, P. et al. (2003) Randomized clinical trial of therapeutic exercise in subacute stroke. Stroke, 34(9): 2173-2180.
2. Boyd, L.A., Winstein C.J. (2001) Implicit motor-sequence learning in humans following unilateral stroke: the impact of practice and explicit knowledge. Neuroscience Letters, 298(1): 65-69.
3. Goldie, P.A., Matyas, T.A., Evans, O.M. (2001) Gait after stroke: initial deficit and changes in temporal patterns for each gait phase. Arch Phys Med Rehabil, 82(8): 1057-1065.
4. Patterson, K.K. et al. (2008) Gait asymmetry in community-ambulating stroke survivors. Arch Phys Med Rehabil, 89(2): 304-310.
5. Patterson, K.K. et al. (2010) Evaluation of gait symmetry after stroke: a comparison of current methods and recommendations for standardization. Gait Posture, 31(2): 241-246.
6. Beauchamp, M.K., et al. (2009) Immediate effects of cane use on gait symmetry in individuals with subacute stroke. Physiother Can, 61(3): 154-160.

Palmar Pressure Thresholds in Grasp and Pinch Functions – Analysis on Patients with Peripheral Nerve Damage

M. Sathish Kumar Paul[1,2], Rekha Vijayakumar[2] and Sudesh Sivarasu[3]

[1] Schieffelin Institute of Health Research & Training Centre, Karigiri, India
[2] School of Bio Sciences and Technology, VIT University, Vellore, India 632014
[3] MRC/MIRU–Division of Biomedical Engineering, Department of Human Biology, Faculty of Health Sciences, University of Cape Town, South Africa 7925

Abstract— **The peripheral nerve damage causes significant impairments if not detected and treated early. The loss of protective sensation in the patients with peripheral neuropathy leads to high pressure areas thus leading to secondary impairments. Impairments further leads to inability in carrying out normal activities of daily living. The study aimed to develop a sensory glove with the tactile sensors embedded in fabrics and to predict the pressure threshold level caused by friction and further prevent the high pressure while carrying out specific hand functions. Customized software was developed and the variations in the superficial pressures of the hand were recorded using tactile sensors while the patients were being involved in their routine daily living activities. The distribution patterns of the pressure in pre defined areas of hand were traced while the patients hand function activities involved the grasp and pinch powers. The study were conducted on (n = 100) patients from different job profile, gender and from different geographical location. The glove embedded with the tactile sensors helped identify pressure variations while the patients were involved in specific hand function activities. The pattern of the result suggests that the pressure were maximal while in the middle of an activity and is minimal at the onset and the end of the activity. The buzzer set along the glove gave an instant auditory feedback to the patient on the activity which causes prolonged high pressures to the skin surface of the hand. The portable device can be used anywhere in the community and will help in prescribing appropriate orthosis and adaptive tools and appliances for the patients and help prevent ulcers while doing their activities.**

Keywords— **Nerve, Pressure, Hand, Functions**

I. Introduction

The peripheral nerve damage causes significant impairments if not detected and treated early in Leprosy [1]. Impairments further leads to inability in carrying out normal activities of daily living. Different reconstructive surgical procedures help in the correction of the impairments and in regaining the functional activities that are lost in leprosy affected patients [2]. Evaluation and assessment of these impairments and the functional activities pre and post interventions are essential for the success. A wide range of standardised tools, simple equipments and tests are developed and available for the assessment [2]. Pre fabricated assistive devices and aids are supplied to patients to help patients enhance their activities of daily living [3].

Common daily living activities like riding bikes, cycling and using work tools like hammer would be difficult if the high precision activities of the hand are impaired. Prehensile functions such as grasp and pinch powers are essential to do activities with the hand. The force exerted on a grasped object before it slips free or from the grasp of the hand is defined as "breakaway strength" [4]. In Leprosy, patients if not detected the peripheral nerve damage early leads to severe sensory and motor impairments causing ulcers and muscle imbalance. Though various tests have been established to assess the effects of the peripheral neuropathies [5-8], there has been very minimal applications to give a feed back on the effects and further prevent new impairments.

The present study aimed to develop a glove with the tactile sensors embedded in fabrics and provide feedback based on the pressure threshold level caused by friction and further prevent the high pressure while carrying out hand functions involving the grasp and pinch prehensile functions.

II. Methods

Customized software was developed and the variations in the superficial pressures of the hand were recorded using tactile sensors while the leprosy affected patients were being involved in their routine daily living activities[9]. Flexi force pressure sensors with an active sensing area of 0.375 diameter were used to identify the pressure variations. The distribution patterns of the pressure in pre defined areas of hand were traced while the subjects were involved in doing their normal hand function activities involving their grasp and pinch powers as in figure 1.

The study were conducted on (n = 100) subjects from different job profile, gender and from different geographical location. The subjects were randomly allocated for the study according to their age group and their level of damage of

J. Goh and C.T. Lim (eds.), *7th World Congress on Bioengineering 2015*,
IFMBE Proceedings 52, DOI: 10.1007/978-3-319-19452-3_29

the peripheral nerves in the hand because of Leprosy. The identified subjects either had only ulnar, median paralysis or were with both the impairments.

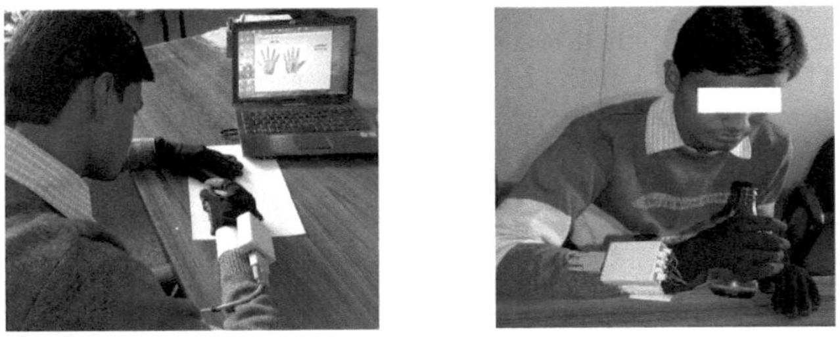

(a) (b)

Figure 1: Tactile Sensor glove used to acquire data (a) Writing – Pinch function, (b) Holding – Grasp function

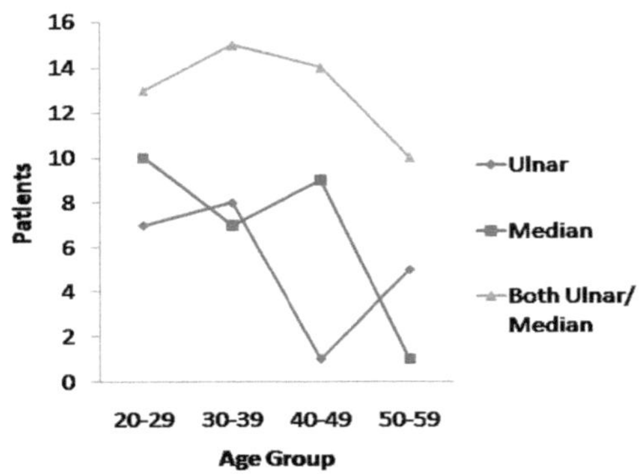

Fig 2: Age group patients tested with the tactile sensor gloves

The nerve paralysis for the subjects on whom the analysis were done have had the paralysis for more than a year and have had adapted to do the functional activties with the impairment. The pressure variations in the hand were recorded while the subjects were involved in certain specific functions like cooking, ploughing, riding bikes and while drinking with bottles [10]. The palmar aspect of the hand was divided into four regions as distal lateral, distal medial, proximal lateral and proximal medial, fig 2 shows the patients with impairments as per their age. The variations in the pressure distribution for each activity were analyzed and plotted according to the regions.

III. RESULTS

The glove embedded with the tactile sensors helped identify pressure variations while the patients were involved in their normal hand functional activities. The pattern of the result in fig 3 suggests that the pressure variations while doing the identfied functional activities were maximal in the the distal lateral and the proximal lateral and were minimal in the disal medial and the proximal medial areas in the palmar regions of the hand. The threshold limit were determined for the palmar regions depending on the maximal pressure that were identified.

The buzzer set along the glove on the highest threshold value gave an instant auditory feedback to the patient on the activity which causes prolonged high pressures to the skin surface of the hand. Fig 4 shows the threshold pressure values set for each of the identified 9 points on the palmar surface of the hand for the subject working as a farmer. A simple glove embedded with sensors and an alarm provided to the patients coming to the hospital for using while doing the functional activities will help them protect themselves from injuries and worsening their impairments.

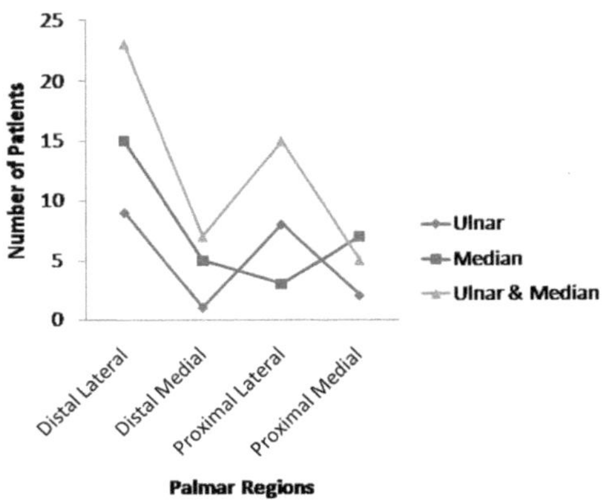

Fig 3: Sites of hand tested with the tactile sensor gloves

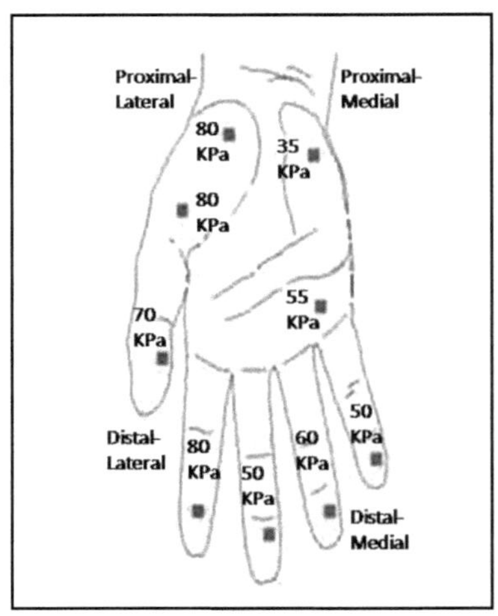

Fig 4: Threshold pressure values set for patients working as farmers

IV. CONCLUSIONS

The robust sensory gloves can be used on patients with peripheral nerve damage irrespective of their profession. The gloves will also help predict the pressure threshold while doing their functional activities using prehensile functions. The cost effective device will help predict the ulcer prone areas in the anesthetic hands and aslo feet in the leprosy affected patients. The portable device can be used anywhere in the community and will help in prescribing appropriate orthosis and adaptive tools and appliances for the patients and help prevent ulcers while doing their activities.

The sensors will help clinicians, therapists and other health professionals with limited engineering skills to use this device and develop a treatment plan for the patient. This device through the telemetry system would help the patient from remote part of any country utilize the best of treatment available through consultants and develop treatment strategies for them

ACKNOWLEDGMENT

The authors would like to thank the Director and staff of the Leprosy Mission Trust India. The authors also acknowledge the patients active participation and cooperation in the study.

CONFLICT OF INTEREST

The authors declare that they have no conflict of interest.

REFERENCES

1. Charosky CB, Gatti JC, Cardama JE: Neuropathies in Hansen's disease. Int J Lepr. 51:576-586, 1983
2. Richard Schwarz, Wim Brandsma: 2004 *Surgical Reconstruction & Rehbailitation in Leprosy and other Neuropathies* Ekta Book Distributors Pvt. Ltd., Nepal, pp: 1-10, Chap:1
3. Ganapati R, Pai V V, Kingsley S. Disability prevention and management in leprosy: A field experience. Indian J Dermatol Venereol Leprol 2003;69:369-74
4. Rajulu, S. L., Klute, G. K., & United States. (1993). A comparison of hand grasp breakaway strengths and bare-handed grip strengths of the astronauts, SML III test subjects, and the subjects from the general population. Washington, D.C: National Aeronautics and Space Administration, Office of Management, Scientific and Technical Information Program.
5. Hammond CJ, Klenerman P. Protective sensation in the foot in leprosy. Lepr Rev, 1988; 59: 347–354.
6. Abbot NC, Beck JS, Samson PD et al. Impairment of fingertip vasomotor reflexes in leprosy patients and apparently healthy contacts. Int J Lepr, 1991; 59: 537–547.
7. Wilder-Smith A, Wilder-Smith E. Electrophysiological evaluation of peripheral autonomic function in leprosy patients, leprosy contacts and controls. Int J Lepr, 1996; 64: 433–440.
8. Wim H Van Brakel, Peter G. Nicholls, Loretta Das, Pramila Barkataki, Pranava Maddali, Diana N. J. Lockwood, Einar Wilder - Simth . The INFIR Cohort Study: assessment of sensory and motor neuropathy in leprosy at baseline Lepr Rev, 2005; 76: 277–295
9. S.K. Paul., V. Rekha., Sudesh Sivarasu., "*Tactile Sensing Fabrics for Detecting Impairments in Leprosy Patients*" Proceedings of the "7th International Conference on Appropriate Healthcare Technologies for Developing Countries AHT 2012."
10. S.K. Paul., V. Rekha., Sudesh Sivarasu., "*Tactile Sensing Fabric glove to predict peak palmar pressure and prevent early impairments in leprosy affected patients - 2 year follow up study*" Proceedings of the "8th International Conference on Appropriate Healthcare Technologies for Developing Countries AHT 2014."

Corresponding author:

Author: Sudesh Sivarasu Ph.D

Institute: MRC/MIRU – Division of Biomedical Engineer-ing, Department of Human Biology, Faculty of Health Sciences,

City: Cape Town

Country: South Africa

Email: Sudesh.Sivarasu@uct.ac.za

Design Evaluation of REMAP Exoskeleton

Preethika Britto[1], Rekha Vijayakumar[2], and Sudesh Sivarasu[3]

[1] Biomedical Engineering Division, School of Bio Sciences and Technology, VIT University, Vellore, India 632014

[2] Environmental Biotechnology Division, School of Bio Sciences and Technology, VIT University, Vellore, India 632014

[3] MRC/MIRU – Division of Biomedical Engineering, Department of Human Biology, Faculty of Health Sciences,

University of Cape Town, South Africa 7925

Sudesh.Sivarasu@uct.ac.za

***Abstract*— Lower extremity paralysis results in significant global morbidity and mortality. In India it is discovered that 3.96 percent of the population is handicapped. Roughly 1% of the world's population relies on wheelchairs for mobility. There are currently about 262,000 spinal cord injured (SCI) individuals in the United States, with roughly 12,000 new injuries sustained each year at an average age of injury of 40.2 years. Of these, at least 44% (at least 5300 cases per year) result in paraplegia. In an effort to restore some degree of legged mobility to individuals with paraplegia, several lower limb orthoses have been developed.**

To solve the above mentioned issues we have developed an exoskeletal suit that could be worn around the waist till the toe which is driven by actuators that enable the user to maneuver without any external assistance. The design is anthropomorphic and utmost care has been taken to ensure that the biomechanically the system is similar to the normal leg. The overall biomimetic approach towards the design has helped to achieve a perfect synergy between the exoskeleton and the wearer. The elaborated design ideology comes out of six iterative re-modifications.

Keywords— **Paraplegia, Exoskeletons, Design analysis, Anthromorphic modelling**

I. Introduction

Lower extremity paralysis results in significant global morbidity and mortality. In India it is discovered that 3.96 percent of the population is handicapped. Roughly 1% of the world's population relies on wheelchairs for mobility. [6].There are currently about 262,000 spinal cord injured (SCI) individuals in the United States, with roughly 12,000 new injuries sustained each year at an average age of injury of 40.2 years. Of these, at least 44% (at least 5300 cases per year) result in paraplegia [7]. One of the most significant impairments resulting from paraplegia is the loss of mobility, particularly given the relatively young age at which such injuries occur. Surveys of persons with paraplegia indicate that mobility concerns are among the most prevalent [2], and that chief among mobility desires is the ability to walk and stand [3]. In addition to impaired mobility, the inability to stand and walk entails severe physiological effects, including muscular atrophy, loss of bone mineral content, frequent skin breakdown problems, increased incidence of urinary tract infection, muscle spasticity, impaired lymphatic and vascular circulation, impaired digestive operation, and reduced respiratory and cardiovascular capacities [1].

In an effort to restore some degree of legged mobility to individuals with paraplegia, several lower limb orthoses have been developed. Although wheelchairs play a high role in enhancing paraplegic mobility, the options for people with mobility disorders have been limited. Humans were not designed to sit for hours on end [4]. Constant sitting might lead to pressure sores, atrophied leg muscles and brittle bones. Wheelchair users are at elevated risk for carpal tunnel syndrome or repetitive strain injury from the constant impact of the hands against the push rims of the wheels [3]. To solve the above mentioned issues we have developed an exoskeleton suit that could be worn around the waist till the toe which is driven by actuators that enable the user to maneuver without any external assistance. The design is anthropomorphic and utmost care has been taken to ensure that the biomechanically the system is similar to the normal leg.

II. Materials and Methods

A. Materials and Designs

The material analysis enabled the selection of proper materials for the facilitation optimized design. Materials are chosen in such a way that it consumes less weight and volume. The identified materials are listed as: Al 2023 is extremely light weight and has high tensile strength it can be fabricated with metal fusion techniques, machinability and it forms the entire exoskeleton structure and rigid casings. (Titanium + Steel alloy) electroplated with Zn or Cr has ability to withstand high shear stress so it can be used for fixtures, fasteners, bolt, nuts and mountings. Cast-iron/Steel is light weight and ideal for power transmission inside the system.

J. Goh and C.T. Lim (eds.), *7th World Congress on Bioengineering 2015*,
IFMBE Proceedings 52, DOI: 10.1007/978-3-319-19452-3_30

B. Structural Elements

The entire exoskeleton is a single functional unit. Based on the human anatomy this integrated unit is discrete into the different functional entities. All the attributes of the individual elements are approximated to reach human anthropomorphology

Figure 1 Complete Exoskeleton design

C. Foot and ankle design

This design has one DOF for foot and Knee joint, with 2 DOF for Hip. It has been facilitated to affix the actuators along the sides of the exoskeleton with the joint couples with all functional elements for power transmission, dampers to act as active/passive suspension for ankle joint. Flaps are provided to take large pressure with minimal contact area. This facilitates the absorption or energy during initial contact and releases it during the heel strike. The foot part has a stud connected to a damper that performs this function. Also the materials are carved out without disturbing the structural stability to ensure that the system has low weight. The ankle joint connects the foot part to the lower shank of the exoskeleton; it has all the power transmission elements that facilitate a smooth rotation between the two.

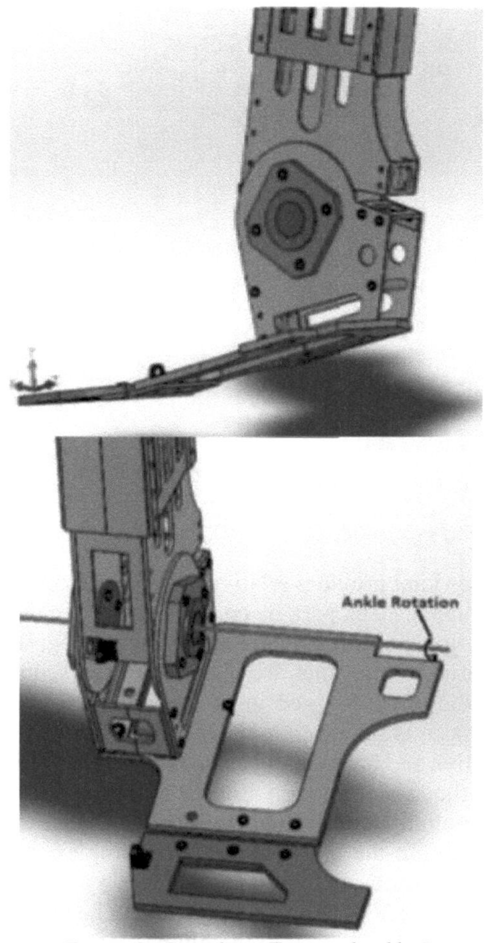

Figure 2: Compliant Foot and ankle Part

D. Shank Part

The shank is constructed out of four sheet metal entities, fastened to form a cubical structure that sides over the other. In the previous designs adjustment of shank according to user height was not possible. But in this design facilitation to adjust to any wearer's height has been provided. The design shows the best possible shank design as a result of iterative adjustments over its former. Joint's motion stoppers provide controlled range of motion. They are designed as mechanical projections over the semi-circular motion path of the two relative parts, when the range of motion ends the motion are constrained by the collision of the projections. Also for safety regulatory stoppers are provided with 35 degrees of flexion beyond which the motion is constrained.

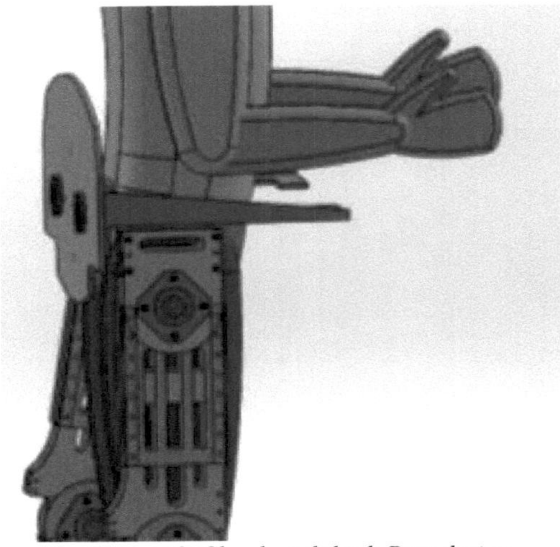

Figure 3: Shank and thigh Part design

E. Knee and hip joint

The knee joint provides relative motion between the lower thigh and upper shank part, by connecting the two using a pin joint. The joint elements include inner coupling, a bearing and a bearing socket that are fastened to the semi-circular edges of the parts. Rugged joint structures are used in the knee joint as it is subjected to the higher torque values. Knee joint is active and is driven by an actuator connected perpendicular to the axis of rotation. Similar to other joints the knee also has motion stoppers that prevent the over flexion and extension within the range of motion. Complex design and functional elements are incorporated into hip joint as it needs to accommodate 3 DOF into a single unit. The hip joint connects the entire lower limb to the hip base via pin joints that can have relative motion between the two. The lower hip part connected to thigh with one DOF which provides the flexion and extension of the lower limb within the range of motion.

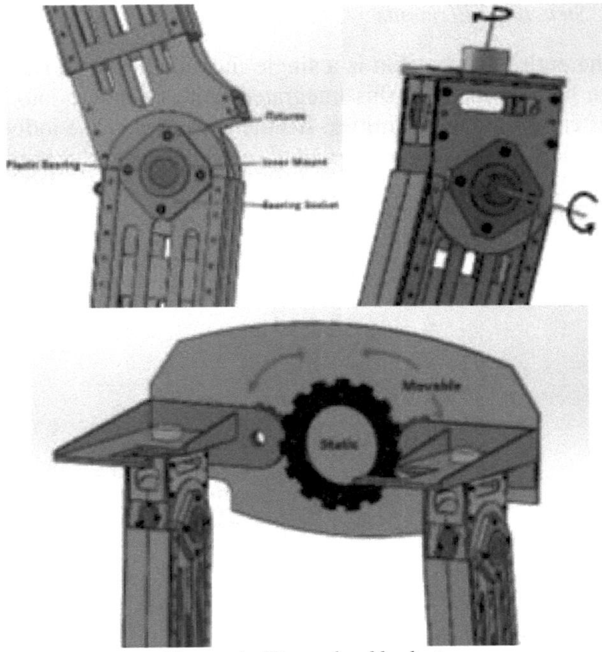

Figure 4: Hip and ankle design

III. CONCLUSIONS

The overall biomimetic approach towards the design has helped to achieve a perfect synergy between the exoskeleton and the wearer. All the joints of the wearer and the robot system lie on the same axis. Straps are provided to adhere the user intact. Also the hip joint accommodates the user comfortability. As this is the preliminary design, more patient comfort levels can be achieved on further alterations over the covers.The elaborated design ideology comes out of six iterative re-modifications. The current design is evident in proving its reliability towards the former designs. Further structural analysis investigates the physical behaviour of a model understand and improve the mechanical performance of a design. Minimizing the size and weight of the structural elements was traded off against maintaining structural stiffness so that the payload could be adequately supported. The strength to weight ratio of the exoskeleton components was maximized using finite element analysis

ACKNOWLEDGMENT

The authors wish to acknowledge Dr.Lazar Mathew for his constant support and guidance throughout the work

CONFLICT OF INTEREST

The authors declare that they have no conflict of interest

References

1. Pons, J. L. (Ed.). (2008). Wearable robots: Biomechatronic exoskeletons (Vol. 338). Hoboken, NJ, USA: Wiley.
2. Spencer Murray, and Michael Goldfarb (2013), Towards the Use of a Lower Limb Exoskeleton for Locomotion Assistance in Individuals with Neuromuscular Locomotor Deficits*
3. Kazerooni, H., & Steger, R. (2006). The Berkeley lower extremity exoskeleton. Journal of dynamic systems, measurement, and control, 128(1), 14-25.
4. Mohammed, S., & Amirat, Y. (2009, February). Towards intelligent lower limb wearable robots: Challenges and perspectives-State of the art. In Robotics and Biomimetics, 2008. ROBIO 2008. IEEE International Conference on (pp. 312-317). IEEE.
5. Chu, A., Kazerooni, H., & Zoss, A. (2005, April). On the biomimetic design of the Berkeley lower extremity exoskeleton (BLEEX). In Robotics and Automation, 2005. ICRA 2005. Proceedings of the 2005 IEEE International Conference on (pp. 4345-4352). IEEE.
6. Ryan J. Farris, Hugo A. Quintero & Michael Goldfarb Preliminary Evaluation of a Powered Lower Limb Orthoses to Aid Walking in Paraplegic Individuals
7. Factsheet on Persons with Disabilities-www.un.org/disabilities
8. Spinal Cord Injury Facts and Figures at a Glance.- www.nscisc.uab.edu
9. Preethika Britto, Rekha V & Sudesh Sivarasu - Conceptual Design of a Paraplegic Walker - Appropriate Healthcare Technologies for Developing Countries – AHT2012. The 7th International Conference – World Health and Wellbeing, 18 – 19 Sep, 2012. London, UK Paper no. 0011
10. Preethika Britto, Rekha V & Sudesh Sivarasu Adaptive Control of Biomechanically Inspired Orthotic Exoskeleton in Paraplegic Rehabilitation, Biomedical Engineering / Biomedizinische Technik, 2014, Vol. 59. (s1); pp. 1034-1036

Corresponding author:

Author: Dr. Sudesh Sivarasu
Institute: MRC/MIRU – Division of Biomedical Engineering, Department of Human Biology, Faculty of Health Sciences,
Street:
City: Cape Town
Country: South Africa
Email: Sudesh.Sivarasu@uct.ac.za

FEM Modeling for Performance Evaluation of Microwave Ablation Applicator When Using T Prong Monopole Antennas

M. Chaichanyut and S. Tungjitkusolmun

King Mongkut's Institute of Technology Ladkrabang Department of Electronics, Faculty of Engineering, Bangkok, Thailand

Abstract— **This paper present three-dimensional finite element method (3D FEMs) simulate of temperature distributions in liver cancer tissue. This study focuses on the evaluation of the performance of a T Prong Monopole Antennas. The configurations of monopole antennas were considered: Two-prong Monopole (2PM), Three-Prong Monopole (3PM) and Equilateral Four-Prong Monopole Antenna (4PM). The 3D FEMs solutions were based on Maxwell and bio-heat equations for studying 2.45GHz T Prong monopole antenna design. We apply the microwave power was 20 W; the duration time for ablation in all case was 300s. From our simulation results, all antennas type have occurred the hotspot at the tip of prong of the antenna. Two-prong Monopole antenna can be induced the highest temperature in cancer tissue. Thermal distributions were obtained and compared. The results indicate that the Two-prong Monopole antenna generated maximum temperature within liver cancer tissue and can be maximum ablation cancer tissue (5.94 cm3). The temperature distribution is determined by the direction of bifurcation of the antenna.**

Keywords— **Microwave ablation, T-prong antenna, bio-heat equation, Coaxial antenna, liver cancer**

I. INTRODUCTION

Microwave coagulation therapy is one such technique where a thin microwave antenna is inserted into the tissue (tumor). The microwave heats up the tumor, producing a coagulation region where temperature above 50°C is reached and the cancer cells are effectively destroyed. The important characteristics for the microwave ablation are relative permittivity ε_r and electric conductivity σ. The heating is more efficient in materials with high conductivity. To have the knowledge of the relative permittivity and the electric conductivity of the tissue in the area to be treated is very important for the antenna design [1].

Thermal ablation procedures are used to destroy unhealthy tissue by way of a very high temperature increase Microwave ablation (MWA) uses an electromagnetic field at microwave frequencies emitted by an antenna inserted into the human body as heat source. MWA uses localized heating to cause tissue necrosis. Water molecules movement in the tissue caused by the varying electric fields; induce frictional heating [2].

Several types of coaxial-based antennas, including the coaxial slot antenna [3], coaxial dipole antenna [4], coaxial monopole antenna [5], coaxial cap-choke antennas [6], [7], and others have been designed for MWA or microwave hyperthermia therapies in an attempt to prevent this backward heating while creating as large an ablation radius as possible. The cap-choke antenna seems to most efficiently prevent backward heating [8].

In this paper, a new simulated interstitial applicator for microwave ablation referred as T Prong monopole antenna has been proposed. We adopt the commercial software COMSOL as a numerical model. We first solve the Maxwell equations and compute the electromagnetic specific absorption rate. Results are then used to solve the bio-heat equation to obtain the temperature distribution in the cancerous tissues.

II. MATERIALS AND METHOD

This section shows the application of computer software to aid in the simulation. The models are evaluated the temperature distribution in the liver by the MWA.

A. *Computational Model*

For the simulation of the proposed T-Prong monopole Antenna, we utilized numerical modeling to solve 3D FE analyses in order to obtain temperature distributions during MW ablation. We solved our FE models using COMSOL Multiphysics 3.5a (COMSOL, Inc., Burlington, MA). The software allows specifying the geometry of antenna design and then solves maxwell's equations and Pennes bioheat equation in the surrounding tissue [9]. The liver is assumed to be of infinite in its extent, which is obtained by using low-reflection boundary conditions. The absorbed electromagnetic energy Q is computed from the electromagnetic field distribution in tissue, and is given by Eq. 1

$$Q = \frac{1}{2}\sigma|E|^2 \tag{1}$$

and the SAR [W/kg] in tissue is calculated as a function of position as follows Eq. 2

J. Goh and C.T. Lim (eds.), *7th World Congress on Bioengineering 2015*, IFMBE Proceedings 52, DOI: 10.1007/978-3-319-19452-3_31

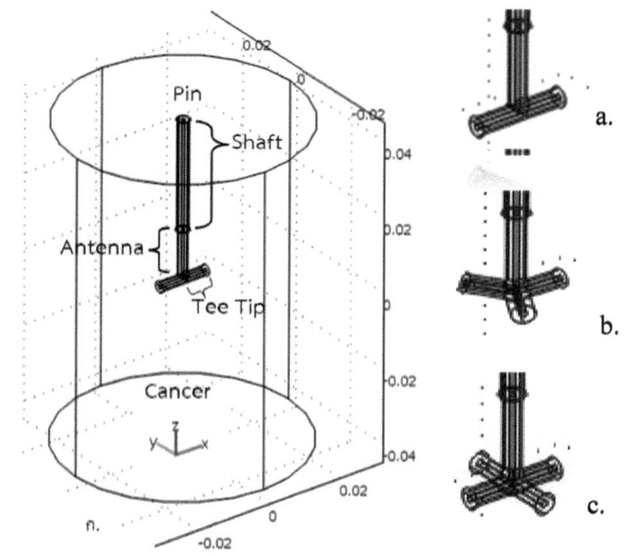

Fig. 1 Configurations of the T-Prong monopole antennas. (a). Two-prong Monopole, (b). Three-Prong Monopole and (c). Equilateral Four-Prong Monopole Antenna.

Where ρ is the density (kg/m^3) , E is the Electric field intensity (V/m).

$$SAR = \frac{\sigma}{\rho}|E|^2 \quad (2)$$

The resistive heating which is calculated from maxwell's equations is used as an input to thermal model. Pennes bioheat equation has been used to model the heat transfer during MWA.

$$\rho c \frac{\partial T}{\partial t} = \nabla.k\nabla T + SAR - h_{b1}(T - T_{b1}) + Q_m \quad (3)$$

$$h_{b1} = \rho_{b1} c_{b1} \omega_{b1}$$

Where c is the Specific heat (J/kg·K), k is the Thermal conductivity (W/m·K), T_{b1} is the temperature of the blood, ρ_{bl} is the blood density (kg/m^3), c_{bl} is the Specific heat of the blood (J/kg·K), ω_{bl} is blood perfusion (1/s) and the h_{bl} is the convective heat transfer coefficient accounting for blood perfusion in the model. The energy generated by the metabolic processes Q_m, (W/m^3).

B. *Geometric T-Prong monopole antennas*

Comsol allows users to create the geometric model as well as assign material properties and necessary boundary conditions. The geometric model is sketched in Fig 1. and

Table 1 Specifications Of The Geometric Model

Type	Shaft (mm)	Antenna (mm)	Tee tip (mm)	cancer (mm)	angle θ	angle ρ
2PM	29	13	6.5	30	90	180
3PM	29	13	6.5	30	90	120
4PM	29	13	6.5	30	90	90

Table 2 Electromagnetic properties at 2.45 GHz

Properties	Cancer	Liver	Antennas
Relative permittivity (ε_r)	43.03	43.03	2.03
Relative permeability (μ_r)	1	1	1
Conductivity (σ) [s/m]	1.79	1.69	PEC*

* PEC (Perfect electric conductor)

Table 3 Thermal properties for thermal analysis

Material	ρ [kg/m³]	c[J/kg.K]	k[W/m.K]	σ [S/m]
Conductor	6450	840	18	10^8
Cancer	1060	3600	0.512	1.79
Shaft	70	1045	0.026	10^{-5}

the dimensions of the T-Prong monopole antennas are listed in Table 1, while its material properties are shown in Table 2 and 3 respectively [10],[11].

The boundary conditions assigned to FEMs in this study complied with the following [12]:

1. The propagation mode for the microwave was assumed to be TEM, where the feeding source was launched inside the cable. The total field was computed by the technique of the scattering problem
2. The tangential electric field is zero by defining all perfect conducting boundaries, such as the conductors in the coaxial cable.
3. The boundary at the coaxial input port is a hard feeding source aperture. For this type of source aperture, the port is treated as a hard source, launching an incident wave but not allowing any reflected wave to be absorbed.
4. The potential difference between inner and outer conductors is 20 W and assigns the time of simulation for microwave ablation to be duration time was 300s.
5. We assign reference temperature for liver and cancer is 37 °C

III. Results

After we created the geometrical model, generated finite element mesh, and assigned all the material properties and boundary conditions, we ran the coupled electromagnetic-thermal simulations using Comsol. We then performed post-processing to obtain the temperature distributions for different duration time shown so Fig. 2 and Table 3. We consider the region temperature exceeds 50 °C, the threshold for successful hepatic ablation or irreversible tissue damage. Table 4 summarizes the maximum temperature, thickness of lesion, average ablation zone dimension (length x diameter mm.) and the lesion volume from microwave ablation.

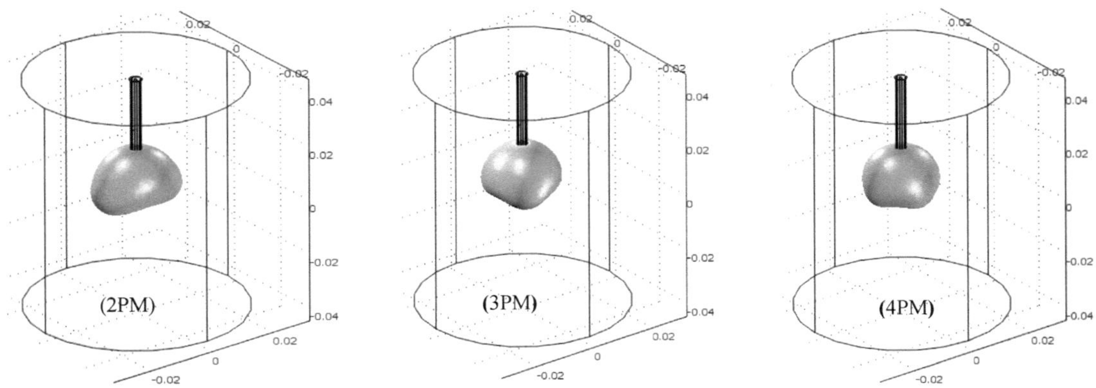

Fig. 2 Lesion zone of T-Prong monopole antenna at power 20W, duration time 300s

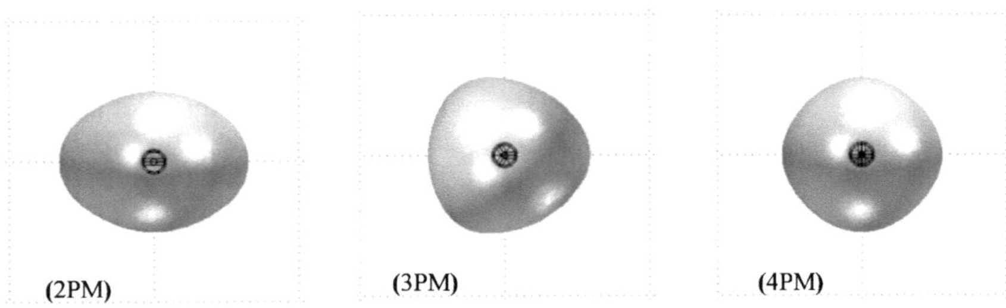

Fig. 3 illustrates the lesion ablation on x-y plane, considered the region temperature exceeds 50 °C

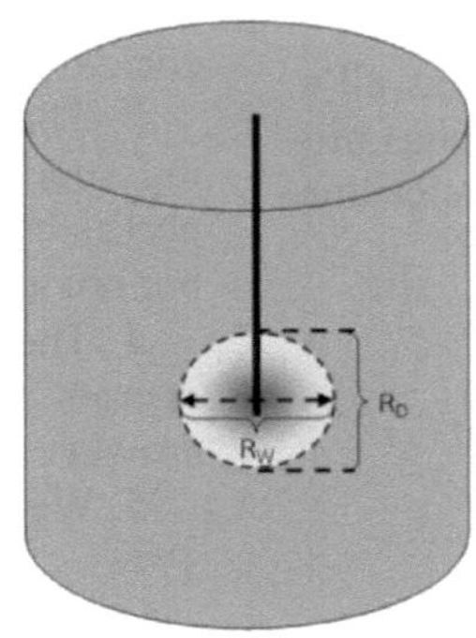

Fig. 4 The position measured of the extent of lesion formed.

Table 4 Simulation result of maximum temperature, thickness of lesion, average ablation zone dimension (length x diameter mm.) and the lesion volume

T Monopole antenna						
Type	**Max Temp(°C)**	**Thickness**	**Lateral**	**Longitudinal**	**Widest x Length (mm.) axial ratio (RW/RD: see fig 4)**	**Volume(cm^3)**
2PM	252.59	10	14	12	14x12(1.16)	5.94
3PM	83.49	11.5	11.5	10	11.5x10(1.15)	5.48
4PM	110.44	10	11	11	11x11(1)	5.09

From our simulation result, it found that the temperature distribution occur along the length of prong of antenna, lesion shape or heating profile was defined in direction of prong conductor of the antenna. From Table 4, the 2PM antenna can be induce the maximum temperature within cancer and its create the widest lesion size. But the 2PM antenna creates the thickness lesion size was minimal. The 4PM antenna creates the lesion was symmetry. While the 3PM antenna can be create the lesion size maximum thickness.

IV. Conclusions

In this research work, the performance of proposed antenna referred as T-Prong antenna has been studied numerically. The optimal dimensions of each proposed antenna, under specific conditions, were obtained. The applicator fed by each antenna delivered the same power level to tissue; however, different shapes of temperature distribution were obtained with each antenna; consequently, each applicator obtained can be used for different cancer therapies according to the size of the tumor.

Acknowledgment

This work is supported by the Thai Research Foundation Grant TRG4580106 and the Biometric for Human Identity Verification Forum, of the Biomedical Engineering Society of Thailand (ThaiBMES).

References

1. C. L. Brace, „Radiofrequency and Microwave Ablation of the Liver, Lung, Kidney, and Bone: What Are the Differences?", Current Problems in Diagnostic Radiology, vol. 38, no. 3, pp. 135–143, May 2009
2. O'Rourke, Ann P; Haemmerich, Dieter; Prakash, Punit; Converse, Mark C; Mahvi, David M; Webster, John G. Current status of liver tumor ablation devices. Expert Review of Medical Devices, Volume 4, Number 4, July 2007, pp. 523-537
3. K. Saito, Y. Hayashi, H. Yoshmura, and K. Ito, "Numerical analysis of thin coaxial antennas for microwave coagulation therapy," in *Proc. IEEE Antennas Propagation Soc. Int. Symp.*,vol. 2, 1999, pp. 992–995.
4. W. Hurter, F. Reinbold, andW. J. Lorenz, "A dipole antenna for interstitial microwave hyperthermia," *IEEE Trans. Microw. Theory Tech.*, vol. 39, no. 6, pp. 1048–1054, Jun. 1991.
5. S. Labonte, A. Blais, S. R. Legault, H. O. Ali, and L. Roy, "Monopole antennas for microwave catheter ablation," *IEEE Trans. Microw. Theory Tech.*, vol. 44, no. 10, pp. 1832–1840, Oct. 1996.
6. S. Pisa, M. Cavagnaro, P. Bernardi, and J. C. Lin, "A 915-MHz antenna for microwave thermal ablation treatment: physical design, computer modeling and experimental measurement," IEEE Trans. Biomed. Eng., vol. 48, no. 5, pp. 599–601, May 2001.
7. J. C. Lin and Y. J. Wang, "The cap-choke catheter antenna for microwave ablation treatment," *IEEE Trans. Biomed. Eng.*, vol. 43, no. 6, pp. 657–660, Jun. 1996.
8. K. Saito, S. Hosaka, S.-Y. Okabe, H. Yoshimura, and K. Ito, "A proposition on improvement of a heating pattern of an antenna for microwave coagulation therapy: introduction of a coaxial-dipole antenna," *Electron.Commun. Japan*, pt. I—Communications, vol. 86, pp. 2351–2357, 2003.
9. Analysis of Tissue and Arterial Blood Temperatures in the Resting-Human Forearm J Appl Physiol, August 1, 1948 1:(2) 93-122
10. K. SAITO, Y. HAYASHI, H. YOSHIMURA and K. ITO, 'Heating characteristics of array applicator composed of two coaxial-slot antennas for microwave coagulation therapy', *IEEE Trans. Microwave Theory and Tech.*, 48, pp.1800–1806,2000
11. D. Haemmerich, S T. Staelin, J Z. TSAI,S. Tungjitkusolmun, D.M. Mahvi, and J. G. Webster 'In vivo electrical conductivity of hepatic tumours,' *Physiol. meas.*, 24, pp.251-260,2003
12. J.D. Kraus and D.A. Flesich, '*Electromagnetics with applications*', 5th Edition, McGraw-Hill Company, pp. 389–419,1999

Automated Infusion Control with Core Body Temperature for Infants under IV Administration

Preethika Britto, Suganthan Veerachamy, Jayasubha Ravi Yathav*, and Habeebunnisha Kaleelur Rahman

Biomedical Engineering, School of Biosciences and Technology, VIT University, Vellore, India

Abstract- **Measuring the body temperature of patients in prolonged infusion shows drop in core body temperature in accordance with the change in atmospheric conditions like room temperature etc. Babies kept inside the intensive care unit are more prone to the fluctuating temperature, resulting in sequential body temperature drop off. Though intravenous fluid warming system adjusts the fluid to appropriate temperature, it does not take feedback temperature from the patient's body. Hypothermic conditions might cause serious illness to epileptic babies and the babies under postoperative care. Designing a system, which would addresses this complication is at the urge of distinction. This paper addresses the automatic sensing and cutoff in infusion to prevent the occurrence of such emergency conditions. Temperature sensor is placed at the extremities of the baby. This sensor measures baby's core body temperature at an hourly interval. The average normal body temperature ['normal' varies from babies to babies] is determined initially. Infusion stops and heater goes on when the temperature goes below the set normal temperature. Once the fluid is warmed to a set optimum temperature, infusion occurs. This automated infusion of warmed fluids helps in bringing back the normal body temperature of the babies, thus avoiding emergencies.**

Keywords- **Infusion, body temperature, babies, fluids, temperature sensor.**

I. Introduction

Infusion pump is a medical equipment mostly used in hospitals and in ambulatory to introduce medications or nutrients in to the blood vessels in very efficient and the most accurate manner. It is mostly used intravenously but subcutaneous, arterial and epidural infusion also can be used. There are many types of infusion systems which can be operated manually or semiautomatic or fully automatic. For better precision fully automatic systems are used [1]. The body temperature of the infants in the Intensive Care Unit under IV administration plunges during the infusion. It provides a route to hypothermia. Hypothermia is a condition when the body temperature plunges below 35 °C. It is categorized as three types depending on the core body temperature. Those types are mild, moderate and severe hypothermia. The symptoms of these three types vary for each type [Table 1]. It is caused by two ways which are continuous exposure to cold and metabolic activities of the body which causes the heat loss [2] [3].

Infants are more prone to hypothermic condition because of their larger surface area than the body weight. So that they tend to lose their body heat faster than children and adults. Hypothermia is not a fatal condition. But it might cause serious illness to the infants who are epileptic. Nowadays most of the Intensive Care Units are built with air conditioner. The human body temperature will vary depending on time of the day. During nights, the temperature of the body will be slightly lower than the daytime. Thus, the set temperature in the air conditioner becomes colder at nights than days. It becomes very difficult for the baby to resist the cold. In this case, even when if the intravenous fluid is maintained at the optimum temperature it is not sufficient for the patient's decreased body temperature. The intravenous fluid heater does not measure and adapt according to the patient's body temperature.

Presently there are several methods to treat hypothermic condition caused due to the flexible body temperature, which is not read by the current intravenous fluid heating systems. All the current available methods are manual which includes covering the infant with warmed blankets, wrapping the extremities (feet & palm) of the infant and infusing warmed intravenous fluid to the baby. The fluid temperature inside the tube is maintained by various methods [4] [5][6][11]. Current systems are not taking any feedback from the infant's body i.e. Changes in their body temperature due to changes in the room or surrounding temperature. The temperature of the fluid is kept as same as the optimal temperature which is 38 °C irrespective of the changes in the body temperature of the patient.

* Corresponding author.

J. Goh and C.T. Lim (eds.), *7th World Congress on Bioengineering 2015*,
IFMBE Proceedings 52, DOI: 10.1007/978-3-319-19452-3_32

Table 1 Hypothermia Classification

Types	Symptoms	Temperature
Mild	Shivering	32-35 °C
Moderate	Drowsiness & No shivering	28-32°C
Severe	Unconscious & No shivering	20-28°C

Therefore, when there is a drop in room temperature due to that the temperature of the fluid will also drop in certain cases where the intravenous fluid heaters are not used. This drop in the temperature of the fluid will lead to hypothermia in the infant [15]. If the problem occurs at night times when the nurses are not alert it will cause serious problem to the infant. . It may lead to death of the patient [11] [13].Manual methods, which are employed in hospitals, are complex and require attention of the nurses all the time. If the proposed system is utilized then the work is reduced along with the human error to decrease the mortality rate So this paper describes an automatic system which controls this problem during such critical conditions.

II. Materials and Methods

This system contains two different modules which are motor and heater. The motor controls the infusion pump and the heater is used to warm the infusion fluid. The control of all these modules is given by controller. The other elements of this system are temperature sensor, LCD (Liquid Crystal Display) and an alarm system. Block diagram is shown in Figure 1.

A. Temperature Sensor

There are several types of temperature sensors which are used to measure the temperature. Those are LM35, thermostat, thermistor etc. To measure core body temperature of the infant the sensor can be placed in few places on the body [14]. Those places are armpit, anus, tympanic membrane, mouth (oral) and forehead [7][11]. But studies say that the placement of the sensor on armpit or anus will give the accurate core body temperature measurement of the infant [8]. There are temperature sensor probe also available, which will be comfortable for the patient and for the doctor to measure.

B. LCD & Alarm System

16x2 Liquid Crystal Display (LCD) is used. Continuous measurement of body temperature of the patient is displayed in LCD. A buzzer is connected in the system which works according to the commands which are given by the controller. This is used for the purpose of alertness of the nurses or staffs of that hospital.

C. Controller

The whole system is controlled by the controller which is programmed by the user to control the body temperature. The microcontroller's I/O pins are used by other elements of the system. This controller can control many modules at a time [9]. The development board can be handled easily. The user controls this controller by programming it using software to control the infusion and heater. The programs are written in C, C++, Java [10] [11].

D. Control system and Block Diagram

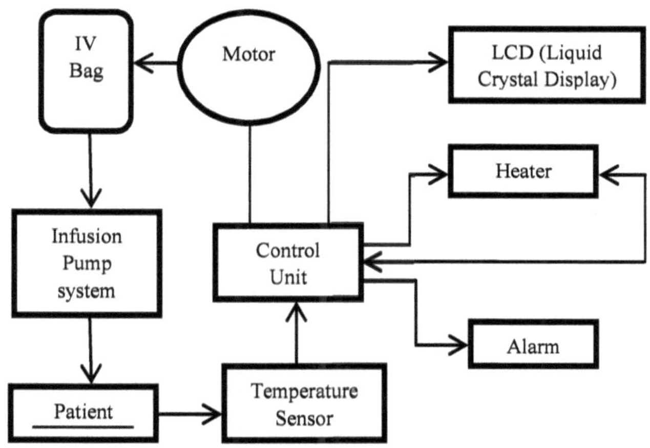

Fig.1 Block Diagram

E. Working

Body temperature of the infant is measured by placing the temperature sensor on the armpit of the infant and it is given to the controller through temperature sensor. Body temperature is displayed in the LCD in degree Celsius. If the input from the temperature sensor i.e body temperature of the infant is normal (37°C) the system will not provide

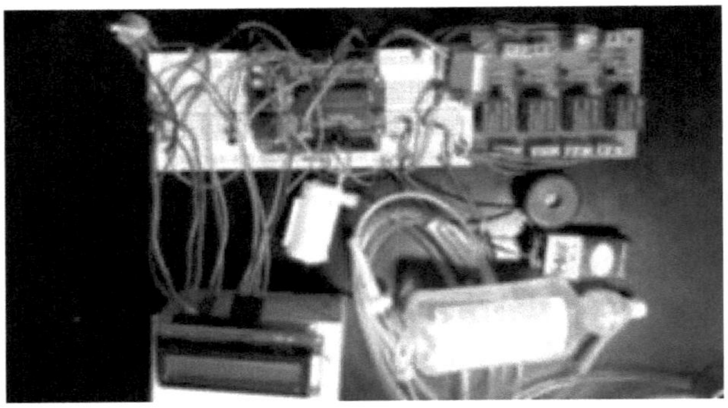

Figure 2 Work set up

any interrupt for the normal infusion. If the body temperature drops below 35°C the controller triggers the alarm. At the same time it will stop the motor which controls the speed of the infusion and it starts the heater to warm the infusion fluid which is to be infused. The motor works at 12V power supply. The temperature sensor and the alarm work at 5-9 V battery. Mostly this 5-9V power supply is given by battery or it can be given from the controller's power supply. There is a feedback system between the heater and the controller unit. The controller will stop the heater once the required i.e. optimum temperature is achieved in the infusion tube. Simultaneously the motor is turned on. Infusion of warmed fluid to the infant, the body temperature of the infant gets back to normal temperature. The alarm will be turned off once the body temperature of the infant is reached normal body temperature. All these processes are controlled by the program which is programmed by the user.

III. Results

This device is designed to measure the body temperature of the infant hourly and controls the infusion to the infants who are admitted in Intensive Care Units under IV administration. The suitable place for the placement of the sensor is the armpit for the babies and infants. Hence, the temperature sensor is placed on the armpit of the infant and the body temperature is measured hourly. Temperature output details are given in table 2.The controller stops the motor of the infusion pump and triggers the alarm. The heater is triggered simultaneously. Once the temperature goes below normal temperature (i.e. <37°C) the motor is stopped and the alarm is triggered. The heater is on and it starts to warm the IV fluid. Once the optimum temperature 38-45 ° C is achieved, the heater is turned off and the motor is on. The alarm is turned off once the body temperature becomes normal i.e. 37°C. Once the heater is triggered

Table 2 Temperature Output

Voltage	Temperature
10mv	1 °C
100mv	10°C
370mv	37°C

temperature of 38°C to 40°C is achieved inside the intravenous fluid and then the infusion is started. Increasing the temperature of the intravenous fluid from 38°C to 40°C automatically provide a drastic change in maintaining the body temperature of the infants .Thus providing a valuable health care solution for this problem. The system incorporation is simple and is used along with the infusion pump. It is of great deed when the infusion system does not have the insulation in their tubing and fluid heater to maintain the heat loss from the infusing fluid.

IV. Discussions

In this paper we have presented an automatic system which does not need doctors or nurses attention always. This system contains temperature sensor, controller, alarm system, heater and motor of the infusion pump. The sensor measures the body temperature of the infant and gives it as input to the controller. The controller stops the infusion to the infant and triggers the alarm & turns on the heater to warm the infusion fluid to be infused to the infant when the body temperature drops below 35°C. Once the optimal temperature is achieved in the infusion fluid the heater is stopped and the infusion starts to flow through the tube. The alarm is turned off once the body temperature of the infant becomes normal.

V. Conclusion

This system is fully automated. There is no need for the presence of the nurses or doctors nearby the patient and no need to check often. The user controls are not required since the device controls the mechanism automatically. It will give alarm when the body temperature drops below normal body temperature. Because of the feedback from the patient's body to the controller unit it is used to maintain the body temperature in more precise manner and the unexpected critical situations and deaths can be avoided. Future work includes the utilization and integration of the system which measures the body temperature of the patient without making any contact.

Conflict of Interest

The authors declare that they have no conflict of Interest.

References

1. R. Assunção, P. Barbosa, R. Ruge, P. S. Guimarães, J. Alves, I. Silva and M. A. Marques, "Developing the control system of the syringe pump", ISEP, REV, 2014.
2. Cameron C. Bangs, M.D. The Mountaineers, "Hypothermia Recognition and Treatment", Robert Douwens, 1986.
3. "Emergency Conditions due to Temperature Change", Chapter 14, Pre-course Material.
4. J. Satoh*, M. Yamakage†, S. I. Wasaki*, A. Namiki§, "Performance of Three Systems for Warming Intravenous Fluids at Different Flow Rates", 2006
5. Rongguo Yan+, Da Ji and Xuefei Han, "Implementation on Medical Infusion Pump with Constant Warm Temperature Flow", 2012 International Conference on Biological and Life Sciences IPCBEE vol.40 ,2012.
6. Khodayar oshvandhi, Fatemeh Hasan Shiri, Mohammed Reza Fazel, Mahmoud Safari, Ali Ravari, "The effect of pre-warmed intravenous fluids on prevention of intraoperative hypothermia in cesarean section", NCBI, Iran J Nurs Midwifery Res. 2014.
7. Johevajile K.N Mazima, Michael Kisangiri, Dina Machuve, "Deign of Low Cost Blood Pressure and Body Temperature interface", International Journal of Emerging Science and Engineering (IJESE), 2013.
8. Carlo Alberto Boano, Matteo Lasagni, and Kay R¨omer, "Non-Invasive Measurement of Core Body Temperature in Marathon Runners", IEEE, 2013.
9. Thomas E. Murphy, Hien Dao (UMD), Caitlin Williams, "Introduction to the Arduino Microcontroller", Hands-on Research in Complex Systems", Shanghai Jiao Tong University, 2012.
10. Charles E. Smith, MD, FRCPC,* Ronak Desai,† Virgilio Glorioso, MD,‡ Antonio Cooper, MD,‡ Alfred C. Pinchak, PE, PhD, MD,‡ Joan F. Hagen," Preventing hypothermia: convective and intravenous fluid warming *versus* convective warming alone"- Presented in part at the 4th American-Japan Anesthesia Congress, San Francisco, 1997
11. Kurz A, Sessler DI, Christensen R, Dechert BA. Heat balance and distribution during the core-temperature plateau in anesthetized humans. Anesthesiology 1995; 83:491-499.
12. Ramesh Agarwal, Ashish Jain, Ashok Deorari, Vinod K Paul Division of Neonatology, "Post-resuscitation management of asphyxiated neonates" Department of Pediatrics All India Institute of Medical Sciences Ansari Nagar, New Delhi – 110029.
13. Margaret wood, D.G. shand and a.j.j. wood," the sympathetic response to profound hypothermia and circulatory arrest in infants", 1980.
14. S.Rossi, E.Roncati Zanier, I Mauri, A Columbo, N Stocchetti "Brain temperature, body core temperature, and intracranial pressure in acute cerebral damage",2001.
15. Igor B. Mekjavic , Ola Eiken, "Contribution of thermal and nonthermal factors to the regulation of body temperature in humans", Journal of Applied Physiology , 2006.

Authors Address:
Author: Prof. Preethika Britto
Affiliation: Assistant Professor
Institute: VIT University
Street: "Baraka Illam", No 25, Second Street,
Bishop David Nagar,
City: Vellore
Country: India
Email: preethika.arockiaraj@gmail.com

Development of Bi-phasic Wheeze Generator for Biomedical Applications

Ravichandran Subbaraman[1], Then Tze Kang[2], Valencia Yeo[1], Siti Aishyah Bte Roslan[1], Nur Sharifiqah Bte Roslee[1], Shirlynn Ng Wan Ern[1], Hamsan Muthayya[1], Javiel Ng Jun Wen[1], Favian Ng Chuan Fang[1], and Shanshan Lin[1]

[1] School of Engineering, Temasek Polytechnic, Singapore
[2] National University of Singapore, Singapore

***Abstract*— This paper discusses the design and development of a bi-phasic pneumatic wheeze generator that has been developed to provide wheeze signal for calibrating biomedical tools for analyzing wheeze without the participation of patients. The bi-phasic wheeze generator essentially consisted of vibrating columns, which are energized by a new airway generator providing inspiratory, and expiratory flow patterns. These vibrating columns have been designed to customize the frequency and the pitch of the low frequency wheeze tones. Wheezes are usually defined as continuous tones with a dominant frequency of 400Hz or more. In some patients, they may be audible at some distance from the patient. The transmission of wheezing sounds through airways is better than transmission through the lung to the surface of the chest wall. Wheezing may result from localized or diffused airway narrowing or obstruction from the level of the larynx to the small bronchi. The airway narrowing may be caused by bronchoconstriction, mucosal edema, external compression, or partial obstruction or by tenacious secretions. The mechanisms underlying wheeze production appear to involve an interaction between the airway wall and the gas moving through the airway. Their mechanisms of production have been compared with musical wind instruments. This new development has provided the scope to test and verify low frequency wheeze tones that are rather difficult to generate. We have designed a new wheeze generation setup based on our previous experience in generating high frequency wheeze tones. To conclude, the newly developed bi-phasic pneumatic wheeze generator has expanded the scope to perform simulated conditions to fine-tune analytical packages for greater accuracies.**

***Keywords*— Wheeze generator, bronchoconstriction, low frequency wheeze tones, bi-phasic pneumatic wheeze, simulated conditions.**

I. Introduction

Breath sounds may be caused by induction of turbulence of the air at the segmental bronchi or lobar. As the gas velocity decreases, the critical velocity that is required to induce turbulence is lesser in the smaller bronchi. Hence, airflow in the smaller airways is perceived to be laminar and silent. Nominal breath sounds that were recorded over the lungs have main frequency bands up to 200 – 250Hz and the main energy level of tracheal sounds are said to extend up to 850 – 1000 Hz. All types of breath sounds are de- pendent on the turbulence induced by different airflow-rate. Breath sounds are not uniform over the lungs and have regional variations in sound intensity. At the apex, the sound is less intense during an inspiration performed from residual volume [1]. There are four types of classification for normal breath sounds; tracheal, bronchial, bronchovesicular and lastly, vesicular. Ultimately, the auscultated breath sounds depends on the area of examination. Vesicular breath sounds are soft, low-pitched sounds, where the inspiratory phase lasts longer then the expiratory phase. Bronchovesicular breath sounds are soft and high pitched while bronchial breath sounds are loud, hollow and harsh. Tracheal breath sounds are high pitched and loud [2-4].

II. Adventitious Lung Sounds

When lung sounds are louder than the breath sounds in certain situations, it usually indicates disease. Wheezes, or continuous adventitious lung sounds, are usually superimposed on the underlying breath sounds. The word "continuous" mentioned earlier does not mean a sound that continues all the way through the respiratory cycle. It instead signifies a period of more than 250ms. Patient with wheeze from asthma may be audible to the ear at some distance away from the patient as wheezes can often be heard from the patient's open mouth. The mechanism responsible for creating the wheezing sounds in the intra-thoracic airways appears to involve the walls in the airway related with the gas travelling through [5]. Researchers have recorded several wheezing sounds at several surface sites in asthmatic patients, and found out that frequency sounds having more than 1,000Hz were more clearly detected over the trachea than at the chest wall [6]. The encounter in narrowing of the airway caliber by mucosal thickening caused by bronchospasm that is responsible for the creation of continuous adventitious lung sounds [7]. Wheezes can also be classified

J. Goh and C.T. Lim (eds.), *7th World Congress on Bioengineering 2015*,
IFMBE Proceedings 52, DOI: 10.1007/978-3-319-19452-3_33

as monophonic (singular pitch), or polyphonic (multiple frequencies perceived simultaneously). The transmission of wheezing sound through the airways is better than transmission through the lung to the surface of the chest wall. The higher-frequency sounds are more clearly detected over the trachea than at the chest [8]. The lung tissues mainly absorb the higher frequency components of breath sounds [9]. Crackles, wheezes, and rhonchi are most commonly classified as adventitious breath sounds. Crackles are discontinuous sounds and can be classified as fine or coarse. Coarse crackles are loud and high pitched while fine crackles are soft and high pitched. A crackle is generated when an abnormally closed airway snaps open during inspiration or closes at the end of expiration [10]. Wheezes and rhonchi are continuous adventitious breath sounds, which are generated by air flowing through a narrowed airway. Wheezes are high-pitched continuous musical sounds with a dominant frequency 400Hz or more. Rhonchi are also continuous adventitious musical sounds. They are, however, lower pitched and have a snoring or gurgling quality with a dominant frequency of 200Hz or less [11].

III. Wheeze Simulations Using Models

In this paper, we will discuss more on the engineering and design aspects of bi-phasic pneumatic wheeze generator that has been developed to provide wheeze signal for calibrating biomedical tools. The architecture of the bi-phasic wheeze simulation is further discussed below.

IV. Bi-phasic Wheeze Simulator Model

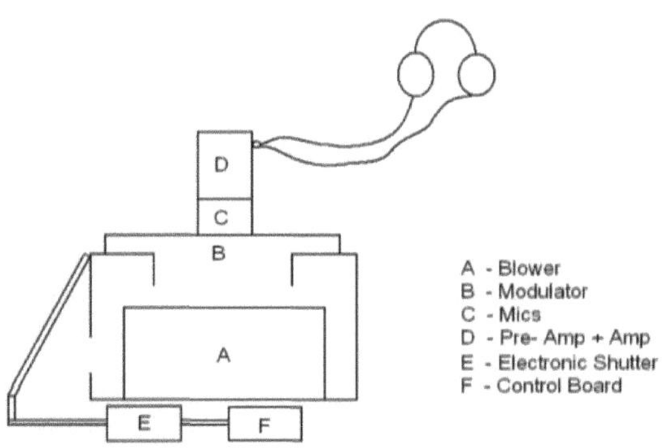

Fig. 1 Block Diagram of Wheeze Simulator

The wheeze simulator designed for generating wheeze signals in the frequency range varying from 200Hz to 700Hz is an integrated structure consisting of a ventilator and other associated engineering modules. The ventilator used is a Bi-level Continuous Positive Airway Pressure (CPAP), which provides two different positive airway pressures for the inspiratory and expiratory cycles.

A. Blower

A positive airway pressure is the starting pointing of the architecture, in any airway model, the ventilator is a Bi-level Continuous Positive Airway Pressure (CPAP), which provides two different positive airway pressures for the inspiratory and expiratory cycles. In the current design, we have not made use of the ventilator to provide our positive airway pressure. However, a delicate blower that provides the same pressure is used, as it is less noisy than the blower system of the ventilator.

B. Modulator

This modulator has two sub-modules, one consisting of the modulator and the second module for calibration. This component of the design provides the right pitch of the wheeze tones of interest.

C. Microphones

A microphone has been positioned right above the modulator to receive the pitch with less external interference. These microphones are sensitive and balanced to record the ambient pitch outside the modulator.

D. Amplifiers

The pre-amplifier and amplifier for picking up the signals from the microphones are positioned outside the modulators to have the ambient signal output of the simulator in the form of electrical signals for analysis.

E. Electronic Shutters

Electronic shutters provide the inspiratory and expiratory pressures that are required for producing the required wheeze tones. They are electromechanical in nature and controlled by the control circuit board.

F. Control Circuit Board

These are digital circuits with timer, flip-flops and drivers for producing the required digital output to activate the electromechanical shutter.

G. Digital Storage Oscilloscope

Though not integrated fully inside the simulator, it is an important accessory for monitoring the signals for research purposes.

H. Monitoring Station

We did not construct a monitoring station exclusively for signal analysis but we made used of the storage oscilloscope for monitoring the signals from the low pass filter. This signal output was used for documenting the audio output of the wheeze model for further analysis.

V. WORKING PRINCIPLES

A. Blower Module

The first highlighted block of the diagram is the electronic shutter as shown in the figure below.

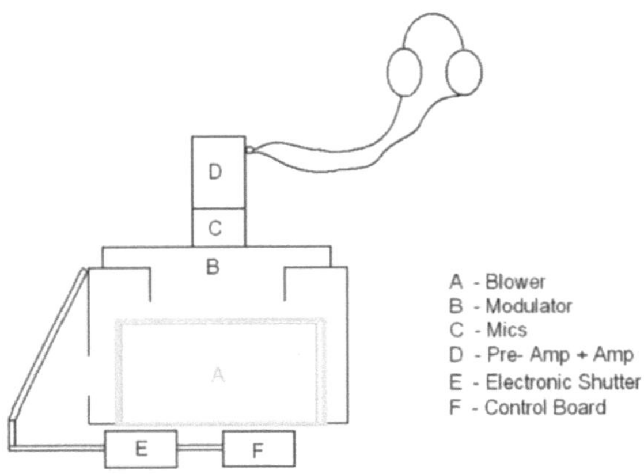

Fig. 2 Blower Module

Bi-level Continuous Positive Airway Pressure (CPAP) provides two different positive airway pressures for the inspiratory and expiratory cycles. A delicate blower provides the same pressure is used, as it is less noisy than the blower system of the ventilator. The speed of the blower is controlled by the control circuit and at a given speed; the aperture opening of the electronic shutter decides the inbuilt pressure of the system. The digital circuits with timer, flip-flops and drivers control the electromechanical shutter.

B. Modulator

The second highlighted block of the diagram is the modulator as shown in the figure below.

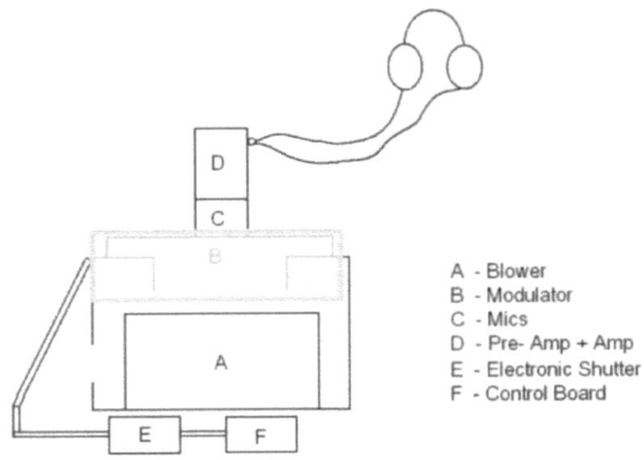

Fig. 3 Modulator

This modulator has two sub-modules, one consisting of the modulator and the second module for calibration. The modulator is basically consisted of reeds seen in some of the winds instruments and the sizes of the reeds are customized to produce the correct pitch and frequency of the wheeze tones. In our current design, the reeds are designed for frequencies between 200Hz and 700Hz. The modulator has several small compartments, which can be selected depending on the required pitch.

C. Electronic Shutter

The third highlighted block of the diagram is the electronic shutter as shown in the figure below.

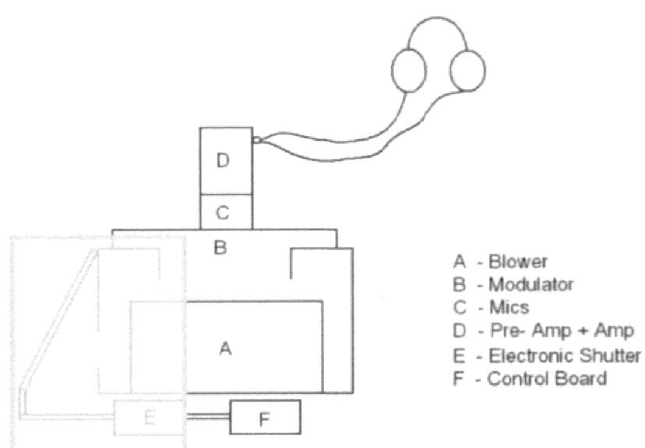

Fig. 4 Electronic Shutter

The electronic shutters module provides the inspiratory and expiratory pressures, and they work in conjunction with the blower mechanism to produce the required positive airway pressure. These shutters are spring loaded and bidirectional shutters that are activated by slider relay mechanism. These are controlled by the electronic circuitry of the system.

VI. Conclusions

To conclude, the newly developed pneumatic wheeze generator has expanded the scope to perform simulated conditions to fine-tune analytical packages for greater accuracy. Our initial research on development of models for generating the wheeze tones has been quite successful and we have managed to develop wheeze tones that are well within the range advocated by the American Thoracic Society (ATS). This development is perhaps the prelude for further more advanced models and had certainly given us the directions for future development. Preliminarily results are quite promising and have a potential application area for calibrating analytical packages for analyzing adventitious lung sounds.

References

1. Ploy-Song-Sang Y, Macklem PT, Ross WR (1978) Distribution of regional ventilation measured by breath sounds. The American review of respiratory disease 117: 657–664
2. Skoda J. (1853) A treatise on auscultation and percussion. Highley & Son, London
3. Hannon RR, Lyman RS (1929) Studies on pulmonary acoustics. II. The transmission of tracheal sounds through freshly exenterated sheep's lungs. Am Rev Tuberc 19:360
4. Gavriely N, Palto Y, Airoy G (1981) Spectral characteristics of normal breath sounds. Journal of Applied Physiology 50.2 50: 307-314
5. Forgacs P (1978) Lung sounds. Cassell and Coller MacMillan-Publishers Ltd, London
6. Takezawa Y, Shirai F, Sawaki S. el al. (1980) Comparison of wheezes over the trachea and on the chest wall. Fifth International Lung Sounds Conference, London, England
7. Fredberg JJ, HolfordSK. (1983) Discrete lung sounds. Crackks (rates) as stress-relaxation quadrupoles. The Journal of the Acoustical Society of America 73:1036
8. Fenton TR, Pasterkamp H, Tal A, Chernick V (1985) Automated spectral characterization of wheezing in asthmatic children. Biomedical Engineering IEEE Transactions 1: 50–55
9. Wodicka GR, Stevens KN, Golub HL, Cravalho EG, Shannon DC (1989) A model of acoustic transmission in the respiratory system. Biomedical Engineering, IEEE Transactions 36.9: 925–934
10. Piirila P, Sovijarvi AR (1995) Crackles: recording, analysis and clinical significance. European Respiratory Journal 8.12: 2139–48
11. Meslier N, Charbonneau G, Racineux JL (1995) Wheezes. European Respiratory Journal 8.11:1942–1948
12. Subbaraman, R., Ern, S. N. W., Muthayya, H., Wen, J. N. J., Fang, F. N. C., Lin, S., & Kang, T. T. (2014) Development of Pneumatic Wheeze Generator for Analysis of Wheeze in Biomedical Applications. In The 15th International Conference on Biomedical Engineering: 809-812.

Author: S.Ravichandran
Institute: Temasek Polytechnic
Street: 21 Tampines Ave 1
City: Singapore
Country: Singapore
Email: dr.sun.moon@gmail.com

In Vitro Practical Evaluations on Heating Characteristics of Thin Microwave Antenna in Maximum Temperature Control Mode

M. Chaichanyut and S. Tungjitkusolmun

King Mongkut's Institute of Technology Ladkrabang Department of Electronics, Faculty of Engineering, Bangkok, Thailand

Abstract— **Microwave thermal therapy is one of the modalities for cancer treatment. There are several schemes of microwave heating. The authors have been studying thin Coaxial Quarter Conductor Antennas (CQCA) for intracavitary microwave heating aiming at the treatment of hepatic cancer. Experimental protocol was composed by a radiation microwave power system and a thermometry system. We apply the microwave power during experiments was 10W, 20W, 30W, 40W, 50W, 60W, 70W and 80W which we set the maximum temperature control at 90°C for all case Experiment, Thermal sensors were placed next to the antenna at 1mm, a large number of experiments on porcine liver are carried out, the temperature distribution within the porcine liver are measured , for cases of different injected microwave power. Experiment for finding a possibility of the treatment. In this study, in order to consider practical situations of the treatment, heating characteristics of the antenna inserted into sample tissue. Moreover, the relation between coagulation size of the tissue, the radiation power from the antenna and the volume of lesion which the hepatic cancer was successful hepatic ablation. From these investigations, some useful results for practical treatments were found.**

Keywords— **Microwave ablation, Microwave antenna, bioheat equation, Coaxial antenna, porcine liver**

I. Introduction

Microwave ablation (MWA) offers an alternative therapy to radio frequency ablation (RFA). MWA may offer key advantages over RFA including higher treatment temperatures, larger active heating zone and reduced susceptibility to ablation zone distortion when used near large vasculature [1].Another major advantage of MWA is its ability to deliver energy to tissue despite inter-ablation temperatures above 100°C. MWA probe efficiency measured as the ratio of reflected power to forward power (scattering parameter) is however affected by the dielectric properties of the tissue within the ablation zone which significantly change over an ablation procedure due to temperature rise and associated tissue dehydration [2],[3].

MW ablation has been considered as a promising way of heating tissue deeper than RF currents. Since MW power deposition inside tissue decays with distance following a second power law, as compared to the fourth-power dependence of RF ablation, deeper lesions can be obtained [4]. Catheter antennas used for MW ablation have included both monopoles [5],[6]and helices [7].

This study demonstrates successful small-signal measurement of MWA probe broadband scattering parameters periodically over the duration of a high power microwave ablation cycle and introduces an electromagnetic modeling technique which accurately represents the broadband scattering parameters of a MWA probe at the onset and completion of an ablation cycle without the need for a multiphysics simulation software package.

In addition, coagulation necrosis can be expected by high radiation power from the antenna for large target tumor. So, in order to correct fundamental data, the relation between coagulation size of the biological tissue and the radiation power from the antenna will be investigated experimentally.

The authors have already been investigated the possibility of the treatment by numerical simulations using a high resolution human body model [8]. Therefore, in this study, heating characteristics of the antenna will be considered under two practical situations for aiming at the actual treatments. This paper presents the principle and geometries of an improved microwave percutaneous coagulator developed by us will be investigated, and some in vitro studies on porcine livers are carried out to demonstrate the performances of the coagulator, including the temperature distribution, ablation shape and size etc.

II. Materials and Method

In the following, we describe the Mathematical models, coaxial antenna designs and In Vitro Experimental Microwave System and setup

A. *The bio-heat equation*

In order to know the change in temperature and potential distributions in the hepatic tissue during microwave ablation, we solved the following bio-heat transfer [9], [10] equation, which governs the heating during hepatic ablation:

$$\rho c\frac{\partial T}{\partial t} = \nabla.k\nabla T + J.E - h_{b1}(T - T_{b1}) - Q_{el} + Q_m \qquad (1)$$

$$h_{b1} = \rho_{b1} c_{b1} \omega_{b1}$$

J. Goh and C.T. Lim (eds.), *7th World Congress on Bioengineering 2015*,
IFMBE Proceedings 52, DOI: 10.1007/978-3-319-19452-3_34

Where ρ is the Density (kg/m^3) , c is the Specific heat (J/kg·K), k is the Thermal conductivity (W/m·K), J is the Current density (A/m^2), E is the Electric field intensity (V/m), T_{bl} is the temperature of the blood, ρ_{bl} is the blood density (kg/m^3), c_{bl} is the Specific heat of the blood(J/kg·K), ω_{bl} is blood perfusion (1/s) and the h_{bl} is the convective heat transfer coefficient accounting for blood perfusion in the model. The energy generated by the metabolic processes Q_m, (W/m^3) and the heat exchanged between the ablation electrode and the tissue, Q_{el} (W/m^3) were neglected since they were small

B. Electromagnetic wave equation

The distribution of the electric field is determined from the electromagnetic wave equation solved here for time harmonic electromagnetic source (with the angular frequency $\omega = 2\pi f$), and expressed in complex form for the electric, respectively for the magnetic field strength

$$\nabla \times H = \sigma E + j\omega\varepsilon E \quad (2)$$

Then used to calculate the distributed heat source $J.E$

$$J = \sigma E + j\omega\varepsilon E \quad (3)$$

Where J is the current density (A/mm),E is the electric field intensity (V/mm), ω is the angular frequency(rad) and ε is the electric permittivity(F/m). And we can compute the electric field intensity by solving the equation

$$\nabla^2 E + \omega^2 \mu\varepsilon \left(1 + j\frac{\sigma}{\omega\varepsilon}\right) E = 0 \quad (4)$$

Coaxial antenna designs (CQCA)

Interstitial ablation antennas are usually made starting from a coaxial cable (or a tri-axial cable). In this study, we first consider the heating characteristics of Coaxial Quarter Conductor Antennas, which operating frequency of the antenna is 2.45 GHz, The antennas used in the experiment were flexible coaxial cable (RG CABLE 4.95mm (0.194inch)/50Ohm, Specification were Mil-C-17/60-RG142) The inner conductor is made from silver-plated copper wire (SPCW),which having a diameter of 0.95mm. and the coaxial dielectric used is a low-loss polytetrafluoroethylene(PTFE)., and were used with BNC-type connector. We modified this coaxial antenna following: the outer conductor was stripped off from the distal end of the antenna) [see Fig. 1].

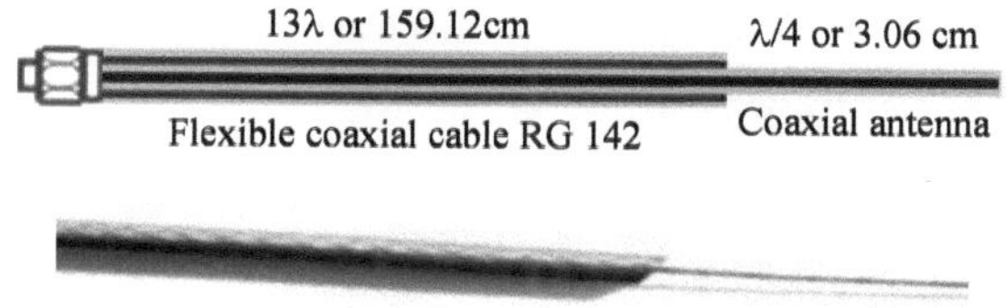

Fig 1. Coaxial antenna for intracavitary thermal therapy

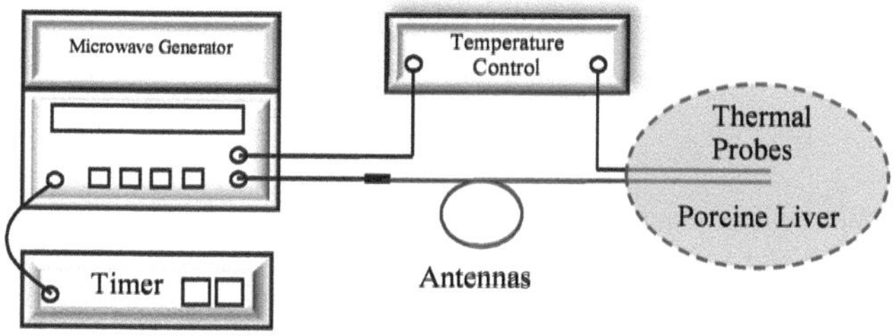

Fig 2. Experimental setup for tissue coagulation.

The active length of the antenna loaded in tissue is nominally (λ/4 or 30.06 mm in length), while the active length of the coaxial feed line are 13λ or 159.12cm. The length of the antenna tip also affects the power reflection and shape of the heating pattern. The CQCA extremity of the inner conductor is in contact with the medium.

In Vitro Experimental Microwave System and setup

This microwave system for liver ablation can be described with the block diagram shown in Fig.2. The system consists of: Microwave generator, timer, Temperature control and tissue sample (Porcine liver)

- Microwave generator: The source has an available microwave power up to 80W continuous wave (CW). It operated at 2450MHz. Some experiments are carried out with microwave coagulation treatment instrument developed in our research group with microwave power ranging from 0~ 80W. The system was equipped with a low-loss. We apply the microwave power during experiments was 10W, 20W, 30W, 40W, 50W, 60W, 70W and 80W, for a period of 4 minutes. The maximum temperature for experiment is controlling at 90°C of all case.
- Temperature control: We used copper-constantan thermocouples (thermo- couple K type) to measure the temperature within porcine liver. The thermocouples were calibrated in the ice water and an accuracy of ±0.1°C as obtained. In this experiment, thermo probe Sensors were placed alongside the antenna, at 1mm. for every experimental test.

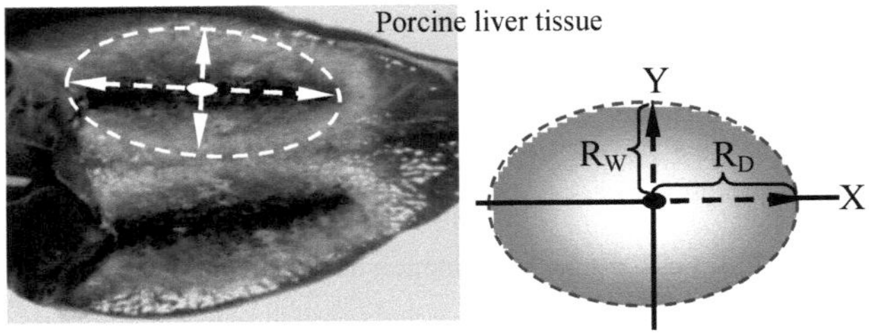

Fig. 3 The position measured of the extent of lesion formed.

III. RESULTS

We will be measure the initial temperature of the porcine liver tissue and record the initial temperature data. Fig. 2 shows the experimental setup. The antenna for experimental manufacture shown in Fig. 1 is inserted into the porcine liver tissue and is connected to microwave ablation system. In order to analyze the relation between radiation power of the antenna and coagulated region, the coagulation size is defined as Fig. 3. Shown Illustrates the distances measured for the extent of lesion formed around the antenna. RW is widest radius lateral directions of coagulation size; RD is longitudinal directions, which we measure along the length of the antenna.

In vitro trials have been conducted to characterize ablation zone shape and dimension as a function of the radiated power and maximum temperature controlled at 90°C, starting with an initial temperature of 25 °C. In additional, each data in Table I represents the average value obtained from four different repetitions of the corresponding measure. A measurement variability of 3mm has been estimated for these trials.

The SAR distribution gives information on the ability of the design to confine the electromagnetic field close to the antenna's active part. However, the final thermal lesion depends not only on the SAR values, but also on the thermal properties of the tissue, which surrounds the antenna, and on the presence of a cooling system. Since the ablation zone shape should be as close as possible to a spherical one, the axial ratio (R_W/R_D), i.e., the ratio between the maximum radial extension and maximum length of the induced thermal lesion will be used as a reference parameter, quite similar to [11].According to the definition, a thermal lesion shape close to a sphere has an axial ratio of 1, while elongated ones will have axial ratios smaller than 1. Shown in table 1 Here, example of a photograph of coagulated tissue is also shown. Fig. 4 and Fig. 5 present the lateral and longitudinal coagulation sizes under various radiations powers (incident power minus reflection power). The reflection powers of the antenna did not change dramatically during the coagulation.

Table 1 Experimental results of average ablation zone dimension (length x diameter mm.) in vitro porcine liver as a function of irradiated power and maximum temperature controlled at 90°c

Power(W)	Time(s)	Ablation Zone
10	812	22x32(0.69)
20	596	18x36(0.50)
30	432	20x36(0.56)
40	327	20x38(0.53)
50	257	24x36(0.67)
60	198	24x36(0.67)
70	153	19x30(0.63)
80	87	18x30(0.60)

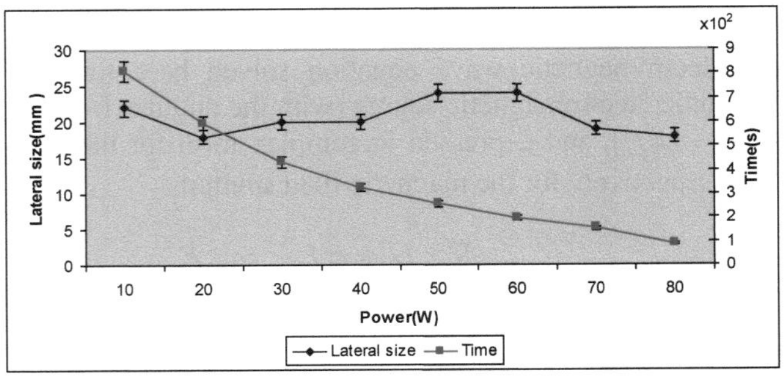

Fig. 4 Coagulation sizes in lateral directions (Error bar 5%).

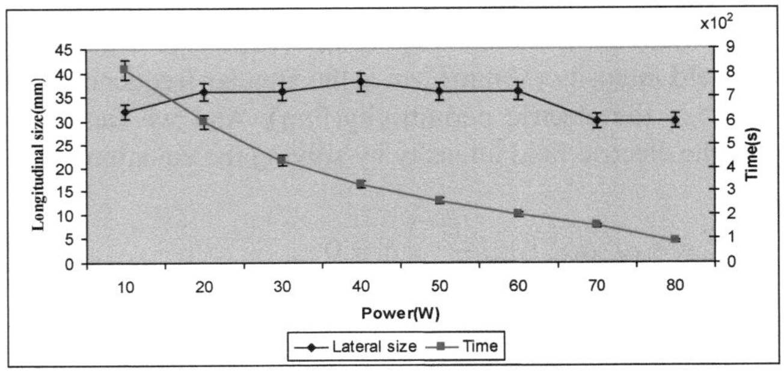

Fig. 5 Coagulation sizes in longitudinal directions(Error bar 5%).

Table 2 Coagulation volume of lesion

Power(watt)	Time (Second)	Volume(mm^3)
10	812	6591.62
20	596	5091.43
30	432	6285.71
40	327	6051.05
50	257	9051.43
60	198	9051.43
70	153	4538.29
80	87	4412.57

An ellipsoidal coagulated region is generated around the antenna. Therefore, the longitudinal sizes are almost twice as large as lateral sizes in all cases. Fig. 6 and Table 2 shows the volume of the coagulated region, when the coagulated regions have been assumed ellipsoidal body.

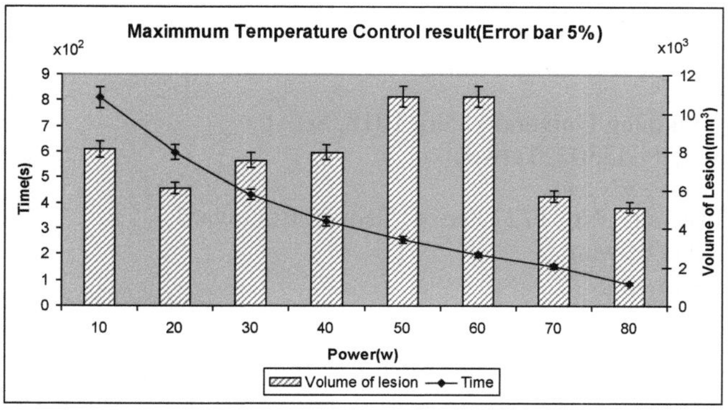

Fig. 6 Coagulation Volume under various radiation powers

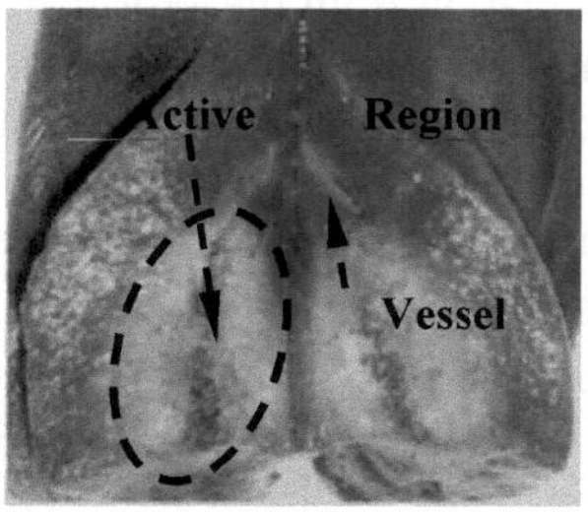

Fig. 7 Shown example of in vitro porcine liver coagulation zone lesion created with 10 W maximum temperature controlled at 90°C

From the results, it is found that the coagulation sizes and volumes are not directly proportional to the radiation power. In addition, there was no serious cable heating due to the ohmic loss of the antenna during microwave radiation.

IV. Discussion

Our study shows that the boundary temperature in normal porcine liver cells. Meanwhile, tissues would look like pink at the edge of the coagulation area, which is probably caused by incompletely necrosis. When the temperature reaches 90°C, the cells would predicatively become necrosis completely, and if this situation lasts long time, the color turns to be sable, which may be called as carbonization.

In additional, Vessels cause inhomogeneity and anisotropy of the antenna medium and, thus, alter the radiation characteristics, reduce the impedance match, and reduce power deposition into the tissue. Vessels *in-vitro* samples are filled with fluid and air, which create material inhomogeneities, increase reflected power, and reduce the energy deposited in the tissue. This likely caused the smaller lesion size.

V. Conclusions

In this paper present the experiments on porcine liver are carried out, the temperature distribution within the porcine liver are illustrated and calculate coagulation volume, for cases of different injected radiation microwave power and set maximum temperature controlled at 90°C

It is obvious to see that the coagulation size become larger with the increase of output power of the magnetron. In our experimental work, the minimum coagulation size of the porcine liver is about 18mm. ×26mm. (0.69)., at case of 80W which the duration time for ablation was 87s, while the maximum size could reach 24mm. ×30mm. (0.80)., at case of 50W×257s and 60W×198s respectively, In the case of microwave output power higher 60W, small tumors with diameter under 2cm can be completely destroyed, implying the possibility to cure the larger tumors. These experimental researches can probably provide some useful reference to the clinic treatment of the liver cancer using microwave percutaneous coagulator system.

References

1. A.S. Wright, F.T. Lee Jr. and D.M. Mahvi, Hepatic microwaveablation with multiple antennae results in synergistically larger zonesof coagulation necrosis. *Annals of Surgical Oncology* 2003; 10:275-283.
2. P. Manley, "Dielectric Changes with Temperature in Ex-vivo BovineLiver," unpublished, November 2007.
3. C.L. Brace, Temperature-dependent dielectric properties of liver tissue measured during thermal ablation: Toward an improved numerical model. *Proc. 30th Annu. Intern. IEEE EMBS Conf.* British Columbia,Vancouver Canada 2008 ; 230-233.
4. J. C. Lin, Y. L. Wang, and R. J. Heriman, "Comparison of power deposition patterns produced by microwave and radio frequency cardiac ablation catheters," *Electron. Lett.*, vol. 30, no. 12, pp. 922–923, June 1994.
5. J. C. Lin, "Catheter microwave ablation therapy for cardiac arrhythmias,"*Bioelectromagnetics*, vol. 20, no. S4, pp. 120–132, 1999.
6. R. D. Nevels, G. D. Arndt, G. W. Raffoul, J. R. Carl, and A. Pacifico,"Microwave catheter design," *IEEE Trans. Biomed. Eng.*, vol. 45, pp.885–890, July 1998.
7. Z. Gu, C. M. Rappaport, P. J. Wang, and B. A. VanderBrink, "Development and experimental verification of the wide-aperture catheter-based microwave cardiac ablation antenna," *IEEE Trans. Microwave Theory Tech.*, vol. 48, pp. 1892–1900, Nov. 2000.
8. M. Chaichanyut, S. Tungjitkusolmun," Temperature Distributions of Hepatic Cancer Tissue in Temperature Controlled Microwave ablation Mode", *International Conference on Cellular & Molecular Bioengineering,Nanyang Technological University*, Singapore,10 to 12 December 2007
9. CHATO, J. (1980): 'Heat transfer to blood vessels', *ASME Trans. Biomech. Eng.*, 102, pp. 110–118
10. KRAUS, J. D., and FLESICH, D. A., (1999): '*Electromagnetics with applications*', 5th Edition, McGraw-Hill Company, pp. 389–419
11. P. Prakash, M. C. Converse, J.G.Webster, and D. M. Mahvi, "An optimal sliding choke antenna for hepatic microwave ablation," IEEE Trans.Biomed. Eng., vol. 56, no. 10, pp. 2470–2476, Oct. 2009.

Effects of Low-Intensity Exercise Training on Tissue Oxygen Saturation of Lower-Extremity in Community-Dwelling Older Adults

Jian-Guo Bau[1], Yu-Fang Chung[2], and Hua-Jian Lin[2]

[1] Department of Biomedical Engineering, Hungkuang University, No. 1018, Sec. 6, Taiwan Boulevard, Taichung City 43302, Taiwan
baujg@sunrise.hk.edu.tw

[2] Department of Electrical Engineering, Tunghai University, No. 1727 Sec. 4, Taiwan Boulevard, Taichung City 40704, Taiwan
yfchung@thu.edu.tw,
ariellc06@yahoo.com.tw

Abstract—**Muscle strength and muscle oxygen saturation are two typical indicators for the evaluation of exercise effects in rehabilitation and sport medicine. The aim of this study was to compare the effects of low-intensity exercise training on muscle strength and muscle oxygen saturation monitored by near infrared spectroscopy (NIRS) in older adults. Eighteen healthy community-dwelling older adults, age mean (SD), 78.7 (7.3) years, were recruited. Ten subjects (exercise group) participated 6-week training program for lower-extremity, and the others were control. The training program was composed of 15-minute exercise course using three machines for lower-extremities training three times a week for six weeks. The intensity of machines were adjusted to 50% of one repetition maximum for each participants respectively. Heart rate, blood pressure, blood oxygen, muscle strength and muscle oxygen saturation, were measured before and after the 15-minute course in week zero (Week 0, before the 6-week training program) and 7th week (Week 7, after the training program). The Student's t-test was used to compare the difference of these variables between groups. The effective fall in quadriceps muscle oxygen saturation, D_{eff}, is defined as the decrease in oxygen saturation from quiet baseline (BL) to the end of one 15-minute course (Post-exercise, PE). After 6-week training program, D_{eff} did not change in the exercise group, while it decreased significantly in control group. Quadriceps muscle strength did not change after 6-week program in the exercise group, but decreased about 6% in the control group (not significant). Heart rate, blood pressure and blood oxygen were unaffected in the exercise group and the control group. The ambient temperatures were 25.8 ℃ and 19.9 ℃ for Week 0 and Week 7 respectively ($p<0.0001$). The training program counterbalance the muscular function decline due to seasonal variation from the end of autumn to winter for subjects participating the exercise program, while the function declined in control group. We suggest the non-invasive measurement of muscle oxygen could be used to assess the effect of physical activity program in community.**

Keywords—**tissue oxygen, physical activity, cold climate, aging, health promotion.**

I. Introduction

Habitual physical activity benefits both body and mental health. Especially in older people, it plays a critical factor not only in the prevention of the incidence of cardiovascular diseases, obesity, osteoporosis, or diabetes [1,2,3,4] but also in the improvements of daily functioning tasks. According to the recommendation from the World Health Organization for health maintenance and promotion [5], adults aged 65 or older should accumulate at least 150 minutes of moderate-intensity physical activity throughout the week. However, upon review of the previous study, it appears that the most sedentary groups in the United States was adults aged ≥60 years, and they spent about 60% of their waking time in sedentary pursuits [6]. Moreover, the levels of physical activity may vary with seasonality, and the effect of cold weather has been identified as a barrier to participation in physical activity [7], which should be associated with the increase of all-cause and cardiac mortalities in most developed societies in the winter.

In order to promote physical activity in older adults, the factors that may influence physical activity participation should be understood. The related factors include the personal characteristics, such as individual's knowledge, skills and psychological attributes, and the environmental factors, such as the supports from family members, friends and other exercise participants. In addition to above factors, a convenient assessment tools that are sensitive to more moderate-intensity or even to low-intensity physical activities favored by older adults should be developed for the physical activity intervention studies [8]. Questionnaire is the typical tool, which collect the personal information such as exercise adherence, self-reported physical activity levels, physical health status and sedentary behavior in leisure time. Accelerometer, which record the amount of physical activity, are the objective tool for the understanding of the energy expenditure in exercise intervention programs. The effectiveness of the physical activity programs are generally be evaluated by the improvement of the physiological functioning variables, such as cardiorespiratory fitness and musculoskeletal function performance.

Muscle strength and muscle oxygen saturation [9] are two typical variables for the evaluation of exercise effects in rehabilitation and sport medicine. Muscle oxygen saturation, with abbreviations SmO_2 refers to the percentages of hemo-

J. Goh and C.T. Lim (eds.), *7th World Congress on Bioengineering 2015*,
IFMBE Proceedings 52, DOI: 10.1007/978-3-319-19452-3_35

globin and myoglobin that are carrying oxygen, and is defined as follows:

$$SmO_2 = \frac{Oxygenated\ hemoglobin\ and\ myoglobin}{Total\ amount\ of\ hemoglobin\ and\ myoglobin} \quad (1)$$

The aim of this study was to compare the effects of low-intensity exercise training on muscle strength and SmO_2 monitored by near infrared spectroscopy (NIRS) in community-dwelling older adults.

II. Materials and Methods

A. Subjects

All volunteers were informed about the training program and the investigation in an information session. Of the 17 subjects, who had given consent, 9 subjects participated in the 6-week training program (Exercise group), while the other 8 subjects were control group. Both groups were matched in age, weight, height and physical activity (Table 1). All participants were asked not to change their physical activity habit during the study period intentionally. The research protocol has been approved by Research Ethics Committee of Taichung Tzu Chi Hospital, with approval number REC103-48.

B. Study design

The whole study was divided into three stages. Stage 1 was the pre-test stage, Stage 2 was the training stage, and Stage 3 was the post-test stage.

Stage1 (pre-test stage, Week 0): (1). Measure the quadriceps and hamstrings muscle strength of each participants, and then adjust the resistance of the training equipment for each participant. The resistance of the training equipment were set to 50% of 1 RM (one repetition maximum) of the corresponding muscle. (2). Measure the quadriceps muscle oxygen saturation in quiet condition (baseline, BL). (3). After the participants were instructed how to use the training equipment, ask them to finish the training course with the training equipment in their personal intensity. One course was composed of two sets of muscle training with 1-2 minutes interval in 15 minutes. (4). Measure the quadriceps muscle oxygen saturation again in post-exercise condition (PE).

Stage2 (Week 1 to Week 6): Each participant was asked to attend the training course three times a week for a period of six weeks. The light exercise training course is same as that in Stage 1. The training equipment used in course was designed for light-intensity training on the quadriceps muscle and hamstrings muscle. During this stage, the training was performed under the instruction by professional sports instructor.

Stage3 (post-test stage, Week 7): same procedures as pre-test.

b) Measurement Procedure

The muscle oxygen saturation was monitored by near infrared spectroscopy (Moxy monitor, Minnesota, USA). The sensor was located on the medial quadriceps. Each measurement was the average of 100 seconds data.

In addition, all subjects were reminded to avoid rigorous exercise one day before the measurement and should be with regular diet and plenty of sleep. The measurement was conducted in non-fasting condition, and at least 60 minutes after meal.

Table 1 Mean and standard deviation values for subject characteristics, anthropometric measures and physical activity

	Exercise group (n=9)	Control group (n=8)
Age (yr)	79.8±7.3	76.8±.7.6
Weight (kg)	56.7±8.3.	58.0±7.6
Height (cm)	155.4±8.3	158.9±8.2
BMI (kg/m2)	23.4±2.0	23.0±2.2
Total PA (min/day)	29.1±39.2	46.3±47.6

Table 2 Heart rate, systolic (SBP) and diastolic (DBP) blood pressure measured in Pretest and Posttest

	Exercise group (n=9)		Control group (n=8)	
	Pre-test	Post-test	Pre-test	Post-test
Heart rate (min-1)	72.2±13.1	71.9±12.6	74.4±11.6	76.8±16.0
SBP (mmHg)	128.4±20.4	121.5±22.8	125.1±21.5	120.3±22.2
DBP (mmHg)	67.1±16.6	62.3±10.9	68.3±14.6	65.6±12.5

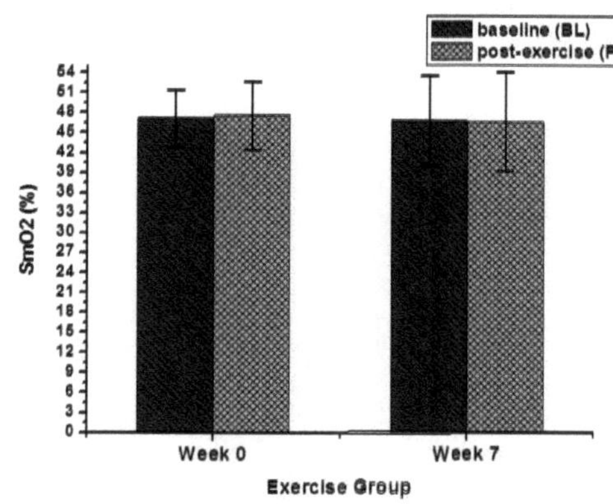

Fig. 1 Quadriceps muscle oxygen saturation, SmO_2, before (Week 0) and after (Week 7) the 6-week training program for the participants in exercise group. SmO2 in quiet baseline (BL) and at the end of one 15-minute course (Post-exercise, PE) were measured.

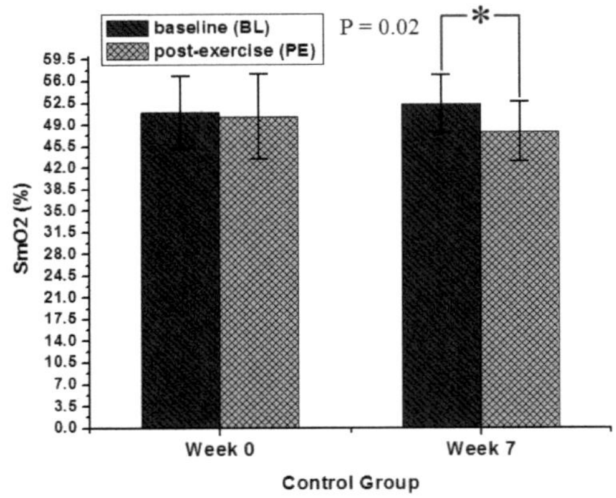

Fig. 2 Quadriceps muscle oxygen saturation, SmO2, before (Week 0) and after (Week 7) the 6-week training program for the participants in control group. SmO2 in quiet baseline (BL) and at the end of one 15-minute course (Post-exercise, PE) were measured.

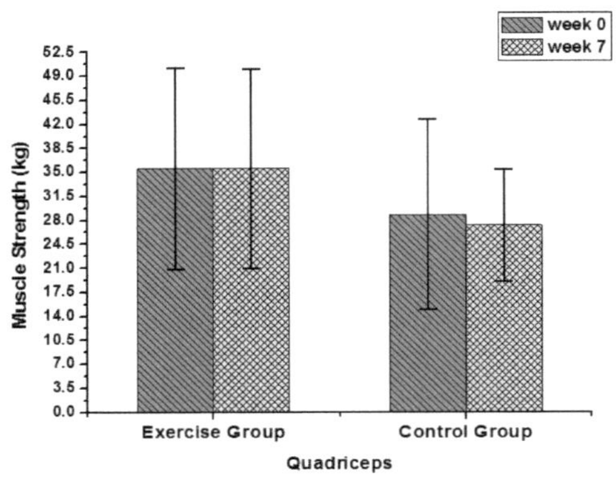

Fig. 3 Quadriceps muscle strength of the participants in exercise and control groups before (Week 0) and after (Week 7) the 6-week training program.

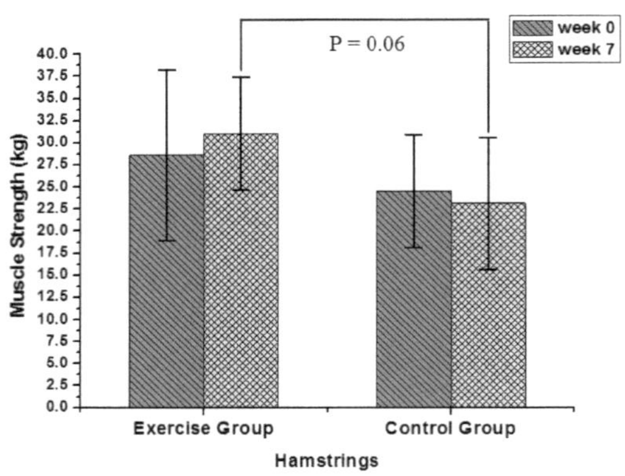

Fig. 4 Hamstrings muscle strength of the participants in exercise and control groups before (Week 0) and after (Week 7) the 6-week training program.

III. RESULTS

The 6-week training program began in November and then ending in December. At this time, the climate change significantly in Taiwan. The ambient temperatures, in which the physiological status of the participants was evaluated, were 25.8 ℃ and 19.9 ℃ for Week 0 and Week 7 respectively ($p<0.0001$). After the 6-week training program (in Week 7), the quadriceps muscle strength did not improve significantly for all participants in exercise group, but decreased about 6% in the control group (not significant), shown in figure 3. The hamstrings muscle strength (in figure 4) also exhibited the similar result. The effective fall in quadriceps muscle oxygen saturation, D_{eff}, is defined as the decrease in SmO_2 from quiet baseline (BL) to the end of one 15-minute course (Post-exercise, PE). In Week 0 (before the program), the PE SmO_2 did not change significantly while compared with BL SmO_2 for both exercise group and control group (figure 1 and figure 2). On the other hand, after 6-week training program (Week 7), D_{eff} did not change in the exercise group, while it decreased significantly in control group ($P=0.017$). Heart rate, blood pressure and blood oxygen were unaffected after the 6-week program in the exercise group and the control group. The heart rate and blood pressure in Week 0 and Week 7 was shown in table 2.

IV. DISCUSSIONS AND CONCLUSIONS

The measurement of muscle oxygen saturation reflects the balance between supply and demand for oxygen in tissue around the capillaries. If the more oxygen is being demanding by the surrounding tissue than is being delivered by the microcirculatory perfusion, such as muscle in exercise condition, the oxygen saturation will go down. However, the oxygen saturation will recover to the pre-exercise level gradually when the tissue stop working. However, if the microcirculation is impaired, the recovery will be insufficient and the local tissue will reveal the low saturation level.

There previous research suggested that the importance of maintaining physical activity during adverse environmental conditions should be underlined [10]. Our study reveal that the SmO_2 of the participants in control group decreased after the 15-minute course in winter, while it recoverd to the baseline sooner in autumn. Oppsitely, this phenomena did not find on the participants of exercise group. In other words, by entering the winter season, if the amount of physical activity is not sufficient, the decrease of the eviromental temperature can result in the decline of the blood supply to the peripheral muscle tissues. Therefore, even the subjects take part in the same training course, the lower blood supply will cause the lower oxygen saturation in the peripheral tissue.

The effectiveness of physical exercise therapy on mobility, physical functioning, physical activity and quality of life in community-dwelling older adults with impaired mobility has been proved [11,12]. Moreover, the decline of the muscle oxygen saturation due to seasonal variation from the end of autumn to winter can be counterbalanced, we suggested that even the improvement of the muscle strength was not significant, the 6-week lignt-intensity exercise program will be helpful for the muscular function of older adults. We recommend older adults should increase their amount of physical activity by exploiting any available indoor facilities in adverse season. With the advantages of non-invasiveness and easy-operation, the variation of muscle oxygen could be used as a microcirculatory index to assess the efficacy of physical activity program in community.

Acknowledgment

We would like to acknowledge Floriey Industry International Co., Ltd. for providing the training equipment.

Conflict of Interest

The authors declare that they have no conflict of interest.

References

1. Hamer, M., Venuraju, S.M., Urbanova, L., Lahiri, A., Steptoe, A., 2012. Physical Activity,Sedentary Time, and Pericardial Fat in Healthy Older Adults. Obesity. http://dx.doi.org/10.1038/oby.2012.61 (Silver Spring, April 5).
2. Kelley, G.A., 1998. Exercise and regional bone mineral density in postmenopausal women: a meta-analytic review of randomized trials. Am. J. Phys. Med. Rehabil.77, 76–87.
3. Paganini-Hill, A., 2011. Lifestyle practices and cardiovascular disease mortality in the elderly: the leisure world cohort study. Cardiol. Res. Pract. 2011, 983764.
4. Santos, D.A., Silva, A.M., Baptista, F., Santos, R., Gobbo, L.A., Mota, J., Sardinha, L.B., 2012.Are cardiorespiratory fitness and moderate-to-vigorous physical activity independently associated to overweight, obesity, and abdominal obesity in elderly? Am. J. Hum. Biol. 24, 28–34
5. Haskell, W. L., I. M. Lee, et al. (2007). "Physical Activity and Public Health." Medicine & Science in Sports & Exercise 39(8): 1423-1434.
6. Older Adult Perspectives on Physical Activity and Exercise: Voices From Multiple Cultures Amount of Time Spent in Sedentary Behaviors in the United States, 2003–2004
7. The effect of season and weather on physical activity: A systematic review
8. J Gerontol A Biol Sci Interventions to Promote Physical Activity by Older Adults Med Sci-2001-King-36-46
9. Non-invasive quantitative assessment of oxidative metabolism in quadriceps muscles by near infrared spectroscopy.
10. Seasonal variations in physical activity and implications for human health
11. Effects of physical exercise therapy on mobility, physical functioning, physical activity and quality of life in community-dwelling older adults with impaired mobility, physical disability and/or multi-morbidity: A meta-analysis
12. A community-based exercise program to improve functional ability in people with Alzheimer's disease: a randomized controlled trial

Use macro [author address] to enter the address of the corresponding author:
Author: Jian-Guo Bau
Institute: Department of Biomedical Engineering HUNGKUANG UNIVERSITY
Street: No. 1018, Sec. 6, Taiwan Boulevard, Shalu District
City: Taichung City
Country: Taiwan (R.O.C.)
Email: baujg@sunrise.hk.edu.tw

Development of Amperometric Biosensor for Creatinine Detection

Pratondo Busono

Center for Communication and Information Technology, BPPT, Jakarta, Indonesia

***Abstract*— A method for fabrication of a disposable amperometric biosensor for determination of creatinine concentration was developed. An enzyme ink was prepared by mixing of 10 mM CaCl2, 1 wt% PEG-400, 4 wt% Cabosil and 2.5 wt% Hydroxyethyl Celluose (HEC) and added with creatinine amidohydrolase, creatine amidinohydrolase, sarcosine oxidase, and horseradish peroxidase. The enzyme ink was then printed on the surface of working electrode. Electropolymerization of pyroll was conducted on the surface of working electrode to prevent the enzyme deteriorate. The resulting biosensor was characterized using cyclic voltametry. Enzymatically generated hydrogen peroxide has amperometrically been detected at a carbon-working electrode.**

The performance of the biosensor was evaluated using standard creatinine. The biosensors have a linear detection range of 0.5 × 10^{-5} – 1.0 × 10^{-3} M with response time of 40 s. The sensitivity of the biosensor was 65 mV/pA. The detection limit of the biosensor is 50 μM. The biosensor works at optimum pH and temperature of 7.25 and 34 ^{0}C, respectively at buffer concentration of 50 μM. The biosensor shows a good stability over a period of 1 months, which is only decreasing to 80 % of its initial activity. The influence of the interfering ascorbic acid and uric acid on the biosensor shows a little effect on its performance. The biosensor had been compared with the measurement conducted with spectrometry techniques. A good agreement was observed between both the measurements.

Keywords— **creatinine, amperometric biosensor, creatinine amidohydrolase. creatine amidinohydrolase**

I. INTRODUCTION

Creatinine is an important clinical analyte for the diagnosis of muscular disfunction and renal failure [1]. An increasing creatinine level in blood or urine is an indication of an onset of kidney disease. The normal clinical range of creatinine in human serum is in between 44 and 106 μM. The pathological value due to kidney failure, the creatinine level may increases more than 1000 μM [1].

Routine creatinine determination in clinical laboratory is mostly using spectrophotometric method. It is based on the Jaffe reaction [2]. However, this method is time consuming, since it needs realtively large amount of blood sample and needs bulky equipment. For this reason, the development of a techniques to creatinine determination is very important. Biosensor is a very good alternative since it is mainly simple, low cost, and fast response. More recently, biosensor has emerged as a promising technology, especially for applications requiring rapid and continuous monitoring.

Most creatinine biosensors are based on the catalytic conversion of creatinine to creatine in the presence of creatinine amidohydrolase enzyme. The fundamental reaction for creatinine detection is based on the following coupled reactions [3],

$$Creatinine + H_2O \xrightarrow[\text{Amidohydro lase}]{\text{Creatinine}} Creatine \quad (1)$$

$$Creatine + H_2O \xrightarrow[\text{Amidinohyd rolase}]{\text{Creatine}} Sar\cos ine + Urea \quad (2)$$

$$Sar\cos ine + H_2O + O_2 \xrightarrow[\text{Oxidase}]{\text{Sar cos ine}} Glycine + Formaldehy\,de + H_2O_2 \quad (3)$$

They ultimately produce hydrogen peroxide, the species that is electrochemically transformed at the electrode to generate an amperometric signal.

Enzyme immobilization is one of the crucial steps in the fabrication of a biosensor. There are various immobilization techniques available for enzyme encapsulation such as adsorption, covalent bonding and cross-linking and sol-gel encapsulation. Currently, researchs in biosensor are mostly focused on resolving the drawbacks of enzyme instability, fragility, sensitivity enhancement and wider response range. Some of these drawbacks of enzyme instability were solved by utilizing a mechanically stable membrane such as polyacrylamide (PAA) and polyvinyl alcohol (PVA) or electrosythesis membranes such as polyaniline (PANi), polypyrrol film and polythiophene. Among them, polypyrrole is one of the most widely used conducting polymers in the fabrication of biosensors, since their conductivity and electroactivity do not strongly depend on pH of the electrolyte. In previous works it was demonstrated the efficiency of polypyrrole as a mediator to the ammonia detection. And its response could be improved using different dopants into the film preparation.

J. Goh and C.T. Lim (eds.), *7th World Congress on Bioengineering 2015*,
IFMBE Proceedings 52, DOI: 10.1007/978-3-319-19452-3_36

II. Material and Method

A. Material

Creatinine Amidohydrolase (EC 3.5.2.10) from Pseudomonas sp, Creatine Amidinohydrolase (EC 3.5.3.3) from Actinobasillus sp and Sarcosine Oxidase (EC 1.5.3.1) from Arthrobacter sp were purchased from Toyobo. Co. Ltd. Creatinine, Polyethylene Glycol (PEG-400), Cabosil, Hydroxyethyle Cellulose (HEC), $CaCl_2$, Uric Acid (Purity was 99.8%), Ascorbic Acid and Horseradish Peroxidase (HRP) were purchased from Sigma-Aldrich (St. Louis, MO, USA). Glutaraldehyde was purchased from Merck. Carbon conductive ink (Electrodag PF-407) and insulation ink (Electrodag 452SS) were purchased from Acheson Colloids, USA. Platinum powder 99.995% was purchased from Aldrich, USA. The reagents are 10 mM Phosphate Buffer Solution (PBS). It was prepared from Na_2HPO_4 and KH_2PO_4 with sodium cloride in 0.135 mM and the pH was adjusted to 7.4. Other reagent was citrate buffer. It was prepared by mixing of 0.1 M citric acid monohydrate and 55.5 ml of 0.1 M trisodium citrate dihydrate (pH = 7.4). Serum was purchased from Merck Chemical. All chemicals were analytical reagent grade. Three-electrode configuration was purchased from Uniscan, Great Britain. Modification was conducted on working electrode by printing with the mixture of platinum powder dan carbon ink.

B. Electrode Preparation

The enzyme ink was prepared by mixing of 10 mM CaCl2, 1 wt% PEG-400, 4 wt% Cabosil and 2.5 wt% Hydroxyethyl Celluose (HEC). This mixture was added into Citrate buffer solution (pH=7.4) and hydrated for 6 h. The mixture was added with 500 units creatinine amidohydrolase, 400 units creatine amidinohydrolase, 300 units sarcosine oxidase, and 500 units horseradish peroxidase.

C. Enzyme Immobilization

Enzyme immobilization is one of the crucial steps in the fabrication of a creatinine biosensor. It is conducted by printing the enzyme ink on the surface of working electrode. Before enzyme immobilization was conducted, the electrode was rinsed with PBS. The amount of enzyme ink over the transducing membrane affects activity, response time and linear dynamic range due to thickness of sensing membrane and amount of enzyme.

Pyroll electropolymerization was carried out using Uniscan PG 580 Potentiostat – Galvanostat. In the electropolymerizaton procces, 30 μL of freshly distilled pyrrol was dissolved in 0.9 mL phosphat buffer 0.1 M., pH 7.4. To this solution, 100 μL of KCl 0.1 M was added. The electropolymerization was conducted galvanostatically by applying the current density of 0.2 mA/cm^2 to the working electrode for 10 minutes. Before substance addition began, the response current base lines were established using a known volume of fresh buffer.

D. Instrumentation

Cyclic voltametry was performed using Uniscan Model PG-580 potentiostat-galvanostat. A pH meter (HI 9124, Hanna Instrument) was used to measure the pH solution. Milton Roy Spectronic 21 was used for comparison in measuring creatinine concentration. Impedance measurement was conducted using HIOKI 3532-50 LCR HiTester.

E. Performance

The biosensor performance can be expressed by the parameters like reproducibility, measuring range, detection limit, response time and selectivity. The reproducibility is an important characteristic for the performance evaluation of a biosensor.

III. Experimental Results

Fig. 1 shows the calibration curve of the amperometric creatinine biosensor. The data were obtaind from cyclic voltametry measurement for creatinine concentration from 10 μM to 1000 μM. The applicable measuring range of the sensor is between 50 μM to 800 μM. By extrapolating the linear part of the curve, it shows that the detection limit of the sensor was 0.1 mM. . The sensitivity of the biosensor was 65 mV/pA. The detection limit of the biosensor is 50 μM.

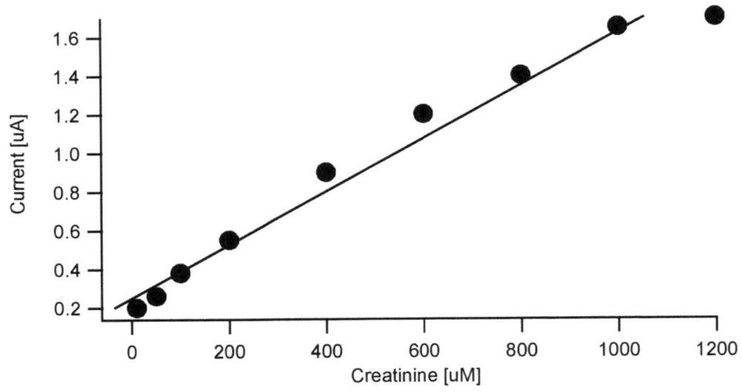

Figure 1 Calibration curve for creatinine concentration between 50-1000 μM.

The biosensor was also checked for the reproducibility. It was checked with 200 μM creatinine solution for upto 10 reused. It was found that its activity decreases to 45 % from initial activity with the relative standard deviation less than 5 %. The measurements were conducted one hour after the biosensor being fabricated. The biosensors were reused 10 times and the delay between measurements were 1 hour. After each measurement, the biosensor was washed with distilled water.

To evaluate the their stability, the 13 biosensors were stored at 4 ^{0}C and periodically tested for more than 2 months. It shows that the biosensors still retained up to 60% of their original activity.

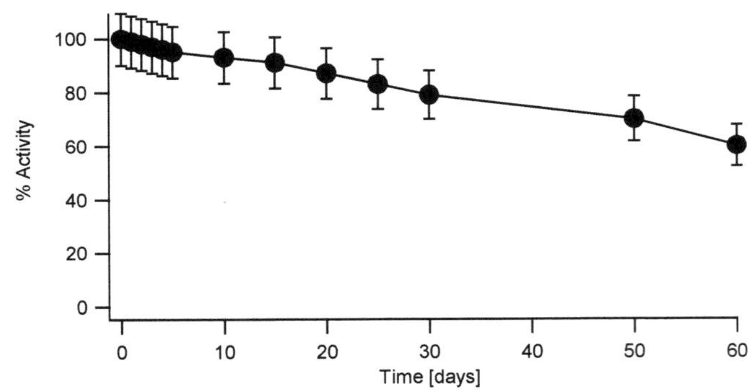

Figure 2 Biosensor stability

Selectivity is the most important characteristic which describes the specificity towards the target ion in the presence of other ions or interfering ions. In this work, the mixed solution method was used to determine the selectivity of the sensors by checking the output response. The most possible interferences present in blood are uric acid, ascorbic acid, and uric acid. Hence, such substances were selected to test the selectivity of the sensor. In the experiment, minor signal changes in sensor response were observed when the creatinine solution of 200 μM was added by 50 μM uric acid, 50 μM ascorbic acid. Such minor change in sensor response was probably due to instability/disturbance caused by the successive addition of the solutions. The similar experiments were conducted several times with other new fabricated biosensors. The results show negligible signal response to interferences which can be ignored in the real application.

The pH dependence of the sensor response was investigated in 100 μM urea solution over the pH range from 6.5 to 9.0. Fig. 3 shows the optimum pH requirement of the biosensor, with the temperature working envirotment of 34 °C. The experimental results indicate that the maximum signal response of the sensor is at pH 7.25. Moreover, it can be noted that at pH value higher than 7.25, the potential decrease due to the decrease in enzyme activity. The broadening of spectrum towards both acidic and alkaline environtment was observed implying that the enzyme becomes less sensitive to pH changes.

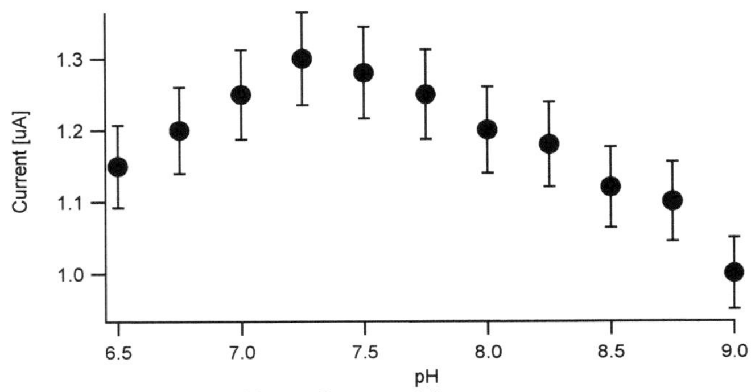

Figure 3 Effect of pH on biosensor response

The effect of temperature on the biosensor response in 10 mM creatinine solution was also investigated at the temperature range between 15 ^{0}C - 70 ^{0}C. The applied potential at working electrode was kept constant of 0.4 V with respect to Ag/Ag Cl reference electrode.

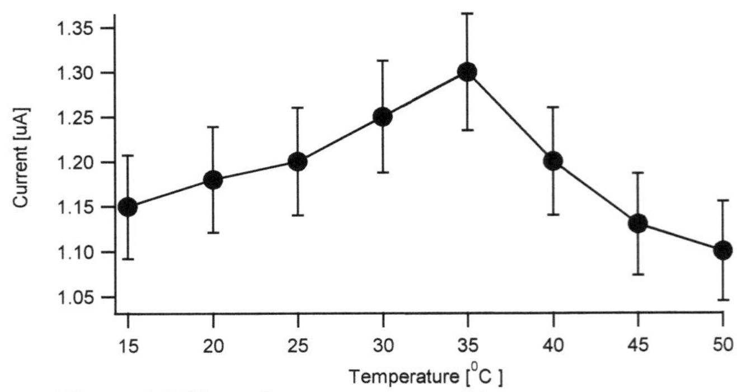

Figure 4 Effect of temperature on biosensor response

Figure 4 shows the effect of operating temperature on the biosensor responses. It shows that the biosensor response increases with increasing of temperature. After reaching to its maximum value at around 35 ^{0}C, the sensor response decreases, because of the denaturation of the enzymes at high temperature. Though, at temperature around 35 ^{0}C the biosensor response reaching maximum value, however it was not well stable for long term performance compared to the sensor response at room temperature. For the experiment shows that when the sensor was tested at 35 ^{0}C, the enzymatic activity degraded drastically and showed poor long term performance. In this work, the room temperature of 18±2 ^{0}C was chosen in order to prevent possible enzyme degradation.

The comparison between the measurement of creatinine concentration in serum using spectroscopy method and biosensor method was also conducted. Close agreements are observed between two measurements. It means that the fabricated creatinine biosensor is valid.

IV. CONCLUSIONS

The creatinine biosensor showed a fast response with less than 10 s and has a quite wide linear range from 50 µM to 800 µM. The creatinine biosensor also exhibited good performances in sensitivity, stability, selectivity, reproducibility and negligible interference to the common interferents. All these advantageous features can make the creatinine biosensor applicable in medical or other areas.

Acknowledgment

This work was supported by the Ministry of Research and Technology, Republic of Indonesia, through Insetif Riset 2011 KP-2011-1758 "Peningkatan Sensitivitas, Akurasi, Selektifitas, Stabilitas dan Reprodusibility Disposable Biosensor Berbasis Teknologi Screen Printing Untuk Kapasitas Produksi Skala Industri". The research was conducted at the Laboratory of Medical Instrumentation, Center for Pharmaceutical and Medical Technology, BPPT, Kawasan Puspiptek Serpong, Indonesia.

References

1. Tiwari A., S.K. Shukla. (2009) Chitosan-g-polyaniline: a creatine amidinohydrolase immobilization matrix for creatine biosensor, eXPRESS Polymer Letters Vol.3, No.9, 553–559.
2. Isildak I., Cubuk O., Altikatoglu M., Yolcu M., Erci V., Tinkilic N. (2012) A novel conductometric creatinine biosensor based on solid-state contact ammonium sensitive PVC–NH2 membrane, Biochemical Engineering Journal 62: 34-38.
3. Tsuchida, T. Yoda, K. (1983) Multi-enzyme membrane electrodes for determination of creatinine and creatine in serum. Clin. Chem., pp. *29*, 51–55.

Author: Pratondo Busono
Institute: BPPT
Street: PTIK, Gedung Teknologi 3 Lt.3
City: Tangerang Selatan
Country: Indonesia
Email: pratondo.busono@bppt.go.id

Finite-Difference Time-Domain Simulation of Localized Surface Plasmon Resonance Adsorption by Gold Nanoparticles

Wen-Chi Lin[1], Wen-Chen Lin[1], Cheng-Lun Tsai[2], and Kang-Ping Lin[1,*]

[1] Department of Electrical Engineering, Chung Yuan Christian University, Chung Li, Taiwan
[2] Department of Biomedical Engineering, Chung Yuan Christian University, Chung Li, Taiwan

kplin@cycu.edu.tw

Abstract— **Using optical sensors to transform light-matter interaction into optical signal has become more and more popular. This is especially true for the fields that require ultra-fast responsibility and remote sensing, such as environmental monitoring, food analysis and medical diagnosis. Among numerous optical sensors, plasmonic nanosensors are of great promise due to their spectral tunability and good adaptability to modern nanobiotechnologies. Localized surface plasmon resonance (LSPR) is the electromagnetic resonance of conducting electrons on metal surface, and it is very sensitive to the variation of environmental refractive index. The LSPR is considered as a useful sensing parameter that provides very good biochemical information. The SPR absorption peak also can be adjusted by changing the nano structure on the LSPR biological sensor chip. In this study, Finite-Difference Time-Domain (FDTD) was applied to simulate the LSPR absorption peak. Four model parameters were modified to study the LSPR sensing sensitivity: (a) the incident light wavelength, (b) the diameter of nanoparticle, (c) the spacing among nanoparticles, and (d) the height of nanoparticle. The simulation results show that 860nm is the best wavelength for the LSPR adsorption measurement. The optimal diameter of nanoparticle is 150nm, and the nanoparticle spacing is 90nm. Higher nanoparticle height provides higher sensitivity, but it also depends on the process capability. The FDTD simulation can be a useful tool to design a LSPR nanoparticle biosensor.**

Keywords—**Localized Surface Plasmon Resonance (LSPR); Biosensor; nanoparticles; Finite-Difference Time-Domain (FDTD)**

I. Introduction

Spectroscopy is applied in the fields of Biophysics / Biochemistry basic research, new medical diagnostic methods development, disease treatment control and structure characteristics identification which monitor the core effect of molecular structure change in Biophysical / Biochemical spectrum. Therefore, a lot of sensing technologies are applied in biological sense such as (Surface Enhanced Raman Scattering; SERS)[1-4] and (Surface Plasmon Resonance; SPR)[5-7], which is a good technology to provide biological sense information.

When a metal layer changes into metal nanostructure, very strong electric field amplification effect will appear between metal nanoparticles that is Localized Surface Plasmon Resonance (LSPR) [8]. The LSPR absorption wavelength positions change with the particle sizes because the metal nanoparticles are disturbed by visible light source to disturb the electron inside particles while instant polarization appeared. The polarized electrons cause harmonic oscillation inside the nanoparticles. When the nanoparticle size is smaller, the oscillation frequency is faster and this means that the range of visible optical wavelength absorption is shorter. When the nanoparticle size is bigger, the oscillation frequency is slower and this means that the range of visible optical wavelength absorption is longer. In addition, the LSPR absorption wavelength are sensitive to the particle shape and surrounded environment such as the temperature [7], particle size and particle space [6] and so on.

In recent years, a lot of metal particle arrays have been used to enhance the SPR sensitivity and there are a lot of methods to produce metal nanoparticles such as e-beam lithography (EBL) [10], nanoimprint lithography (NIL) , anodic aluminum oxide (AAO) [12], nanosphere lithography (NSL) [1-4] and oblique angle deposition (OAD)[6, 13],etc. However, how to find the optimal condition to increase the biosensor sensitivity is important. The FDTD simulation technology was applied to change 4 different conditions, which are incident wavelength, nanoparticle diameter, nanoparticle space and height that the optimal condition could be found to apply in biological sensing.

II. Simulation

In this study, Rsoft was applied for simulation. The FDTD simulation parameters with incident light wavelength of 300~900 nm and the CW excitation light source. The metal was setting to gold. The combination of both the Drude model and two Lorentz poles was used in the FDTD calculation. The uniaxial perfectly matching layer was applied as an adsorbing boundary. Fig. 1 showed the LSPR simulation conditions diagram. The LSPR absorption peak

* Corresponding author.

J. Goh and C.T. Lim (eds.), *7th World Congress on Bioengineering 2015*,
IFMBE Proceedings 52, DOI: 10.1007/978-3-319-19452-3_37

was different due to change the nanoparticle shape, size, space and height.

Where D is the nanoparticle diameter and s is the space between nanoparticles, t is nanoparticles height and λ is the wavelength of incident light source. The optimal conditions were simulated as the followings.

1. Changes of the wavelength of incident light source.
2. Changes of the gold nanoparticle diameters.
3. Changes of the gold nanoparticle space.
4. Changes of the gold nanoparticle heights.

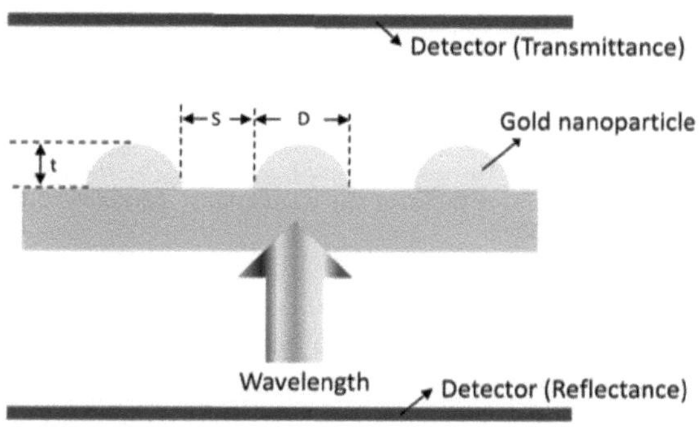

Figure 1. LSPR Simulation Condition Diagram

III. Results and Discussion

This study, initially investigated the wavelength position of the LSPR absorption peak when the nanoparticle sizes were different. In Figure 2, it shows that the LSPR absorption peak appeared in the wave band of 800~900nm when the nanoparticle sizes changed from 100nm to 500nm. In reality, LED incident light source is difficult to be captured when the wavelength is lower than 400nm; therefore, the LSPR absorption wavelengths between 800nm~900nm were simulated and the incident wavelength changed every 10nm displayed in Figure 3.

In Figure 3, it shows that the absorption could reach 12.5% when the nanoparticle diameter was 150nm and produce 860nm LSPR resonance wavelength which cause the maximum absorption. In addition, when the nanoparticle diameter changed to 100nm, the LSPR would move to short wavelength with absorption wavelength of 840nm, which showed that the LSPR absorption wavelengths would change with the nanoparticle sizes and were sensitive.

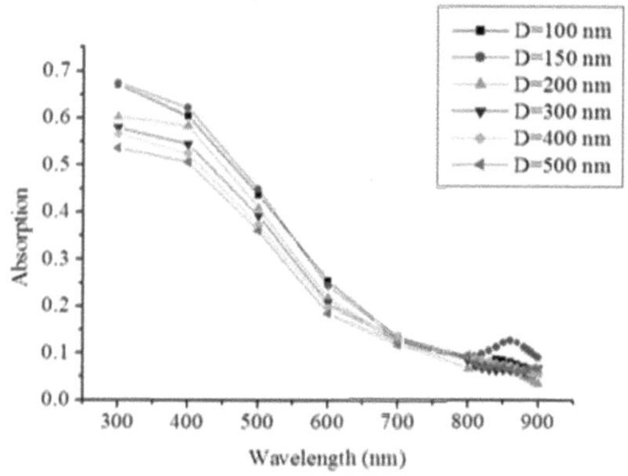

Figure 2. Diameter Change in Different Nanoparticle Sizes and the LSPR Absorption Wavelength Change Diagram.

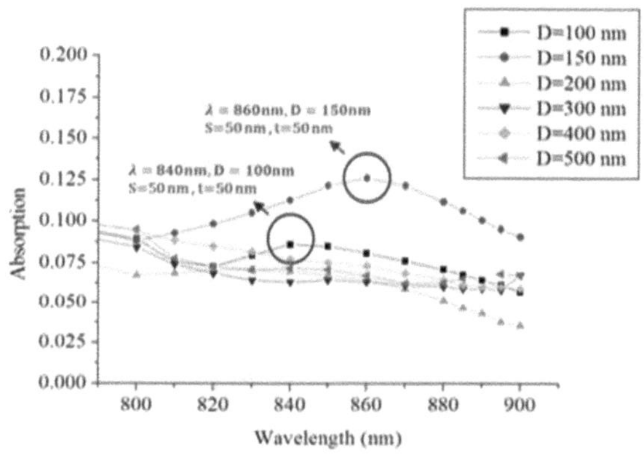

Figure 3. Diameter Change in Different Nanoparticle Sizes and the LSPR Absorption Wavelength Change Diagram and the wavelength change is 10nm.

When the LSPR absorption wavelength closed to 800~900nm was confirmed, the nanoparticle diameter sizes and the changes of LSPR absorption wavelength were further discussed. From Figure 4, it shows that when the nanoparticle diameters changed with 10nm and the diameter was 150nm, the LSPR absorption was the optimal; therefore, the nanoparticle diameter of 150nm was optimal.

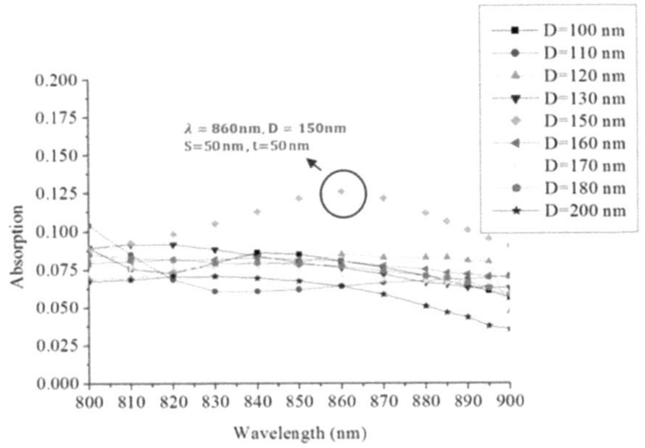

Figure4, Diameter Change in Different Nanoparticle Sizes and the LSPR Absorption Wavelength Change Diagram when the Nanoparticle diameter sizes are 100nm~200nm and the change space is 10nm.

As mentioned previously, the LSPR absorption wavelength position and the space between particles changed with the particle size. In Figure 5, when the gold nanoparticle space changed with 10nm and the space was 90nm, the optimal LSPR absorption appeared when the wavelength was 860nm that shows the optimal result appeared on the gold nanoparticle with space of 90nm. In addition, to apply the LSPR absorption peak dada with the wavelength of 860nm from Figure 5 to Figure 6 that shows when the nanoparticle diameter was 150nm, thickness was 50nm and space was 90 nm, the optimized LSPR absorption peak appeared on the wavelength of 860nm.

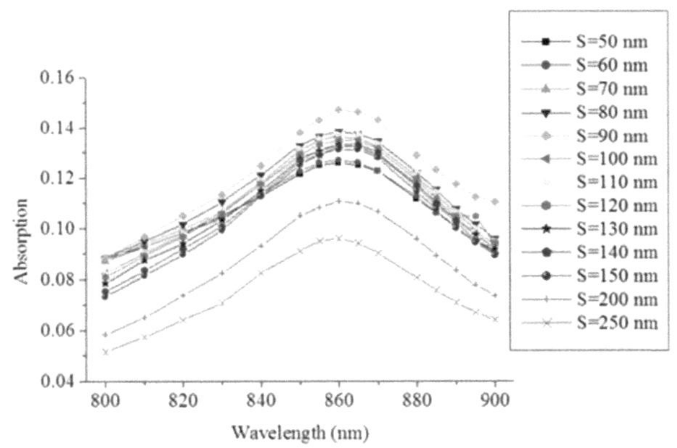

Figure5. Different gold Nanoparticle Space and the LSPR Absorption Wavelength Change Diagram when the Nanoparticle size is 50nm~250nm and the space change between 50nm~150nm is 10nm.

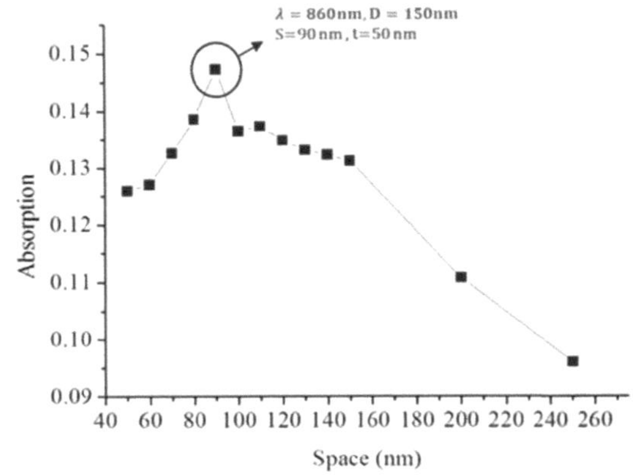

Figure6. Different gold Nanoparticle Space and the LSPR Absorption Wavelength Change Diagram when the Nanoparticle diameter is 150nm and the thickness is 50nm

Finally, to change the nanoparticle height, it was found that the higher the gold nanoparticle height was, the higher the LSPR absorption was. However, on account of the Nano process, the higher the height of the gold nanoparticle was, the higher the Nao process instability was that was the result of shadow effect which limited the height during the back end process to deposit the metal on the chip surface [3]. Therefore, the highest height to produce the biochip during process was considered.

IV. CONCLUSIONS

By the application of FDTD simulation technology, it was found that the optimal absorption peak appeared when the LSPR wavelength was 860nm with nanoparticle diameter of 150nm and space of 90nm. In addition, the higher the gold nanoparticle was, the better the result was that depends on the process capability. Its Index of refraction was different with different detected target. The simulation result could be a reference to design relevant chip format when the LSPR detected target and chip structure changed.

ACKNOWLEDGMENT

The authors are grateful for the financial support from the Hsinchu Science Park, Ministry of Science and Technology under grant number 103MG15, and the Southern Taiwan Science Park, Ministry of Science and Technology under grant number BY-23-08-54-103.

References

1. W. C. Lin, H. C. Jen, C. L. Chen, D. F. Hwang, R. Chang, J. S. Hwang, and H. P. Chiang (2009). SERS Study of Tetrodotoxin (TTX) by Using Silver Nanoparticle Arrays, Plasmonics 4: 178-192.
2. W. C. Lin, S. H. Huang, C. L. Chen, C. C. Chen, D. P. Tsai, and H. P. Chiang (2010). Controlling SERS intensity by tuning the size and height of a silver nanoparticle array, Appl. Phys. A 101: 185-189.
3. W. C. Lin , L. S. Liao, Y. H. Chen, H. C. Chang, D. P. Tsai, H. P. Chiang (2011). Size dependence of Nanoparticle-SERS Enhancement from Silver Film over Nanosphere (AgFON) Substrate, Plasmonic 6: 201-206.
4. W. C. Lin , T. R. Tsai, H.L. Huang, C.Y. Shiau, H.P. Chiang, 2012, "SERS Study of Histamine by Using Silver Film over Nanosphere Structure", Plasmonics 7: 709-716.
5. T. C. Peng, W. C. Lin , C. W. Chen, D. P. Tsai, H. P. Chiang (2011). Enhanced Sensitivity of Surface Plasmon Resonance Phase-Interrogation Biosensor by Using Silver Nanoparticles, Plasmonics 6: 29-34.
6. H. Y. Chung, C. C. Chen, P. C. Wu, M. L. Tseng, W.C. Lin, C. W. Chen, H. P. Chiang (2014). Enhanced sensitivity of surface plasmon resonance phase-interrogation biosensor by using oblique deposited silver nanorods, Nanoscale Research Letters 9:476.
7. C.W. Chen, H. P. Chiang, P. T. Leung and D. P. Tsai, (2008). Temperature dependence of enhanced optical absorption and Raman spectroscopy from metallic nanoparticles, Solid State Communications 148, 413.
8. K.A. Willets and R. P. Van Duyne (2007). Localized Surface Plasmon Resonance Spectroscopy and Sensing. Annu. Rev. Phys. Chem. 58: 267-297.
9. J. N. Anker, W. P. Hall, O. Lyandres, N. C. Shah, J. Zhao, R. P. Duyne (2008) Biosensing with plasmonic nanosensors. Nat. Mater. 7, 442.
10. E.C. Le Ru, P.G. Etchegoin, J. Grand, N. Fe'lidj, J. Aubard, G. Le'vi, A. Hohenau , and J. R. Krenn (2008). Surface enhanced Raman spectroscopy on nanolithography-prepared substrates, Curr. Appl. Phys. 8, 467–470.
11. S. Y. Chou, P. R. Krauss, and P. J. Renstrom (1996). Nanoimprint lithography. J. Vac. Sci. Technol. B 14(6), 4129-4133.
12. H.H. Wang, C.Y. Liu, S.B. Wu, N.W. Liu, C.Y. Peng, T.H. Chan, C.F. Hsu, J.K. Wang, Y. L. Wang (2006). Highly Raman-enhancing substrates based on silver nanoparticle arrays with tunable sub-10 nm gaps. Adv Mater 18:491–495.
13. H. Chu, Y. Liu, Y. Huang, and Y. Zhao, "A high sensitive fiber SERS probe based on silver nanorod arrays", Opt. Express 15, 12230-12239 (2007).

Use macro [author address] to enter the address of the corresponding author:

Author: Kang-Ping Lin
Institute: Department of Electrical Engineering, Chung Yuan Christian University,
Street: 200 Chung Pei Road
City: Chung Li
Country: Taiwan
Email: kplin@cycu.edu.tw

Preliminary Study on Frequency Based Parameters of Myoelectric Signal Using Single Channel Myoelectric Module

I. Priadythama and S. Susmartini

Laboratory of Product Planning and Design, Industrial Engineering Department, Universitas Sebelas Maret, Indonesia

Abstract— **Application of myoelectric module already wide spread in some recent years. Low cost prosthetic hand has been studied to improve its functionality to be an active prosthetic using myoelectric control. Compare to advanced prosthetic, its control can be much simpler such as using only few channels. A problem for capturing surface myoelectric signal with single channel for multi finger control is distinguishing signal for several different type of grasping. The signals should be unique because its electrodes have captured signal combination from many finger muscles crosstalk. Some research has been conducted to study in finding the best frequency based parameter and best filtering technique for signal recognizing, but it was not directly implemented to distinguishing grasping type. This paper aims to show frequencies spectral and their parameters related to two different types of grasping, power grip and precision grip. The study involved 5 male subjects which were treated with two different grasping tasks. A custom made AD620 based bi-potential single channel myoelectric module was used to capture the signals. Fast Fourier Transformation (FFT) based real time analyzer software has been used to show the frequencies spectral of the myoelectric signal. From the preliminary study analysis it can be inferred that spectral pattern is a promising frequency based parameter for distinguishing grasping type.**

Keywords— **frequency based parameter, myoelectric signal, prosthetic hand, grasping.**

I. Introduction

Application of myoelectric module already wide spread in some recent years, especially in rehabilitation devices. In a wheelchair, myoelectric control is a potential alternative to replace control lever or joystick when extremities are no longer functional. For an advanced prosthetic hand or leg, myoelectric signal plays very important role to produce natural active movement. Since smarter but more affordable instrumentation amplifiers were found, myoelectric control are no longer limited for advanced device.

Low cost prosthetic hand has been studied to improve its functionality to be an active prosthetic using myoelectric control. The mechanics of the hand is simplified so that some fingers or finger segments can simultaneously be activated with a single trigger. For example in TBM Hand [1], LARM Hand [2], or UNS LPPD Hand [3], each finger has only 1 DOF so that a movement of proximal phalanx will be followed by its medial and distal. Especially for TBM Hand, all 5 fingers, include thumb, are linked to a distribution plate which allow full fingers activation with a single pulling trigger. Thus compare to advanced prosthetic, its control can be much simpler such as using only few channels or even only single channel.

A problem for capturing surface myoelectric signal with single channel for multi finger control is distinguishing signal for several different type of grasping. According to taxonomy of grasp [4] there are more than 16 grasping type for human hand. However, not all of those types are required for a low cost prosthetic hand or it will be more complex and expensive. In addition, some types are functionally can be replaced by other types, such as disk and sphere, or hook and prismatic wrap. Basically, all of the hand movement can be divided into six grasping type, and then they can be classified again into two basic grasping, power grip and precision grip [5].

In this preliminary report, we tried to study the differences of myoelectric signal characteristics during cylindrical grasping (power grip) and lateral grasping (precision grip). We thought that these two grasping type has much different both from the grasping power and thumb position. The signals should be unique because its electrodes have captured signal combination from many finger muscles crosstalk for those different configurations [6]. For low resolution frequency analyzer as in this preliminary study, that condition would be an advantage.

Some research has been conducted to study in finding the best frequency based parameter and best filtering technique for signal recognizing [7] [8], but it was not directly implemented to distinguishing grasping type. This paper aims to show frequencies spectral and their parameters related to two different types of grasping, power grip and precision grip. In this study we evaluate mean frequency (MNF), median frequency (MDF), and our proposal parameter, frequency spectral pattern (FSP) as characteristic parameters of the spectral.

J. Goh and C.T. Lim (eds.), *7th World Congress on Bioengineering 2015*,
IFMBE Proceedings 52, DOI: 10.1007/978-3-319-19452-3_38

II. Methods

This study involved 5 normal male subjects (aged 23±1 years old, height of 169±3, and weight of 68±4) who were treated with two different grasping tasks, cylindrical and lateral. For cylindrical grasping, the subjects were directed to grasp and release a vertical cylindrical object with Ø of 30 mm with constant speed, one grasp per second, using their left hand. Similar activity was also directed for lateral grasping, but using 5,5h x 8,5w mm standard sized name card paper. Each task consists of 30 trials.

A custom made (-3V – neutral – +3V from 4 AA sized battery) battery powered AD620 based bi-potential single channel myoelectric module was used to capture the signals. The amplification was set to 100 times (±500Ω of gain resistance), which is calibrated using 50 mV 70 Hz sinus signal from function generator and then checked with an oscilloscope. This module was equipped with two Ag-AgCl positive and negative surface electrodes, and silver plated wire as neutral electrode. The positive electrode was attached on flexor carpi radialis muscle while the negative electrode was on extensor digitorum sublimis. The two electrodes were placed around 15 cm from wrist. This position considered can capture all signals from all finger flexor and extensor muscle. The neutral electrode was wrapped to the forearm, around 20 cm from the wrist or 5 cm from the two electrodes.

Fast Fourier Transformation (FFT) based real time analyzer software has been used to show the frequencies spectral of the myoelectric signal. For the measurement, we set the software to 1/24 octave frequency resolution, -80dB to -10dB amplitude range, 10Hz to 50KHz frequency full range, sampling speed of 96KHz, tradeoff speed of 80KHz, and peak hold function was activated. This software was operated in notebook with battery powered mode to prevent noise. The module output was connected to 3.5 mm microphone jack terminal of the notebook and then calibrated to equalize the amplitude.

The measurement was conducted continuously for the 30 trials of each grasping type of a subject. After the 30 trials accomplished, then we collected the frequency and its amplitude peak value. From that data, we then determine the MNF, MDF, and FSP. For MNF, we calculate the average frequency with considering the amplitude of all frequency. For MDF, we determine the middle frequency that can divide the power spectrum equally for low and high frequency. Especially for FSP, we tried to present them via pattern charting with linear and 6th order polynomial trend. We also showed their trend line equations.

III. Result and Discussion

A. MNF Result

It can be shown from Table 1 that MNF is very good for distinguishing between cylindrical and lateral grasping type. Although their mean are close, the magnitudes are different and the data ranges are completely distinct. Total muscle works is strongly suspected to give a contribution in this condition. Cylindrical is classified as power grip which need all finger to move and wrap the object, while in lateral, even though the thumb is more dynamic, the other fingers only move slightly and slowly. The dynamic finger movement needs lots of muscle contraction and increase the myoelectric power. We are still studying whether the MNF can also distinguish almost similar grasping type such as between cylindrical and spherical.

Table 1 MNF Result

Grasping Type	MNF [Hz]	Mean	St. Dev.	Limit
	22.962			23.398
	23.068			
Cylindrical	23.301	23.195	0.203	
	23.162			22.992
	23.482			
	22.203			22.736
	21.435			
Lateral	22.455	22.248	0.487	
	22.451			21.761
	22.698			

B. MDF Result

MDF, as seen in Table 2, seems to be weaker distinguishing parameters compare to MNF since the upper limit of lateral MDF can reach the mean of cylindrical MDF. Similar to in MNF, lateral grasping seems to be more inconsistence since it has lower mean but with higher deviation. Further, we need to study this issue towards other precision grips, whether less or smoother finger movement can cause spectral inconsistency.

Table 2 MDF Result

Grasping Type	MDF [Hz]	Mean	St. Dev.	Limit

	957.846			1006.068
	941.636			
Cylindrical	942.799	963.766	42.302	
	938.315			921.464
	1038.236			
	895.058			964.373
	828.033			
Lateral	933.710	911.550	52.823	
	936.458			858.726
	964.489			

Compare to lateral, cylindrical grasping seems can give more equal power in low and high frequency. MDF can show this characteristic better than MNF. It can be inferred from the MDF result that both cylindrical and lateral has similar characteristics in low frequency region. The strength extension of grasping may give more effect in high frequency region. Many study such as in [9] state that functional range of myoelectric signal is below 1 KHz, but this study present an evidence for functional signal of more than 1 KHz, which is very useful in distinguishing grasping.

Both cylindrical and lateral grasping shows declining trend line at their spectral (Fig. 1 and Fig. 2). This is confirming many literature of electromyography, such as in [10], that beyond 1 KHz, the signals become weaker.

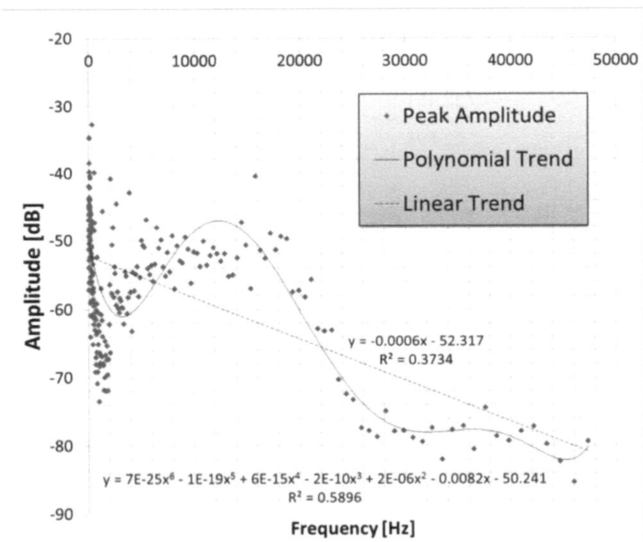

Fig. 1 FSP of Cylindrical Grasping

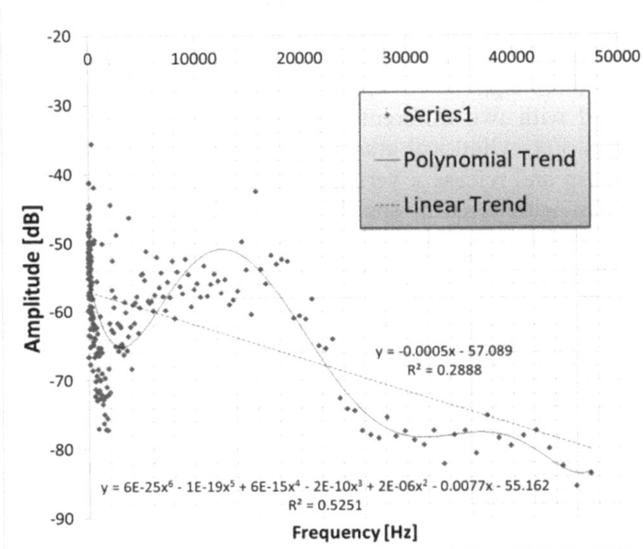

Fig. 2 FSP of Lateral Grasping

Visually, their pattern seems similar, even some datum over 25 KHz seems forming identical pattern for both grasping type. This information can be useful for future grasping distinguishing studies that over 25 KHz the signals are less useful. Future study should set the frequency limit around this frequency.

In order to analyse more detail about the pattern, we present linear and polynomial trend for the spectral. For the linear pattern, even though the curve fitting are still poor with R^2 around 0.2-0.4, they may provide important information. From linear trend line, we can clearly see the slope. Compare to cylindrical grasping, lateral grasping has less steep slope by 16% difference. This information is difficult to read from both the MNF and MDF result. From 6th order polynomial trend line, we can get better curve fitting with $R^2 > 0.5$. Both grasping type give maximum amplitude at 1.25 KHz but with different maximum value. Cylindrical grasping showed superior maximum amplitude over lateral.

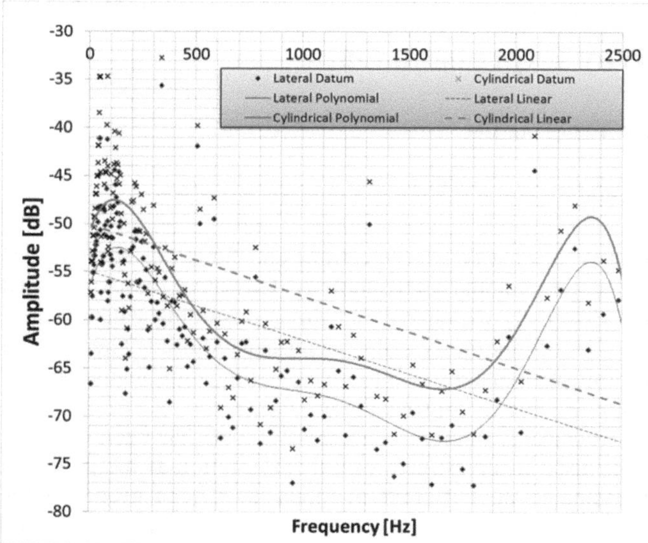

Fig. 3 FSP of Lateral & Cylindrical Grasping Up to 2.5 KHz

At more narrow range (Fig. 3), we can show bigger differentiation at 700 Hz to 1.6 KHz. Cylindrical grasp now gives more moderate slope at this range. With higher spectrum analyser resolution, many more regions to split, and better noise filtering technique, we believe there are many more information we can get. Fig. 3 illustrate that deeper studies in low frequency region are needed since in this area, variability is higher. Future study, especially for distinguishing grasping type, should give more focus on narrow range in low frequency region.

IV. Conclusions

This paper shows that MNF is better frequency based parameters than MDF for grasping type distinguishing purpose. From this preliminary study analysis, it can be inferred that spectral pattern is also a promising frequency based parameter for distinguishing grasping type it can provide many information of signal characteristics which is not provided by MDF and MNF.

References

1. Dechev N, W.L. Cleghorn WL, and Naumann S (2001) Multiple finger, passive adaptive grasp prosthetic hand. Mechanism and Machine Theory 36:1157-1173
2. Rodriguez NEN, Carbone G, and Ceccarelli M (2006) Optimal design of driving mechanism in a 1-DOF anthropomorphic finger. Mechanism and Machine Theory 41:897-911
3. Putranto A, Priadythama I, Susmartini S, and Laksono PW (2014) Conceptual Design of 3 Independent Motor Actuation Transmission Systems for Low-Cost Anthropomorphic Prosthetic Hand. International Journal of Materials, Mechanics and Manufacturing 2(2):121-125
4. Jones LA, Lederman SJ (2006) Human Hand Function, Oxford University Press, New York
5. Fukaya N, Toyama S, Asfour T, and Dillmann R Design of a Humanoid Hand for Human Friendly Robotics Applications, Proc. Human Friendly Mechatronics: Selected Papers of The International Conference on Machine Automation ICMA2000, Elsivier, 2001, ch. 6, pp. 273-278.
6. Kong Y, Hallbeck MS, and Jung M (2010) Crosstalk effect on surface electromyogram of the forearm flexors during a static grip task. Journal of Electromyography and Kinesiology 20:1223–1229
7. S. Thongpanja, A. Phinyomark, P. Phukpattaranont, C. Limsakul (2011) Time-Dependent EMG Power Spectrum Parameters of Biceps Brachii during Cyclic Dynamic Contraction, IFMBE Proc. vol. 35, 5th Kuala Lumpur International Conference on Biomedical Engineering, Malaysia, 2011, pp. 233–236
8. Clancy EA, Bertolina MV, Merletti R , and Farina D (2008) Time- and frequency-domain monitoring of the myoelectric signal during a long-duration, cyclic, force-varying, fatiguing hand-grip task. Journal of Electromyography and Kinesiology 18:789–797
9. McDonald AC, Sanei K, and Keir PJ (2013) The effect of high pass filtering and non-linear normalization on the EMG–force relationship during sub-maximal finger exertions. Journal of Electromyography and Kinesiology 23:564–571
10. Weiss L, Silver JK, and Weiss J (2004) Easy EMG. Butterworth-Heinemann, Wltham, Massachusetts

Author: Ilham Priadythama
Institute: Universitas Sebelas Maret
Street: Ir. Sutami No. 36A
City: Surakarta
Country: Indonesia
Email: ilham@megaspin.net, priadythama@gmail.com

A Bioinformatics Approach to Investigating Significant Genes of Hepatocellular Carcinoma from Microarray Expression Data

Liang-Tsung Huang and Yun-Ting Lai

Department of Biotechnology, Mingdao University, Changhua, Taiwan

Abstract—**Hepatocellular carcinoma is one of most common cancers worldwide and also the top three most common causes of death from cancer in Taiwan. Hepatocellular carcinoma is often diagnosed after clinical deterioration and then the survival rate is low. Therefore, it is crucial to develop early diagnosis method in asymptomatic individuals. Accordingly, the study aimed to analyze, find and evaluate significant genes of hepatocellular carcinoma from microarray expression data by a systematic bioinformatics approach. The study clearly demonstrated the approach to investigate significant genes of hepatocellular carcinoma through significance analysis algorithm and various online databases. We firstly constructed a dataset from gene expression profiling of Affymetrix GeneChip on ArrayExpress database of functional genomics experiments. The dataset consisted of 455 samples in tumor and non-tumor tissues of hepatocellular carcinoma patients. Secondly, a significance analysis algorithm was employed to calculate candidate genes efficiently. Finally, we integrated the information obtained from COSMIC, catalogue of somatic mutations in cancer, and OncoDB.HCC, oncogenomic database of hepatocellular carcinoma, databases to evaluate the gene significance of hepatocellular carcinoma. From 22277 genes, 100 candidate genes were selected by the significance analysis algorithm. Further, we analyzed the studies of the candidate genes in COSMIC database and 98% of these candidates were related to cancer. In particular, 83% of these candidates were directly related to hepatocellular carcinoma. From the other hand, we analyzed the evidence of significant genes in OncoDB.HCC database and 41% of these candidates were supported by different types of evidence, including Stanford microarray, experiments and microarray/proteomic reports. The results showed that the proposal approach can effectively find and evaluate the gene significance of hepatocellular carcinoma from microarray expression data. Thus we suggested that the framework may be applied to other cancer studies based on microarray expression data and provide more possible clues to significant genes.**

Keywords—**Bioinformatics, Significant Genes, Hepatocellular Carcinoma, Microarray.**

I. INTRODUCTION

The successful completion of the Human Genome Project (HGP) advances our knowledge rapidly in genomics research. In post-genome era, one of the most important tasks is considered as genome annotation. Genome annotation is to identify gene locations and functions so as to develop gene regulation technology. However, these processes are extremely expensive and time consuming when a large number of laboratory experiments are required. Therefore, a bioinformatics approach to investigating significant genes is quite feasible to improve the processes.

Nowadays, microarray expression data has been successfully used to investigate useful information at the gene expression level. Different methods have been proposed for tumor classification and relevant gene identification. One of the earliest methods for tumor classification is the weighted voting machine which is based on a linear model [1]. Other methods include hierarchical clustering [2], machining learning [3; 4], compound covariate [5], shrunken centroids [6], partial least square [7], principal component analysis disjoint models [8], factor mixture models [9] etc. Our earlier study presented an integrated method for accurate tumor classification and relevant gene identification from microarray express data [10], and then the relationships between tumor classes and genes may be established by association rule algorithm or decision tree algorithm [11-14]

Hepatocellular carcinoma is one of most common cancers worldwide and also the top three most common causes of death from cancer in Taiwan. Hepatocellular carcinoma is often diagnosed after clinical deterioration and then the survival rate is low. Therefore, it is crucial to develop early diagnosis method in asymptomatic individuals.

Accordingly, the study aimed to analyze, find and evaluate significant genes of hepatocellular carcinoma from microarray expression data by a systematic bioinformatics approach. The study clearly demonstrated the approach to investigate significant genes of hepatocellular carcinoma through significance analysis algorithm and various online databases.

II. MATERIALS AND METHODS

We firstly constructed a dataset from gene expression profiling of Affymetrix GeneChip on ArrayExpress database of functional genomics experiments. The dataset consisted of 455 samples in tumor and non-tumor tissues of hepatocellular carcinoma patients. Secondly, a significance analysis algorithm was employed to calculate candidate

J. Goh and C.T. Lim (eds.), *7th World Congress on Bioengineering 2015*, IFMBE Proceedings 52, DOI: 10.1007/978-3-319-19452-3_39

genes efficiently. Finally, we integrated the information obtained from COSMIC, catalogue of somatic mutations in cancer, and On-coDB.HCC, oncogenomic database of hepatocellular carcinoma, databases to evaluate the gene significance of hepatocellular carcinoma. The main flow chart is shown in Fig. 1.

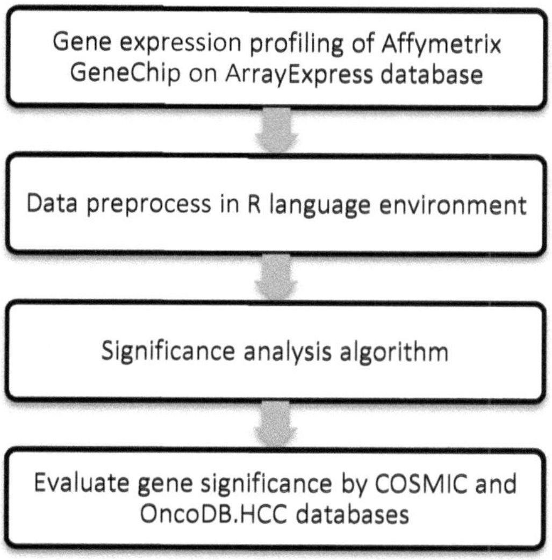

Fig. 1 The flowchart of investigating significant genes of hepatocellular carcinoma from microarray expression data

A. Gene expression profiling of hepatocellular carcinoma

We firstly constructed a dataset from gene expression profiling of Affymetrix GeneChip on ArrayExpress database [15]. ArrayExpress database is one of the important repositories to archive functional genomics data and support reproducible research. The database includes 56771 experiments and 1,668,935 assays (updated to 2015) and describes microarray data by MIAME (Minimum Information About a Microarray Experiment).

ArrayExpress database is maintained by the Functional Genomics group at EMBL-EBI (European Molecular Biology Laborator) and available at https://www.ebi.ac.uk/arrayexpress/ (as shown Fig 2).

In this study, we compiled a dataset of 488 samples from no. E-GEOD-14520 in tumor and non-tumor tissues of hepatocellular carcinoma patients.

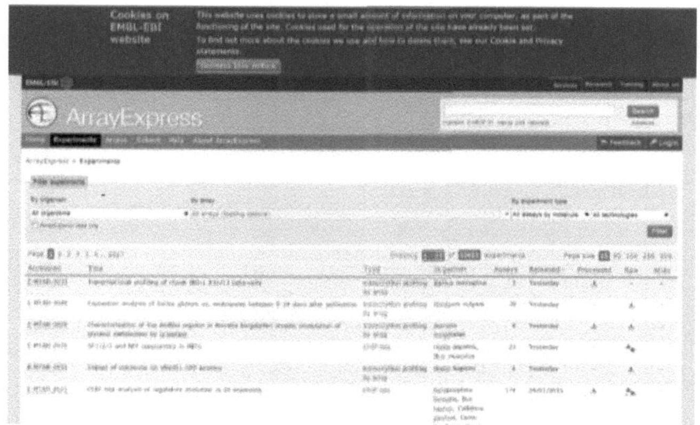

Fig. 2 A snapshot of ArrayExpress database of functional genomics experiments

B. Data analysis environment

For data preprocess and analysis, we used R/qtl [16; 17] as the programming language and development environment. R is a well-developed, simple and effective programming language and provides a wide variety of statistical and graphical packages. R is an official part of the GNU project and available at http://www.r-project.org/.

In this study, we used R 3.1.1 and Bioconductor package, which is developed for the analysis and comprehension of high-throughput genomic data and available at http://www.bioconductor.org/).

C. Significance analysis algorithm

Microarray gene expression data may be used to identify changes of thousands of genes different biological states. However, algorithms are needed to determine whether changes in gene expression are significant.

In this study, we perform SAM significance analysis algorithms. Significance Analysis of Microarrays (SAM) [18] assigns a score to each gene on the basis of change in gene expression relative to the standard deviation of repeated measurements, and then uses permutations of repeated measurements to estimate the false discovery rate.

D. Oncogenomic database of hepatocellular carcinoma

To evaluate significant genes, we processed existing hepatocellular carcinoma databases to find direct evidence.

OncoDB.HCC database [17] is the first comprehensive oncogenomic database for hepatocellular carcinoma (HCC). The database integrates three datasets from public references to provide multi-dimension view of current hepatocellular carcinoma studies, including chromosome aberration studies, gene expression studies and model organisms.

OncoDB.HCC database is maintained by the Institute of Biomedical Sciences, Academia Sinica and available at http://oncodb.hcc.ibms.sinica.edu.tw/index.htm (as shown in Fig 3).

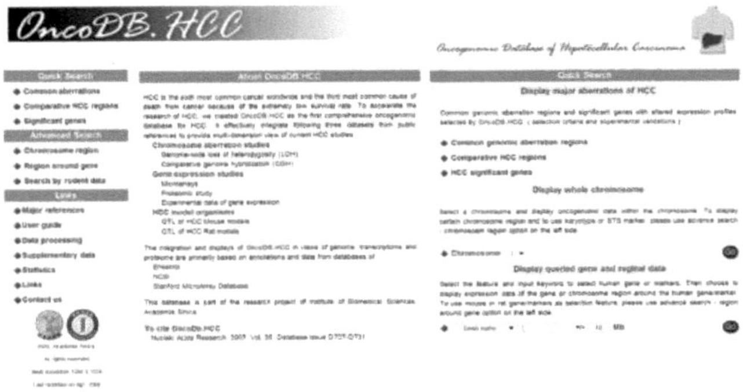

Fig. 3 A snapshot of OncoDB.HCC database of hepatocellular carcinoma

III. Results and Discussion

A. Statistical analysis of gene expression profiling

From 2007-2014, there are total 138 experiments of hepatocellular carcinoma from ArrayExpress database and the sample size is between 1 and 620. We examined the conditions of 138 experiments and then focused on no. E-GEOD-14520 of 488 samples, where probe HT_HG-U133A consists of 455 samples with tumor and non-tumor tissues of hepatocellular carcinoma patients.

B. Analysis of significant genes

Since each samples consists of 22277 genes, we performed SAM algorithm to calculate 100 significant genes. Table 1 lists top 10 significant genes with related information and Table 2 shows high frequency genes.

C. Evaluation of significant genes

In OncoDB.HCC database, there are three catalogues of evidence: (1) Genes significantly up- or down- regulated in at least three independent HCC microarray/proteomic reports; (2) Genes were selected with consistent expression level changes for at least 2 folds in more than 70% patients after reprocessed Stanford HCC microarray data; (3) Genes with wet-lab experimental data from previous reports.

Table 1 Top 10 significant genes by significance analysis of microarrays

No	Name	Symbol	d.value
1	209365_s_at	ECM1	42.9
2	205019_s_at	VIPR1	41.7
3	218002_s_at	CXCL14	41.1
4	205866_at	FCN3	36.4
5	220114_s_at	STAB2	36.3
6	207609_s_at	CYP1A2	35.9
7	204428_s_at	LCAT	35.5
8	203554_x_at	PTTG1	-32.9
9	211762_s_at	KPNA2	-32.4
10	205225_at	ESR1	32.3

Table 2 High frequency genes out of top 100 significant genes

Symbol	No		Occurrence
CYP1A2	6	13	2
KPNA2	9	18	2
CLEC4M	17	78	2
RRM2	28	38	2
MT1F	35	64	2
EGR1	36	87	2
PSMD4	45	83	2
CXCL12	52	66	2
TOP2A	58	59	2
PER1	74	96	2
N4BP2L1	89	99	2

In Table 3, top 5 significant genes and corresponding evidence are listed. For the most significant gene of ECM1, there are more than 94% patients after Stanford HCC microarray experiments and gene significantly up- or down-regulated in independent HCC microarray/proteomic reports.

Table 3 Top 5 significant genes with different types of evidence

Symbol	Chromosome	Array	Stanford	Experiments
ECM1	1	1	94	0
VIPR1	3	1	91	0
CXCL14	5	0	0	0
FCN3	1	3	0	0
STAB2	12	0	0	0

Conclusions

From 22277 genes, 100 candidate genes were selected by the significance analysis algorithm. Further, we analyzed the studies of the candidate genes in COSMIC database and 98% of these candidates were related to cancer. In particular, 83% of these candidates were directly related to hepatocellular carcinoma. From the other hand, we analyzed the evidence of significant genes in OncoDB.HCC database and 41% of these candidates were supported by different types of evidence, including Stanford microarray, experiments and microarray/proteomic reports. The results showed that the proposal approach can effectively find and evaluate the gene significance of hepatocellular carcinoma from microarray expression data. Thus we suggested that the framework may be applied to other cancer studies based on microarray expression data and provide more possible clues to significant genes.

Acknowledgment

The work is funded mainly by the research projects of Grant Numbers: NSC 102-2221-E-451-001 and MOST 103-2221-E-451-004. The authors would like to acknowledge the support from Biomedical Engineering Program, National Science Council and Ministry of Science and Technology, Taiwan.

Conflict of Interest

The authors declare that they have no conflict of interest.

References

1. Golub TR, Slonim DK, Tamayo P, et al (1999) Molecular classification of cancer: class discovery and class prediction by gene expression monitoring. Science 286:531-537
2. Alon U, Barkai N, Notterman DA, et al (1999) Broad patterns of gene expression revealed by clustering analysis of tumor and normal colon tissues probed by oligonucleotide arrays. Proc. Natl. Acad. Sci. U. S. A. 96:6745-6750
3. Zhang H, Yu CY, Singer B, et al (2001) Recursive partitioning for tumor classification with gene expression microarray data. Proc. Natl. Acad. Sci. U. S. A. 98:6730-6735
4. Olshen AB, Jain AN (2002) Deriving quantitative conclusions from microarray expression data. Bioinformatics 18:961-970
5. Hedenfalk I, Duggan D, Chen Y, et al (2001) Gene-expression profiles in hereditary breast cancer. N Engl J Med 344:539-548
6. Tibshirani R, Hastie T, Narasimhan B, et al (2002) Diagnosis of multiple cancer types by shrunken centroids of gene expression. Proc. Natl. Acad. Sci. U. S. A. 99:6567-6572
7. Huang X, Pan W (2003) Linear regression and two-class classification with gene expression data. Bioinformatics 19:2072-2078
8. Bicciato S, Luchini A, Di Bello C (2003) PCA disjoint models for multiclass cancer analysis using gene expression data. Bioinformatics 19:571-578
9. Martella F (2006) Classification of microarray data with factor mixture models. Bioinformatics 22:202-208
10. Huang L-T (2009) An integrated method for cancer classification and rule extraction from microarray data. Journal of Biomedical Science 16:25
11. Huang LT, Gromiha MM, Hwang SF, et al (2006) Knowledge acquisition and development of accurate rules for predicting protein stability changes. Comput Biol Chem 30:408-415
12. Huang LT, Gromiha MM, Ho SY (2007) Sequence analysis and rule development of predicting protein stability change upon mutation using decision tree model. Journal of Molecular Modeling 13:879-890
13. Huang LT, Gromiha MM, Ho SY (2007) iPTREE-STAB: interpretable decision tree based method for predicting protein stability changes upon mutations. Bioinformatics 23:1292-1293
14. Huang L-T (2014) Finding simple rules for discriminating folding rate change upon single mutation by statistical and learning methods. Protein and Peptide Letters 21:743-751
15. Kolesnikov N, Hastings E, Keays M, et al (2015) ArrayExpress update-simplifying data submissions. Nucleic acids research 43:D1113-1116
16. Arends D, Prins P, Jansen RC, et al (2010) R/qtl: high-throughput multiple QTL mapping. Bioinformatics 26:2990-2992
17. Broman KW, Wu H, Sen S, et al (2003) R/qtl: QTL mapping in experimental crosses. Bioinformatics 19:889-890
18. Tusher VG, Tibshirani R, Chu G (2001) Significance analysis of microarrays applied to the ionizing radiation response. Proceedings of the National Academy of Sciences of the United States of America 98:5116-5121

Corresponding author:

Author: Liang-Tsung Huang
Institute: Department of Biotechnology, Mingdao University
Street: 369 Wen-Hua Rd., Peetow
City: Changhua
Country: Taiwan
Email: larry@mdu.edu.tw

Wireless System for Monitoring Body Temperature

D. Tolar[1] , R. Bischofová[1], L. Hynčík[1], and J. Gallo[2]

[1] New Technologies - Research Centre, University of West Bohemia, Plzen, Czech Republic
[2] Department of Orthopaedics, Faculty of Medicine and Dentistry, Palacky University Olomouc, Czech Republic

Abstract— **Wireless communication technologies together with physiological signals measurement provide a wide range of capabilities for monitoring, recording and analysis of the physical and health status of individuals. Continuously monitored data can provide very useful information for evaluating the physical state of the human body. The data can also help to optimize the process of monitoring and treatment of patients, e.g. in intensive care units or elderly care facilities. The system presented consists of wireless sensors, data transceivers and an information system for the visualization and analysis of measurement data. The developed sensor continuously measures body temperature at a given point, and wirelessly transmits the data to the central device that evaluates the data, based on the needs of the operating personnel. The sensor is miniature and battery-powered; the battery lasts for about one year. Also, the sensor is easy to sterilize and store for purposes and needs not only in health care. The system is designed to be implemented in hospitals, elderly care facilities, rehabilitation units, spas and even households. Sensors with extended functionality, which will focus on other different physiological signals like pulse rate, EMG or motion activity of monitored the person are planned for development as part of the system in future.**

Keywords— **sensor, wireless, temperature, health care**

I. Introduction

The system presented here is based on a wireless sensor network operating in the ISM band 2.4 GHz. The system provides the capability of wireless monitoring of patients′ body temperature in large facilities like hospitals.

The user interface of the system provides information on actual patient temperature together with patient temperature trends over time. This is clinically useful information, which can help a surgeon with early detection of health complications. The system presented is now tested at the Department of Orthopaedics, Faculty of Medicine and Dentistry, University Hospital Olomouc, Czech Republic. The aim is to validate the data from continual monitoring of body temperature by matching it with wound healing and other potential complications.

The fate of total hip (THR) and knee (TKR) replacements is determined during the surgery and in the early postoperative period. This is a critical period not only in terms of infection but also in terms of adaptation of host tissues on the biomaterial surface (implant).

Some patients experience increased temperature or fever, which is not easy to explain. Usually antibiotics are administered blindly with a hope that these induce resolution of wound problems and contribute to the development of an aseptic interface between the host and the implant. However, this early postoperative period can be also symptomless; so at least some patients are discharged into aftercare without any evidence of an emerging infectious complication. This is the main reason why we are trying to continuously monitor temperature in the area of the surgical wound and from the measured data predict near future complications.

II. System for measurement, transmission and visualization of physiological signals

The installed system shown in Fig. 1 consists of two separated wireless sensor networks. The first network is located at the intensive care unit. The second network is located at the physiotherapy department, in a different building, a few hundred meters distant. The connectivity of these two subsystems is implemented over a local area network.

Measured data from the physiotherapy department are transferred to the system located in the intensive care unit, which contains the application server. The application server stores and processes the data and allows users to access the data.

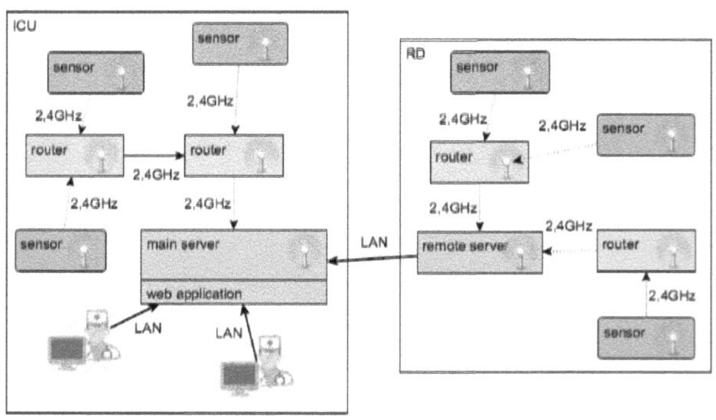

Fig. 1 System installation structure

J. Goh and C.T. Lim (eds.), *7th World Congress on Bioengineering 2015*,
IFMBE Proceedings 52, DOI: 10.1007/978-3-319-19452-3_40

The wireless sensor network implemented uses the ISM 2.4 GHz band. This band is intended to be used for industrial, scientific and medical purposes. It is a free band, which means that it can be used freely, if the communication device fulfills regulatory requirements. On the other hand, using this band it provides no guarantees regarding radio interference levels existing in this band. A part of our wireless sensor network communicates in compliance with the IEEE 802.15.4 standard [1], which defines the physical and media access layer for low-rate wireless networks.

The wireless transmission uses the direct sequence spread spectrum technique, which is the modulation technique where the transmitted signal takes up more bandwidth than the information signal that modulates the carrier or broadcast frequency. This technique takes every single bit which has to be transmitted and replaces it with a sequence of bits. These sequences have a pseudo-random nature. The sequences are then modulated on the carrier frequency. This results in controlled introduction of redundancy into the transmitted signal. The signal is spread into a wider part of the radio bandwidth and is more resistant to interference, which increases the reliability of the transmission. Other users of the same frequency band see this signal as random noise, and without the knowledge of the algorithm for generation of the pseudo-random sequence, it is complicated to reconstruct the transmitted data. This method allows data communication in a noise environment and provides good results when coexisting with other technologies using the same bandwidth [2].

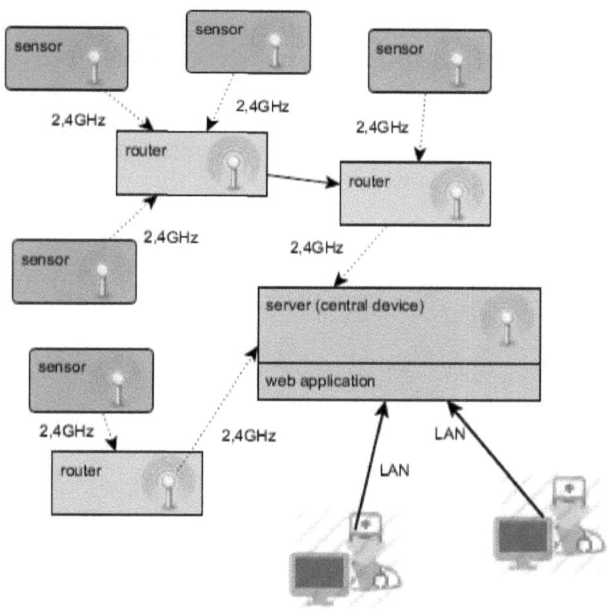

Fig. 2 Basic system component diagram

Components of the system communicate with a transmission power of 2.5 mW or less.

Fig. 2 shows a component diagram of the system. The source of information on physiological signals is wireless sensors. Data from sensors are transferred to the server, which handles, stores and provides user access to data through the web application. Other devices participating in wireless data transfer are routers.

The path of data packets from sensor to server is not fixed and can be updated on each transmission when the condition in the communication channel or the physical position of the sensor within the network covered area changes. Each transmission of a data packet from a sensor to the network is confirmed by the receiving side. This ensures that the sensor is informed if data has successfully reached its destination. The sensor has internal memory, which is used to store data if transmission of the data is not possible for various reasons. This minimizes the risk of losing measured data, but can cause data to be delivered to the system with some delay. This mechanism is useful especially in situations where the sensor location is outside the communication range of the network for some time, as the data are not lost, but transferred to the system whenever the sensor is back in communication range of the network [3].

There may be situations where it is required to cover a large area with the wireless sensor network, or more locations which are distant and direct communication of wireless sensor networks is not possible. In such cases it is not possible to use only a single wireless sensor network, but multiple networks must be configured. However, they can all share the same server and the data can be transparent to users. An example of such a situation is our test setup shown in Fig. 1. Each sensor network has its own server, but only one server is the main server which handles data from all remote servers located in distant sensor networks. The connection between the main and remote servers is implemented through a LAN connection. The remote server prevents data loss by storing measured data in its own storage in case that data can't be delivered to the main server for various reasons.

III. Temperature sensor

The sensor used in the installation is designed to sense the temperature of the skin surface. The sensor also allows detection of patient movements. The location of the sensor on a patient's body is selected by medical staff with regard to targets of the measurement, which can be monitoring of body temperature, monitoring of local temperature near the surgical wound, etc.

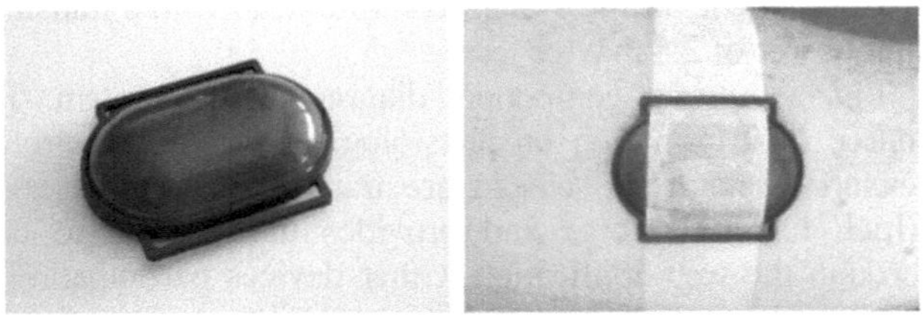

Fig. 3 Temperature sensor and its fixation (right)

The sensor is shown in Fig. 3. Sensor dimensions are 39x25x10 mm. The time period of the measurement can be selected from 1 second to 60 minutes.

In our testing we used a measurement period of 60 seconds. The sensor allows storage of up to 5000 temperature measurements for a situation where it is not possible to transfer the data.

Fixation of the sensor on the measured surface is shown on Fig. 3. Fixation is implemented by using standard 2.5 cm wide medical adhesive tape.

The communication range of the sensor depends on RF interference on the communication channel and also in the surrounding environment. The usual communication range is from 10 to 50 meters.

IV. User interface of the system

The user interface of the system is implemented as a web application, which can be accessed from a web browser. The user interface of the system was designed with emphasis on simplicity and intuitive usage. It was designed with respect to its intended use, which is the hospital environment. Therefore, user roles in the interface are doctor, nurse and administrator. A doctor can view measured patient data, which means actual data for all patients, patient history and the history of measurements. A nurse can view patient data and also add new patients and delete patients (end measurement of the patient). An administrator has access to all functions and settings of the system.

Fig. 4 shows the main screen of the application (values shown in the figure are not real data from measurement). This screen shows current data about monitored patients. There is a tile for each patient showing the actual temperature of sensors associated with the patient and a small chart showing the progression of temperature over time. If a patient has more than one sensor attached, then the highest temperature is shown in big letters. If the temperature exceeds the predefined threshold of 37 °C, then the temperature test color changes from blue to red. The progression chart shows the area of increased temperature in light red color, which allows simple evaluation of the patient´s temperature history.

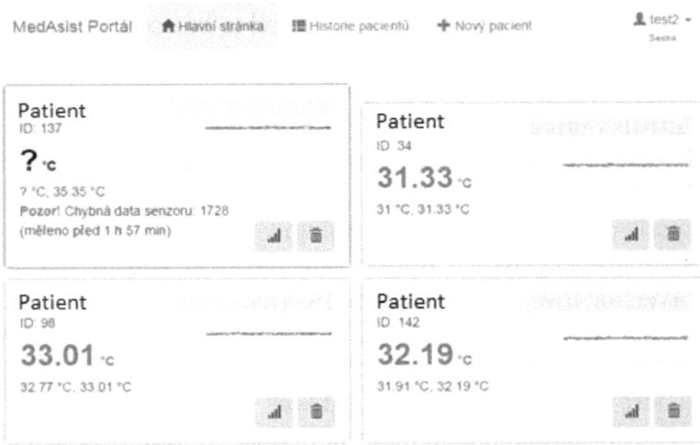

Fig. 4 Main screen of application

This simply allows the evaluation of the effect of applied measures for lowering body temperature. Tiles with patients are sorted with defined rules to first show patients who require special attention, then patients whose sensors did not provide data for a defined period of time, then patients whose temperature entered the area of increased temperature in the last 2 hours.

The remaining tiles are sorted by patient temperature. Situation where a patient´s temperature is not current is a special situation, which under normal conditions should not occur. This situation requires the attention of medical staff, because the reason for the temperature being unknown should be investigated and fixed, for example by replacing a temperature sensor in the case of its malfunction. These rules make identification of patients´ temperature-related problem very quick and effective.

Fig. 5 shows a detailed view of a patient´s temperature over time, which allows a detailed investigation of the patient´s temperature history.

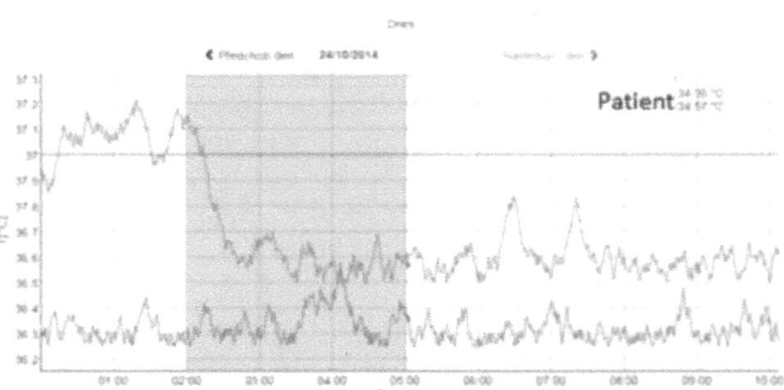

Fig. 5 Patients' temperature detail screen

V. Additional Measurement Issues

Suitable placement of sensors around the surgical wound was important for getting usable measurement results and for meeting patient comfort, medical and hygiene requirements.

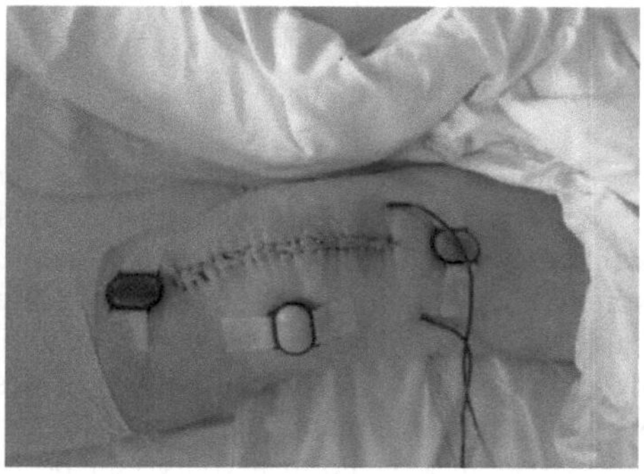

Fig. 6 Placement of sensors near surgical wound

The basic premise was to assess the functionality of the degree of reliability of the measured data. Data obtained from the system was compared to reference measurements taken by standard thermometers used on the intensive care unit. Additional recorded information was on patient activities, changes of position in bed, applying cold packs, etc. The total number of evaluated patients was 44. All patients signed the informed consent and the process was approved by the ethical committee.

Another important part of the research was a sociological survey. The survey examined a representative sample of respondents regarding the social and psychological acceptability of the proposed continuous measurement system [4].

Despite the high level of resources used to monitor the position and activities of a person (GPS, cell phones, fitness bracelets, etc.), it cannot be automatically anticipated that potential candidates will agree to be included in the program of continuous monitoring of selected somatic functions without any reservations [4]. The aim was to assess the opinions of a sample of potential users about the tested system. Further research should determine if the tested system could be beneficial for healthcare providers too.

VI. Conclusions

The first evaluation of temperature data measured by the system in comparison to temperature measurements performed by reference thermometers shows that the system is reliable in detection of host body temperature. The system is appropriate for determining the temperature trends in time. With regard to the measurement of wound temperature the system depends strongly on the configuration of sensors around a wound. The critical part of the whole system is the attachment of the sensor to the skin. There are technical and patient parameters of this step. A disturbance or intolerance can significantly affect the measurement.

Acknowledgment

The work is co-financed by the Technology Agency of the Czech Republic within the project TD020094 "Sensor technology for health care for aging and disabled population" and within the project SGS-2013-026 "Development of biomechanical models for medicine".

Conflict of Interest

The authors declare that they have no conflict of interest.

References

1. IEEE norms at https://standards.ieee.org/findstds/standard/802.15.4-2006.html
2. Atmel AT02845, Coexistence between ZigBee and Other 2.4GHz Products
3. Štork M, Tolar D (2014) Wireless electronic systems for physiological parameters measuring, International Conference on Applied Electronics, Plzeň, Czech Republic, pp. 281–284
4. Šimek D, Vávrová P, Kryštof D. et al (2015) Opinions of clients and health professionals on the use of remote monitoring of basic functions for elderly and patients in the Czech Republic (*in Czech language*). Florence. In press

Author: Luděk Hynčík
Institute: University of West Bohemia
Street: Univerzitní 8
City: Plzeň
Country: Czech Republic
Email: hyncik@ntc.zcu.cz

Application of Appropriate Technology in Ankle Joint to Below Knee Prosthetic Reduce Working Pulse on Transtibial Amputee

L. Herdiman[1,2,*], S. Susmartini[2], N. Adiputra[3], and K. Tirtayasa[3]

[1] Ergonomics Department of Postgraduate Program, University of Udayana, Bali, Indonesia
[2] Department of Industrial Engineering, University of Sebelas Maret Surakarta, Indonesia
[3] Department of physiology, School of Medicine, University of Udayana Bali, Indonesia

Abstract— **The purpose of this study is determining the physiological condition difference between 2 treatment group of below knee prosthetic user, the first group was using prosthetic without ankle joint and another group with multi axis energy store-return ankle joint. Each group consist of 7 male below knee subjects, which were tested with 2 period in cross-over design, between the two period we insert 1 day washing out. The measured physiology response were resting pulse and working pulse when the subject walked as far as 80 meters for 6 minutes with 1.2±0.3 m/s walking speed. The pulse was measured using palpation method with 10 pulse technique. The result of preliminary test shows each group is normally distributed and homogeny with $p > 0.05$. After the two group had been compared, there were no significant period and carry over effect ($p > 0.05$), and significant difference for both treatment ($p < 0.05$). Conclusion of this study is the existence of appropriate technology which providing articulation ankle ability and improvement in walking energy management can significantly reduce cardiovascular load.**

Keywords— **Transtibial amputee; working pulse; design of the ankle; 6MWT, below knee prosthetic.**

I. Introduction

The main problem of local prosthetic foot products in Indonesia is the lack of flexible connection between segments of the limbs. Barriers to technological innovation and knowledge of materials in making a prosthetic foot, cause the local prosthetic makers face difficulties to make a more modern prosthetic.

The purpose of this paper is how to determine the needs of the ankle design of below knee endoskeletal prosthetic foot with seven criteria of appropriate technology. The result is prototype of the ankle that can reduce cardiovascular load. Success in designing the appropriate technology will give an impact in reducing the cardiovascular load of amputee while walking which has never been previously evaluated.

Designing ankle on endoskeletal prosthetic foot should be done through the application of appropriate technology with reference to the seven criteria [1]. The seven element of appropriate technology characteristic include several aspects; technically feasible, economical, ergonomic/ healthy, socio-culturally acceptable, energy saving, environmental friendly, and trendy [2]. The ankle on endoskeletal prosthetic foot should have a relatively simple design that is based on the availability of technology and local raw materials and possible to developed with local designer and operators with affordable manufacturing cost.

The workload in a walking activity can be represented through cardiovascular load as resulted from load which is received by human body [3] [4]. External load (stressor) is load that comes from outside of the body in the form of a task, organization, and environment. For organization, it includes the time, processes and systems. While environment, it consists heat, light intensity, and humidity. Internal load (strain) is a load that comes in a body that is closely related to the desire, satisfaction and others. Assessment cardiovascular load by measuring the increase of pulse should be conducted while walking or finished walk, assessing cardiovascular load using 10 pulses [5].

The pulse is an indicator for evaluating the amputee during walking. The intensity of physical activity show a linear relationship with the pulse, it means an increase in intensity during walking will be followed by an increase in pulse rate. The measurement of the pulse may be conducted through pulse palpation method. That increased pulse can be distinguished based on the work which is classified into the category of cardiovascular load levels [6]. The pulse is also depending on age and gender. Pulse before the work is defined as resting pulse, the amount of the average pulse before starting a job. While working pulse or cardiovascular load is the average number of pulses for doing the job. Then the work pulse is the pulse difference between during work and before work [7]. Six minute walk test (6MWT) might be used to measure the amputee's pulse as a response towards the activity while using a prosthetic foot [8].

The emergence of the cardiovasculer load as a response towards walking activity with prosthetic foot can be shown by an increase in pulse rate. That amputee's activity is a blend of physical and mental workload which is affected by external and internal load [9]. A proper appropriate technology will be indicated by a decrease in pulse rate and cardiovascular load of the subject that makes the body becomes

* Corresponding author.

J. Goh and C.T. Lim (eds.), *7th World Congress on Bioengineering 2015*,
IFMBE Proceedings 52, DOI: 10.1007/978-3-319-19452-3_41

healthy and fit. Design of the ankle joint in endoskeletal prosthetic foot which always consider to the level of suitability of the tools towards the user will give maximum benefit.

II. METHODS

The determination of the ankle joint technical requirements of the appropriate technology was done through focus group discussion (FGD). The generation of technical needs of the ankle design involved designers expert, machining and assembly expert, prosthetic and orthotic expert, rehabilitation expert, ergonomics expert, and the subject (transtibial amputee). The result was a recommendation for action plan on technical issues, manufacturing, and feasibility of the ankle joint design.

The design was tested to 14 male subjects. The test was using a cross-over design that is divided into two groups for Period I and Period II. Washing period time between the two periods was conducted during the first day, as described in Figure 1.

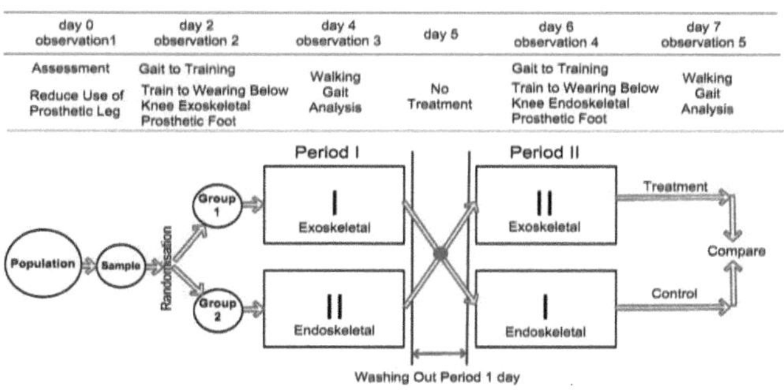

Fig. 1 The cross-over design

Tests walk on transtibial amputee was done with speed of 1.2 ± 0.3 m/s using 6MWT. Measurement of cardiovascular load on the subjects includes resting pulse, working pulse, cardiovascular load, and cardiovascular load percentage (%CVL) which were measured in beat per minute (bpm). The overall execution time of testing in each group was 6 hours.

III. RESULT & DISCUSSIONS

A. Appropriate technology in ankle joint

The design requirements include the technical aspects and cosmetics were resulted trough focus group discussion (FGD) which involves several expert and users of transtibial amputee. The results was described in Table 1.

Table 1 Appropriate technology application through 7 criteria

7 Criteria	Ankle Joint with Design Attributes		Recomendation of Materials and Technology
	Designers and Experts	Users	
Techni-caly feasible	a. Being able to be manufactured in the manual machining process with precision ± 0.01 mm. b. Has a high technical functionality to meet the proper biomechanics. c. The use of local material as far as possible. d. Cosmetically real.		a. Materials using HDPE plastic rod blue or gray. b. Guide the direction of motion of the ankle joint use of spherical plain bearing c. Lock two component parts of the ankle joint with shaft dome. d. Using the double spring which is placed in the anterior and posterior to the ankle joint. e. Using a double-ring locking pressure on the axis as the dome on the upper joint.
Econo-mical	e. Made with low manufacturing costs. f. Use locally available materials. g. Durable, simple, easy to repair. h. Using the capabilities of local technicians. i. Reproduced by local workers.		
Ergono-mics / Healthty	j. Application of store-return energy technologies with multi-axis motion that support the SACH foot to be directed. k. PTB-SC socket wear comfortable. l. Considering the local climate. m. Considering to the situation and working conditions.		
Socio-culturally Accept-able		n. Psychosocially accepted by society of appearance prosthetic foot	
Energy Saving	o. Technology does not require electric power.		
Environ-ment Friendly	p. Material is not harmful for the environment.		
Trendy		q. Technology and design more advanced than ever before.	

The results from focus group discussions provide technical recommendations for the design of the ankle joint: (1) spherical plain bearing function generates motion dorsi flextion–plantarflexion, inversion-evertion and rotation,

(2) the ball joint to the axis as function synergistically to control the direction of movement between fields lateral and medial, (3) double locking ring on the axis of the dome serves as the force created to withstand the weight of the body to the ground and the force generated from the ground surface, (4) a spring that is placed in the posterior and anterior as much as 2 pieces by 2 pieces functioning energy store and return, (5) provide easy assembly for the adapter lower part of the movement with SACH foot, (6) the upper adapter serves to connect the movement with foot shank portion, and (7) the upper depths adapter serves as a lock shank and foot, as described in Figure 2.

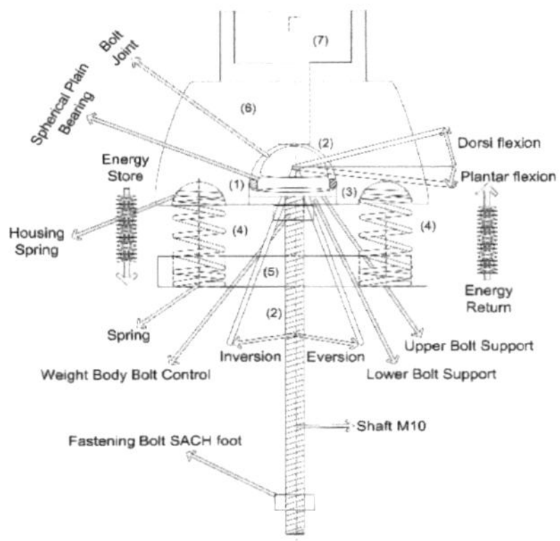

Fig. 2 Ankle joint multi axis energy store-return

The ankle joint in the below knee endoskeletal prosthetic foot is useful for facilitating the direct SACH foot and improve posture while amputee to walk. Prosthetic foot is designed for activity daily living (ADL) with the ankle joint that provide walking assisted energy store-return capability during the stance phase and swing phase. Energy store-return can cause the reduction in the pulse rate while walking, as described in Figure 3.

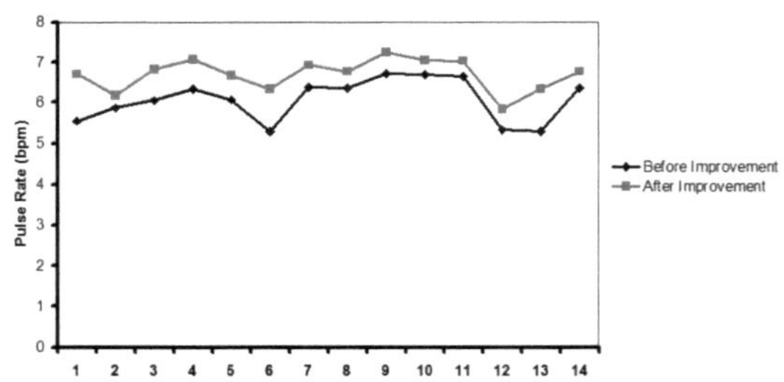

Fig. 3 Pulse rate while walking

B. Subject characteristics

Test of walking with ankle joint attached on the below knee endoskeletal prosthetic foot for a medium stump 20% - 50% with prosthetic foot height 41 cm, short stump < 21% with prosthetic foot height 44 cm, and the long stump < 51% with prosthetic foot height 41 cm 45 cm, as described in Table 2.

Table 2 Transtibial amputee anthropometric data (n = 14)

Parameters	Means	SD	Min	Max
Age (yr)	27.50	7.37	19	41
Height (cm)	163.71	7.63	145	176
Estimation of body weight (cm)	66.02	7.78	57	86
Body mass index (kg/m2)	20.19	1.62	19	24
Wearing experience (yrs)	8.14	6.48	1	20
High prosthetic foot (cm)	45.11	2.71	40	48
Weight prosthetic foot (kg)				
a. Exoskeletal	1.86	0.27	1	2
b. Endoskeletal	1.65	0.13	1	2

C. Testing between groups (n = 7)

Testing between treatment groups consists of the stages of a statistical test of comparability, the effects of the period, residual effects, and the effects of treatment. Comparability performed on the first period before the test conducted on the subject. Effects period carried on between groups by comparing the difference in Period I and Period II, and the residual effects on inter group done by comparing the amount in Period I and Period II.

Each stage testing Shapiro-Wilk test that indicates the data are normally distributed both groups ($p > 0.05$) and test Levene's test data show both homogeneous groups ($p > 0.05$). Furthermore, independent t-test on the pulse test to determine differences between treatment groups in Period I and Period II, the results of which are described in Table 3.

Table 3 Test statistics on the pulse (n = 7)

Pulse	Group	*n*	Means	SD	Value *t*	Value *p*
Comparability (Pulse Rest Period I)	I	7	78.66	3.54	1.562	0.169
	II	7	76.87	2.00		
Effects of the period ($P_0 \rightarrow P_1$); ($P_1 \rightarrow P_0$)	I	7	10.43	4.89	1.415	0.207
	II	7	7.49	5.22		
Residual effects ($P_0 \rightarrow P_1$) ;($P_1 \rightarrow P_0$)	I	7	35.61	11.26	0.236	0.821
	II	7	34.55	15.46		

The test results showed that ($p > 0.05$) means that there is no significant difference of the pulse in the group using the exoskeletal prosthetic foot and groups using endoskeletal prosthetic foot and can also be expressed is the same. A decrease in pulse rate was caused by the difference in treatment.

D. Test treatments in period I and period II (n = 14)

To determine differences in treatment in Period I (before improvement) and Period II (after improvement) test performed pair-t test on the pulse. The results of which are described in Table 4.

Table 4 Test pulse treatment (bpm)

Variables	n	Before Improvement		After Improvement		Value t	Value p
		Means	SD	Means	SD		
Pulse before walking	14	77.76	2.92	76.97	2.87	5.213	< 0.0001
Pulse after walking	14	99.78	8.86	90.03	5.66	7.036	< 0.0001
Cardiovascular load	14	22.02	7.95	13.06	5.87	6.577	< 0.0001
% CVL	14	45.89	16.61	27.27	12.27	6.535	< 0.0001

The test results showed that ($p < 0.05$) means that the pulse Period I to Period II significantly different. Ankle joint on the below knee endoskeletal prosthetic foot can reduce load of cardiovascular while walking. Percentage of cardiovascular loads (%CVL) indicates $p < 0.05$ by t-value = 6.535 and p = 0.0001, that means is %CVL significant difference in Period I to Period II. Differences pulse before improvements, after improvements, cardiovascular load and %CVL, as described in Figure 4.

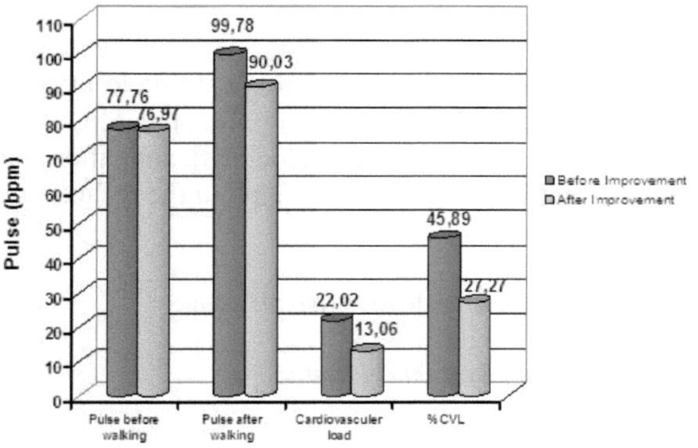

Fig. 4 Comparison of the pulse at the transtibial amputee

Testing the ankle joint for the below knee endoskeletal prosthetic foot with the age range of 19 years to 14 years in age. Use of endoskeletal prosthetic foot with height range of 40 cm to 48 cm. An indication of the difference in pulse rate decrease after the walking of 9.75 bpm and the cardiovascular load of 8.62 bpm, meaning that the ankle joint is designed based on the appropriate technology can function either as a replacement of the amputated limb. Difference in percentage cardiovascular (%CVL) based classification at 18.62 which indicates a change from being relieved, it shows the user fatigue is reduced.

IV. Conclusions

Ankle joint with appropriate technology in the below knee endoskeletal prosthetic foot is able to perform multi-axis motion and energy stores-return. Decrease the load of cardiovascular difference after walking and cardiovascular load of percentage (%CVL) which indicates that the ankle joint is functioning in accordance of technical recommendations on a focus group discussion (FGD).

References

1. Manuaba I. B. A. (2005) Total Ergonomics Enhancing Productivity, Product Quality and Customer Satisfaction. Indonesian Journal of Ergonomics. vol. 6: 1-6.
2. Manuaba I. B. A. (2007) A Total Approach in Ergonomics is a Must to Attain Humane, Competitive and Sustainable Work Systems and Products. Journal Human Ergology. vol. 36: 23-30.
3. Kilbom R. (2003) Occupational Health Conditions in Extreme Environments. British Occupational Hygiene Society, Published by Oxford University Press. Ann. Occupational Hygiene Jour. vol. 47(3): 241–52.
4. Adiputra N. `(2004) Ergonomi. Artikel Pelatihan Kesehatan Kerja Tenaga Kesehatan. Denpasar: Fakultas Kedokteran Udayana Bali.
5. Louhevaara V and Kilbom Å. 2005. Dynamic Work Assessment. on: John R. Wilson & Nigel Corlett (eds.) Evaluation of Human Work. 3rd edition. London: Taylor & Francis ltd. pp. 520-542.
6. Kroemer, K. H. E. (2008) Fitting The Human : Introduction to Ergonomics. 6th edition. Boca Raton: Taylor & Francis Group.
7. Miao-Ju Hsu M., Nielsen D H, Lin-Chan S J, Shur D (2006) The Effects of Prosthetic Foot Design on Physiologic Measurements, Self-Selected Walking Velocity, and Physical Activity in People with Transtibial Amputation. Arch. Phys. Med. Rehabilition Journal. January; vol 87(86): 123-9.
8. Suh-Jen L. and Nisha H. B. (2008) Six-Minute Walk Test in Persons with Transtibial Amputation, Journal Arch Phys Med Rehabil, vol. 89: 2354-59.
9. Paul L. E. (2003) The Six-Minute Walk Test. Respiratory Care of Journal. vol. 48(8): 783-5.

Author: Lobes Herdiman
Institute: Sebelas Maret University
Street: Ir. Sutami No. 36A Road
City: Solo
Country: Indonesia
Email: lobesh@gmail.com

Modelling of Peristaltic Bile Flow in the Papilla Ampoule with Stone and in the Papillary Stenosis Case: Application to Reflux Investigation

A.G. Kuchumov[1], Y.I. Nyashin[1], and V.A. Samartsev[2]

[1] Perm National Research Polytechnic University, Department of Theoretical Mechanics and Biomechanics, Perm, Russian Federation

[2] Perm State Medical University, Department of General Surgery#1, Perm, Russian Federation

Abstract— **Bile is a biofluid produced by the liver and responsible in the duodenum. The object of present research is an investigation of the peristaltic bile flow in the papilla of Vater, which is the narrowest place in the biliary system (responsible for bile production, storing, and transport of the bile into the duodenum for fat digestion). Recently, the pathological bile was revealed to be the non-Newtonian thixotropic fluid. The present paper contains results of solution of problems devoted to the peristaltic flow of bile as Carreau's fluid. Two cases were considered: bile flow in the duct with a stone and bile flow in the stenosed duct. Adopting the perturbation method, we obtained the analytical solutions for axial velocity, flow rate, and pressure. Moreover, the dependences of pressure drop on the flow rate were obtained. Developed mathematical model may be used for studying of such pathology as choledochopancreatic reflux (i.e. pathological flow of the gallbladder bile coming out the common bile duct through the Vater's papilla ampoule into the pancreatic ducts instead of the duodenum). Adopting an assumption that the reflux occurrence corresponds to the case when the flow rate becomes less zero, the critical pressures were computed. The pressure rise magnitudes corresponding to reflux occurring conditions depending on various parameters (amplitude ratio, Weissenberg number, etc.) were obtained.**

Keywords— **bile, Vater's papilla, reflux, pressure drop, perturbation method.**

I. INTRODUCTION

Bile is a one of 32 bio-fluids in human body. Biliary system is responsible for bile production, storing, and transport into the duodenum, where the fat digestion is being proceeded [1]. Biochemical (liver metabolism disorders) and biomechanical (bile stasis in the different regions of the biliary system) reasons are thought to be main for a concrement (calculus) formation. Biliary calculi may lead to a blocking the flow of bile to the intestine and causing obstructive jaundice. Moreover, it may cause severe abdominal pain, inflammation, and even cancer.

The medical practice showed that even after the surgical stone removal, the problems remain.

Papillary stenosis [a disturbance of the sphincter of Oddi, a muscular valve, that prevents the opening and release of bile or pancreatic fluids into the duodenum] caused by stone passage and consequent scar tissue formation is an example of the biliary system dysfunction, which may develop even if the stone is surgically removed.

A little is known about bile reflux reasons related with gallstones presence. Choledochopancreatic reflux (i.e. pathological flow of the gallbladder bile coming out the common bile duct into the pancreatic duct instead of the duodenum) is one of the reason of the pancreatitis (inflammation of the pancreas).It is believed that hypomotility of sphincter of Oddi has an influence on the bile flow and reflux occurring [2].

The given paper is intended to model the process of bile flow in the duct with papillary stenosis and to obtain quantative factors for the reflux formation taking into account the pathological character of flow and peristalsis influence. The duct is modeled as a finite-length tube with narrowing permeable walls. The pathological bile is assumed to be the Carreau's fluid [3]. After that, it is planned to suggest treatment methods to avoid this dysfunction.

The task of present paper is to develop mathematical models of peristaltic flow of pathological bile to understand the biomechanical reasons of such pathophysiological phenomenon [find pressures and velocities] and after that to suggest the methods of treatment to avoid the reflux.

II. MATERIAL AND METHODS

A. Bile and biliary system anatomy

Bile is a bio-fluid produced by the liver and responsible for a fat digestion in the duodenum. Normal daily secretion amount of bile is about 1.5 liters in adults [4].

After the biliary tree passing, a bile together with a pancreatic juice through the channel called the Vateri's papilla ampoule enters the duodenum. The bile accelerates the absorption of D, E, K and A vitamins containing in the fats in the duodenum [5].

An anatomy of common bile duct and pancreatic duct bifurcation is shown in Fig. 1 [6].

When contractive ability of Oddi's sphincter is disordered (due to basal pressure changes, gallstones presence in the ampoule, etc.) [7], the pathological refluxes (i.e. upward

J. Goh and C.T. Lim (eds.), *7th World Congress on Bioengineering 2015*, IFMBE Proceedings 52, DOI: 10.1007/978-3-319-19452-3_42

flow of bio-fluid in another hollow organ oppositely to normal flow) may occur [8].

The choledochopancreatic reflux is a one of the chronic pancreatitis development reason. The choledochopancreatic reflux is bile flow from the common bile duct to the pancreatic duct instead of the duodenum [9].

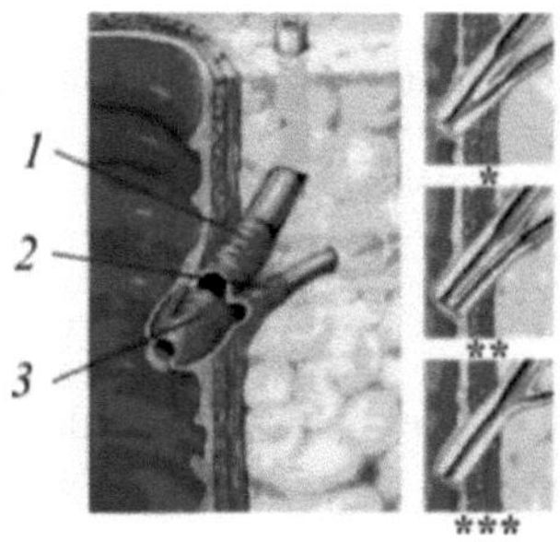

Fig. 1 Anatomy of Oddi's sphincter: a – common anatomy (*1* – sphincter of the common bile duct, *2* – sphincter of pancreatic duct, *3* – sphincter of Vateri's papilla ampoule), b – various anatomy of papilla ampoule (*– bifurcation in the papilla ampoule, ** – parallel situation of ducts, *** – bifurcation before an ampoule)

The investigation and modeling of reflux occurring conditions in the channels with various geometry to provide an biomechanical assistance to surgeons in the treatment of the current disorder is a serious challenge [10, 11]. The present paper is devoted to modeling of choledochopancreatic reflux in the case of bile flow in the duct with a stone and bile flow in the stenosed duct.

B. Problem statement of bile flow in the duct with a stone

The peristaltic bile flow as non-Newtonian fluid (Carreau's fluid) is considered in the Vateri's papilla ampoule with a stone. The problem geometry is shown in Fig. 2. The bile flows in the annulus between the stone, which is considered as an internal immovable cylinder with radius $r_1 a$ and bile duct with permeable wall, which is considered as an external cylinder with radius a. The peristaltic wave is propagating along the external cylinder wall with velocity of c. The flow is considered in the longitudinal and transversal directions in the fixed frame $(\bar{X}, \bar{Y})$.

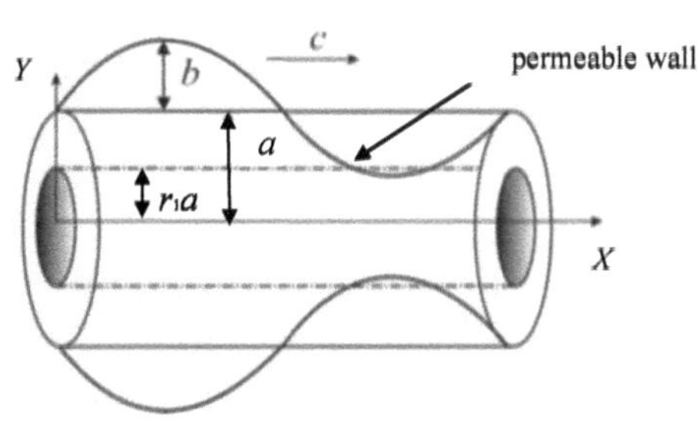

Fig. 2 Geometry of peristaltic bile flow in duct with a stone

Carreau's fluid equation is

$$\eta = \eta_\infty + (\eta_0 - \eta_\infty)(1 + (\Gamma^2 \dot{\gamma}^2))^{\frac{m-1}{2}}, \tag{1}$$

where η_∞ is the infinite shear rate viscosity, η_0 is the zero shear rate viscosity, Γ is the time constant, m is the power law index.

A geometry of external cylinder wall can be described as

$$\bar{h}(\bar{X}, \bar{t}) = a + b\sin\frac{2\pi}{\lambda}\left(\bar{X} - c\bar{t}\right), \tag{2}$$

where b is a wave amplitude, λ is a wave length, $\bar{t}$ is a time. Let us consider flow in the wave frame $(\bar{x}, \bar{y})$ moving with velocity c away from the fixed frame $(\bar{X}, \bar{Y})$.

$$\bar{x} = \bar{X} - c\bar{t},\ \bar{y} = \bar{Y},$$
$$\bar{u} = \bar{U} - c,\ \bar{v} = \bar{V},$$
$$\bar{p} = \bar{P}.$$

Let us introduce non-dimensional variables:

$$x = \frac{\bar{x}}{\lambda},\ y = \frac{\bar{y}}{a},\ v = \frac{\lambda\bar{v}}{ac},\ t = \frac{c\bar{t}}{\lambda},\ \delta = \frac{a}{\lambda},\ r_1 = \frac{\bar{r}_1}{a},\ p = \frac{a^2\bar{p}}{c\lambda\eta_0},\ We = \frac{c\Gamma}{a},$$

$$\tau_{xx} = \frac{\lambda\bar{\tau}_{xx}}{c\eta_0},\ \tau_{xy} = \frac{a\bar{\tau}_{xy}}{c\eta_0},\ \tau_{yy} = \frac{a\bar{\tau}_{yy}}{c\eta_0},\ \dot{\gamma} = \frac{a\bar{\dot{\gamma}}}{c},\ Re = \frac{ca\rho}{\eta_0},\ Da = \frac{k}{a^2},$$

$$h = \frac{\bar{h}}{a} = 1 + \varphi\sin 2\pi x,$$

where $\varphi = \frac{b}{a}$ is the amplitude ratio, We is the Weissenberg number, Da is the Darcy number, Re is the Reynolds number.

After non-dimensionalization, the governing equations in the wave frame written in the terms of the stream function Ψ ($\Psi = \frac{\bar{\Psi}}{ac}$, $u = \frac{\partial\Psi}{\partial y}$, $v = -\delta\frac{\partial\Psi}{\partial x}$) are as follows:

$$\frac{\partial u}{\partial x} + \frac{\partial v}{\partial y} = 0, \tag{3}$$

$$Re\cdot\delta\left[\left(\frac{\partial\Psi}{\partial y}\frac{\partial}{\partial x} - \frac{\partial\Psi}{\partial x}\frac{\partial}{\partial y}\right)\frac{\partial\Psi}{\partial y}\right] = -\frac{\partial p}{\partial x} + \delta^2\frac{\partial\tau_{xx}}{\partial x} + \frac{\partial\tau_{xy}}{\partial y}, \tag{4}$$

$$-Re\cdot\delta^3\left[\left(\frac{\partial\Psi}{\partial y}\frac{\partial}{\partial x} - \frac{\partial\Psi}{\partial x}\frac{\partial}{\partial y}\right)\frac{\partial\Psi}{\partial x}\right] = -\frac{\partial p}{\partial y} + \delta^2\frac{\partial\tau_{xy}}{\partial x} + \delta\frac{\partial\tau_{yy}}{\partial y}. \tag{5}$$

Under the assumptions of long wavelength $\delta \ll 1$ and low Reynolds number, neglecting the terms of order δ and higher and after introduction of boundary conditions, the final problem statement will look like:

$$\frac{\partial^2}{\partial y}\left[\left(1+\left(\frac{m-1}{2}\right)We^2\left(\frac{\partial^2\Psi}{\partial y^2}\right)^2\right)\frac{\partial^2\Psi}{\partial y^2}\right]=0,$$

$$\Psi=0 \text{ at } r=r_1,\ \frac{\partial\Psi}{\partial y}=0 \text{ at } r=r_1, \tag{6}$$

$$\Psi=\frac{q}{2} \text{ at } r=h,\ \frac{\partial\Psi}{\partial y}+\beta\frac{\partial^2\Psi}{\partial y^2}=-1 \text{ at } r=h,$$

where $\beta=\frac{\sqrt{Da}}{\alpha}$ is the permeability parameter, $\bar{Q}=q+1$ is the dimensionless mean flow, where $q=\int_{r_1}^{h}\frac{\partial\Psi}{\partial y}y\,dy$.

More elaborately about boundary condition statement on permeable wall one can find in [12–14]. More elaborately about bile duct wall permeability one can find in [15, 16].

C. Problem statement of bile flow in the stenosed duct

During the long-term stay of the concrement in the papilla ampoule, the scar tissue formation starts so the lumen of the channel decreases. This disorder is called the papillary stenosis. Changes in the duct geometry may also lead to a reflux occur.

Let us consider a peristaltic bile flow as Carreau's fluid in the ampoule of the Vateri's papilla modeled as the tapered tube with length of *L*. The wavelength is comparable with channel length ($L\approx\lambda$), thus the wave number is small, and Reynolds number is neglible. The sinusoidal wave is considered. Wall geometry is described as

$$H'(x',t')=b_0-kx'+g\sin\frac{2\pi}{\lambda}(x'-ct'), \tag{6}$$

where b_0 is inlet radius; k is a slope coefficient, c is a wave velocity, x' is a longitudinal coordinate, g is an amplitude of peristaltic wave. Schematically, the problem geometry is presented in Fig. 3.

Parameters used in the model are: b_0 is 3 mm, *L* is 5 mm, k is 0.5.

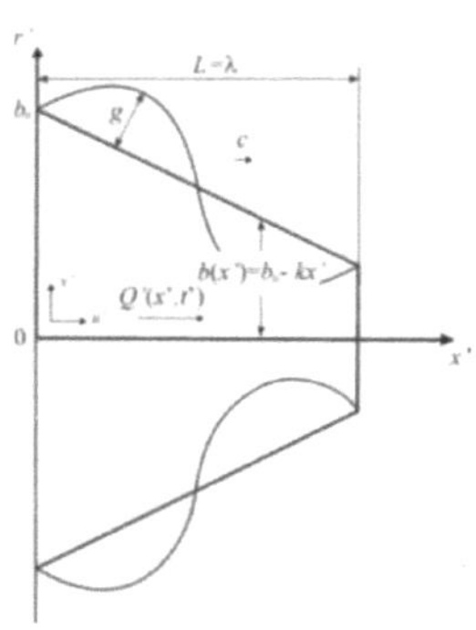

Fig. 3 Geometry of peristaltic bile flow in a stenosed duct

The final statement after simplifications is

$$\frac{\partial p}{\partial x}=\frac{1}{r}\frac{\partial}{\partial r}(r\tau_{rx}),$$

$$\frac{\partial u}{\partial r}=0 \text{ at } r=0, \tag{7}$$

$$u=0 \text{ at } r=h=1-\frac{\lambda kx}{b_0}+\varphi\sin(2\pi(x-t)),$$

where u is the axial velocity.

III. PERTURBATION METHOD

To solve above problems, let us adopt the perturbation method and consider solution in the series form by Weissenberg number (We) expansion:

$$\Psi=\Psi_0+(We)^2\Psi_1+O(We^4),\qquad u=u_0+(We)^2u_1+O(We^4),$$

$$q=q_0+(We)^2q_1+O(We^4),\qquad p=p_0+(We)^2p_1+O(We^4),$$

The dimensionless pressure drop is

$$P_\lambda=\int\frac{\partial p}{\partial x}dx. \tag{8}$$

IV. RESULTS

A. Bile flow in the duct with a stone

Computational results for the bile as Carreau's fluid are presented in this section. Parameters of Carreau's fluid are taken from [3]. Fig. 4 contains the dependence of dimensionless mean flow rate $\bar{Q}$ on pressure drop ΔP_λ at different Weissenberg number (We) $\left(\beta=0.01,\ \varphi=\frac{\pi}{3}\right)$. It should be noticed that ΔP_λ increases with increasing of Weissenberg number. At $\bar{Q}=0$, ΔP_λ is maximal and pressure decreases with increasing of flow rate. Moreover, it should be noticed that flow rate may have negative values corresponding to the case of a reflux occur. Thus, the values of ΔP_λ corresponding to $\bar{Q}=0$ can be considered as criteria of the reflux occur, i.e. the reflux occurs when $\Delta P_\lambda > \Delta P_{\lambda\,crit}$.

B. Bile flow in the stenosed duct

Dependence of ΔP_λ on *t* for a sinusoidal wave at various values of $\bar{Q}$ is shown in Fig. 5. It can be seen that the pressure drop reaches its maximum value at $t=0.27$. At this moment the vessel inlet collapse starts, because the cross section is minimal at the entrance, and thus ΔP_λ is maximal. Subsequently, the pressure drop falls to zero.

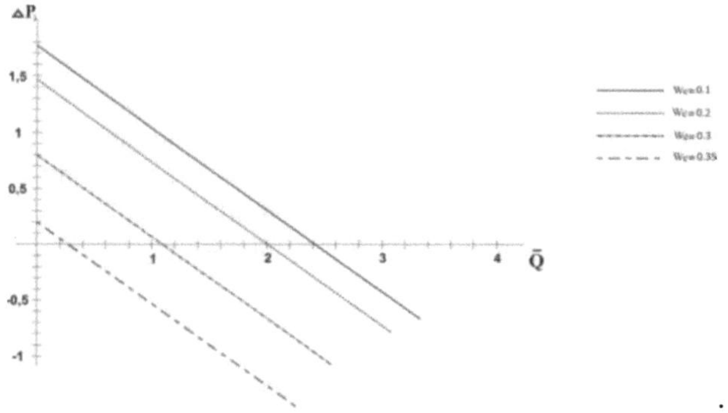

Fig. 4 Dependence of ΔP_λ on $\bar{Q}$ at various Weissenberg numbers

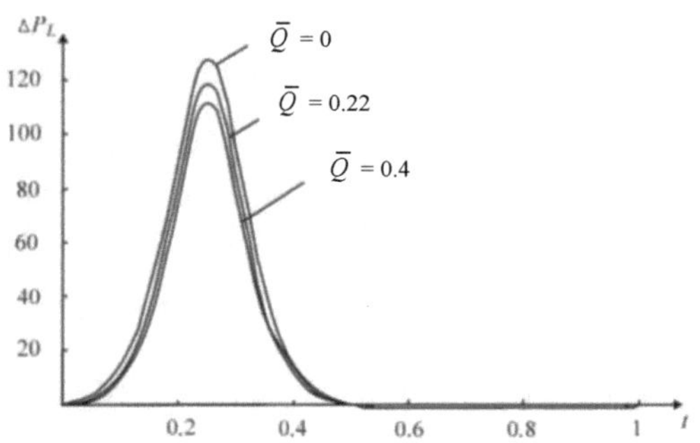

Fig. 5 Dependence of ΔP_λ on t for sinusoidal wave (φ=0.6)

V. CONCLUSIONS

The present paper contains results of solution of problems devoted to the peristaltic flow of bile as Carreau's fluid. Two cases were considered: bile flow in the duct with a stone and bile flow in the stenosed duct. Developed mathematical model may be used for studying of such pathology as choledochopancreatic reflux. Adopting an assumption that the reflux occurrence corresponds to the case when the flow rate becomes less zero, the critical pressures were computed.

ACKNOWLEDGMENT

The reported study was supported by RFBR, research project No. 14-01-31027-mol_a.

CONFLICT OF INTEREST

The authors declare that they have no conflict of interest.

REFERENCES

1. Jian C, Wang G (1991) Biomechanical study of the bile duct system outside the liver. Biomed Mater Eng 1(2):105-113
2. Zhang Z-H, Wu S-D, Wang B et al. (2008) Sphincter of Oddi hypomotility and its relationship with duodenal-biliary reflux, plasma motilin and serum gastrin. World J Gastroenterol 14(25): 4077-4081
3. Kuchumov A, Gilev V, Popov V et al. (2014) Non-Newtonian flow of pathological bile in the biliary system: experimental investigation and CFD simulations. Korea-Australia Rheology Journal 26: 81–90.
4. Hofmann A (2007) Biliary secretion and excretion in health and disease: current concepts. Ann. Hepatology 6: 15-27
5. Buchner A, Sonnenberg A (2002) Factors influencing the prevalence of gallstones in liver disease: the beneficial and harmful influences of alcohol. Am. J Gastroenterol 97: 905-909
6. Horiguchi S, Kamisawa T (2010) Major duodenal papilla and its normal anatomy. Dig Surg 27: 90-93.
7. Pitt H, Roslyn J, Kuchenbecker S et al. (1981) The role of cystic duct resistance in the pathogenesis of cholesterol gallstones. Journal of Surgical Research 30: 508-514.
8. Otto W, Scott G, Rodkiewicz C. (1979) A comparison of resistances to flow through the cystic duct and the sphincter of Oddi. Journal of Surgical Research 27: 68-72.
9. Liang T, Liu Y, Bai X et al. (2010) Sphincter of Oddi laxity: an important factor in hepatolithiasis, World J Gastroenterology 16: 1014-1018
10. Misra M, Rao A (2005) Peristaltic transport in a channel with a porous peripheral layer: model of a flow in gastrointestinal tract. Journal of Biomechanics 38: 779-789.
11. Takabatake S, Ayukava K, Mori A. (1988) Peristaltic pumping in circular cylindrical tubes: a numerical study of fluid transport and its efficiency. J Fluid Mech 193: 269-283.
12. Saffman P (1971) On the boundary condition at the surface of a porous media. Stud Appl Math 50: 93–101.
13. Beavers G, Joseph D (1967) Boundary conditions at a naturally permeable wall. J Fluid Mech 30: 197–207.
14. Vajravelu S, Sreenadh K et al. (2012) Peristaltic transport of a Williamson fluid in asymmetric channels with permeable walls. Nonlinear Analysis: Real World Applications 13: 2804–2822
15. Smith N , Boyer J (1982) Permeability characteristics of bile duct in the rat. American Journal of Physiology Gastrointestinal and Liver Physiology 242: G52-G57
16. Chenderovitch J (1973) Transmembrane potentials and permeability of the wall of the common bile duct. Lille Med 18(3): 297-301

Author: Alex G. Kuchumov
Institute: Perm National Research Polytechnic University
Street: Komsomoskiy Prospect 29
City: Perm
Country: Russian Federation
Email: kychymov@inbox.ru

SPHCOFEM: Solver for Coupling SPH and FE

L. Hynčík

New Technologies – Research Centre, University of West Bohemia, Plzeň, Czech Republic

Abstract— **Numerical simulations in biomechanics and bioengineering consider highly deformable tissues in interaction with fluid flows. The solver SPHCOFEM for investigating fluid flow coupled problems with 1D, 2D or 3D calculations is presented. The fluid domain is represented by smoothed particle hydrodynamics and the boundary is modeled using finite elements. The interaction is done via a continuous contact model. Several material models useful for various hydrodynamics matters and simple finite elements for representing the boundary are implemented. Pre- and post-processors serve for task definition and result analyses.**

Keywords— **meshless, SPH, coupling, FE, solver.**

I. Introduction

Modeling of fluid flow is important in many fields including not only energetics and automotive industry, but also biomechanics, bioengineering and clinical applications.

Laminar fluid flow is easy to describe with standard grid methods, but if any turbulence or complex flow, including multi-phase flows, appears, the grid methods are complicated to control. Additional contact problems with a highly deformable boundary, such as biological tissues, even increase the complexity of the problem.

Standard methods modeling fluid exploit the Eulerian approach using discretization of the domain by a grid. If a complex flow or highly deformable boundary appear, there is an option to implement the Lagrangean approach called smoothed particle hydrodynamics (SPH) [1]. SPH is a meshless method useful for flows, where the domain geometry changes considerably, because the fluid is discretized using the moving separated particles.

The boundary is modeled using the finite element method (FEM) [2]. There are several methods dealing with the interaction between SPH and FEM. The continuous contact model based on direct particle to wall coupling is an optimal choice [3] due to the moving particles.

II. Materials and methods

This work presents an in-house solver coupling SPH and FEM using the continuous contact model. The solver called *SPHCOFEM* (*SPH COupled to Finite Element Method*) investigates 1D, 2D and 3D problems. Since the solver is implemented using the C++ language, the pre- and post-processing is done via GUI (*Graphical User Interface*), developed in the MATLAB environment.

A. Modeling the fluid

The basis of SPH is interpolation theory. Fluid mechanics conservation laws are transformed into integral equations using an interpolation function, the kernel $w(\mathbf{r},h)$, where $\mathbf{r}$ is a position vector and h is the smoothing length [4]. The kernel approximates the delta-function and refers to a weighting function that defines how much of each field variable contributes to the field variable at a given point in space.

Having the values of the physical field $A(\mathbf{r})$ known on a set of N discrete points given by position vectors $\mathbf{r}_i$ and masses m_i, one can design the basic interpolation equation.

$$\langle A(\mathbf{r})\rangle = \sum_{j=1}^{N} \frac{m_j}{\rho_j(\mathbf{r}_j)} A(\mathbf{r}_j)\, w(\mathbf{r}-\mathbf{r}_j, h) \tag{1}$$

Analogically the gradient can be expressed.

$$\langle \nabla_r A(\mathbf{r})\rangle = \sum_{j=1}^{N} \frac{m_j}{\rho_j(\mathbf{r}_j)} A(\mathbf{r}_j)\, \nabla_r w(\mathbf{r}-\mathbf{r}_j, h) \tag{2}$$

The momentum equation, the continuity equation and the heat transfer equation are discretized using equation (2) to the final form.

$$\frac{d\mathbf{v}_i}{dt} = -\sum_{j=1}^{N} m_j \left(\frac{p_j}{\rho_j^2} + \frac{p_i}{\rho_i^2} + \Pi_{ij}\right) \nabla w_{ij} \tag{3}$$

$$\frac{d\rho_i}{dt} = -\sum_{j=1}^{N} m_j (\mathbf{v}_i - \mathbf{v}_j)\, \nabla w_{ij} \tag{4}$$

$$\frac{du_i}{dt} = -\sum_{j=1}^{N} m_j \frac{p_j}{\rho_j^2} (\mathbf{v}_i - \mathbf{v}_j)\, \nabla w_{ij} \tag{5}$$

The second derivative term describing the viscosity of the fluid flow cannot be easily designed using the equations (1) and (2) above. That is why here Π_{ij} which refers to so called artificial viscosity term, with the numerical constants α and β to improve the stability, is added to the equations of motion (EOM) [5].

In contrast to the grid methods, where the interaction of the neighboring cells is processed via the direct connection, SPH is dependent on fast techniques for finding the neighboring particles that contribute to the summation (2). This is

J. Goh and C.T. Lim (eds.), *7th World Congress on Bioengineering 2015*,
IFMBE Proceedings 52, DOI: 10.1007/978-3-319-19452-3_43

called the *N*earest *N*eighbor *S*earch (NNS) and it is the main issue related to the complexity and time consumption of the SPH calculation. The simplest NNS approach is to assess all pairs of particles, but it has a computational demand equal to N^2. SPHCOFEM benefits from pre-distributing the neighboring particles into independent sub-sets to be managed separately. This approach decreases the calculation time considerably; however, for complex turbulent flow, where the domain geometry changes too much, this approach is not always applicable.

There are several equations of state to be implemented for the fluid flow. The developed algorithm easily manages ideal gas due to the compressibility.

$$p = \rho(\kappa - 1)u \tag{6}$$

Here p denotes pressure on any particle in equations (3) and (5), κ is the ratio of the thermal capacities by constant pressure and constant volume, respectively, and u is the internal energy from equation (5). Incompressible flows are simulated within the quasi-incompressible limit [6].

$$p = K\left[\left(\frac{\rho}{\rho_0}\right)^\gamma - 1\right] \tag{7}$$

Here K governs the relative density fluctuation and γ is the constant usually set to 7. K refers to the liquid bulk modulus and is set in order to ensure the density variations under 1% for quasi-compressible fluid [6].

B. Modeling the boundary

The developed algorithm accommodates standard 1D, 2D and 3D finite elements (FE) to be in contact with the fluid domain. The Lagrange equations are used to develop the EOM for the boundary elements.

$$\frac{d}{dt}\left(\frac{\partial L}{\partial \dot{q}_i}\right) - \frac{\partial L}{\partial q_i} = \boldsymbol{Q} \tag{8}$$

Here $L = E_k - E_p$ is the Lagrange function, $\boldsymbol{q}$ is the vector of generalized coordinates and $\boldsymbol{Q}$ is the vector of external forces including the contact forces caused by coupling the boundary FE with the fluid domain particles.

Introducing the kinetic energy E_k and the potential energy E_p, the EOM describing the boundary are developed.

C. Coupling

The interaction between the fluid domain and the boundary is designed via direct coupling using the penalty-based continuous contact model. The approach also utilizes the automatic force calculation. The contact thickness, which is an area with a defined thickness along each element, is defined. For each couple of particle and element, a particle entering that area is detected and the penalty force starts can be calculated taking into account the possibility to push the particle back within a time step [3].

$$F_n = \frac{m_p m_e}{m_p + m_e}\frac{p}{\Delta t^2} \tag{9}$$

Here F_n denotes the normal contact force, m_p and m_e are the penetrating particle and penetrated element masses, respectively, p is the penetration depth and Δt is the current time step. The tangential friction force is easily added by defining the friction coefficient.

D. The algorithm and the software

EOM for SPH and FE are coupled to a single algorithm. Whilst the smoothed particle motion is integrated using the central acceleration method, the FE motion is driven by the conjugate gradient method [3].

The stable time step scheme, taking into account the Courant-Friedrichs-Lewy conditions for SPH stability and the stability condition for FEM, is applied.

To improve the results, the so-called X-SPH variant is also implemented [7]. The X-SPH variant updates the velocities with the constant ε between 0 and 1.

$$\frac{d\mathbf{x}_i}{dt} = \hat{\mathbf{v}}_i = \mathbf{v}_i + \varepsilon\sum_{j=1}^{N} m_j \frac{\mathbf{v}_{ij}}{\rho_{ij}} w_{ij} \tag{10}$$

There is a structured input text file describing the particle coordinates and initial conditions as well as boundary FE. Due to the pre- and post-processors developed as GUI in the MATLAB environment, the MATLAB input file structure using standard MATLAB functions to design the structure is also defined [8].

III. VALIDATION TESTS

SPHCOFEM is validated using the 1D, 2D and 3D tests by comparing the calculation results to available published experimental or numerical data. The validation concerns the shock tube problem in 1D, the broken dam problem in 2D and the crashing ball problem in 3D.

A. Shock tube

The shock tube validation [9] deals with a stationary gas in a tube divided by a diaphragm into two regions, each characterized by a given pressure and density. At time $t = 0$, the diaphragm is released.

The SPHCOFEM input file accommodates 1000 particles. The ideal gas EOM (6) is used. The artificial viscosity coefficients α and β are set both to unity and the specific heat ratio $\kappa = 1.4$. The time to be simulated is 0.230768 s. The boundary is not modeled in the shock tube problem.

B. Broken dam

A free surface flow is investigated and validated here. A rectangular column of water in hydrostatic equilibrium is fixed by the vertical wall. The water column is 0.05715 m high and 0.05715 m wide. At time $t = 0$, the gate is removed instantaneously and the water is allowed to collapse [10].

The SPHCOFEM code is tested using 361 (coarse), 3249 (fine), 12,996 (very fine) and 51,984 (complex) particles, also with and without the X-SPH variant. The EOM is modeled for quasi-compressible fluid using equation (7) with the parameter $\gamma = 7$.

The water has density $\rho = 996$ kg/m^3. Coefficients α and β are chosen as 1.2 and 1.5, respectively. Contact between the water and the boundary is defined by equation (9).

C. Crashing ball

The task is chosen in order to assess the behavior of a water filled elastic ball crashing against a rigid wall at the initial velocity 1 m/s [11]. It concerns the full 3D problem.

The SPHCOFEM input file concerns 9889 particles. The water has density $\rho = 996$ kg/m^3. The EOM is modeled for quasi-compressible fluid using equation (7), with the coefficients $\alpha = 1.2$, $\beta = 1.5$ and $\gamma = 7$.

IV. Results

The results of all simulation including sensitivity analysis for the most used 2D case are evaluated below. Whilst the 1D and 2D cases are compared to experimental results, the 3D case is compared to numerical output.

A. Shock tube

After releasing the diaphragm, a refraction wave starts travelling from the region of higher density to the region of lower density and the shock wave is generated.

Fig. 1 shows the time dependent density (left) and pressure (right) after releasing the diaphragm.

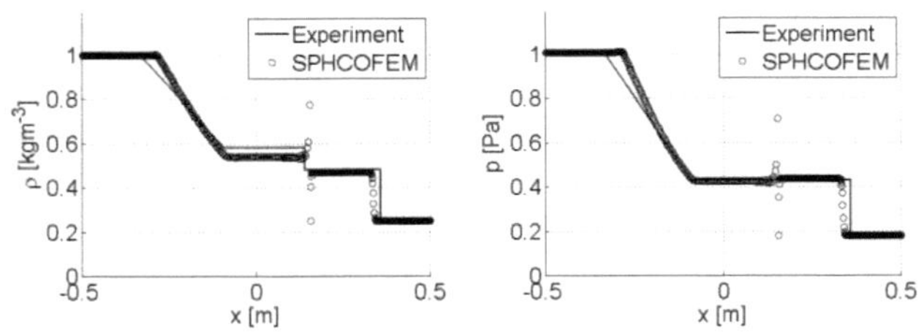

Fig. 1 Density (left) and pressure (right) distribution

B. Broken dam

Due to the complexity of any SPH calculation, the majority of the problems are investigated in 2D as planar flows. The broken dam is a typical problem for the 2D approach.

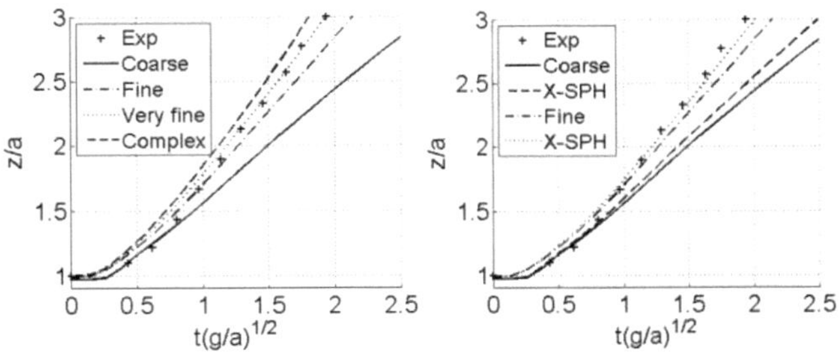

Fig. 2 Broken dam front surge propagation

The task is also used to evaluate the influence of the number of particles, the influence of the chosen method for NNS and the influence of the X-SPH variant.

Fig. 2 (left) shows the propagation of the surge front of the water domain after releasing the dam, taking into account the complexity of the model. Dimensionless time and water front position are defined with the gravitational acceleration g and the water front position z.

Fig. 2 (right) shows the influence of the X-SPH variant. Table 1 summarizes the time demand for particular calculations run on the table PC with Intel® Core™ i5-4570 CPU @ 3.20 GHz 3.20 GHz, 16 GB, 64 bit.

C. Crashing ball

The crashing ball is the only spatial problem, where the results are compared to the FEM computational environment in a different code using full 3D finite elements.

Table 1 Time demand for particular calculations

Model	Particles	Variant	NNS	Calculation [s]
Coarse	19 x 19	SPH	N^2	1.34
		SPH	Sub-sets	1.32
		X-SPH	N^2	5.40
		X-SPH	Sub-sets	5.33
Fine	57 x 57	SPH	N^2	80.65
		SPH	Sub-sets	56.00
		X-SPH	N^2	779.04
		X-SPH	Sub-sets	754.14
Very fine	114 x 114	SPH	N^2	1 850.75
		SPH	Sub-sets	974.21
Complex	228 x 228	SPH	N^2	79 243.73
		SPH	Sub-sets	14 290.08

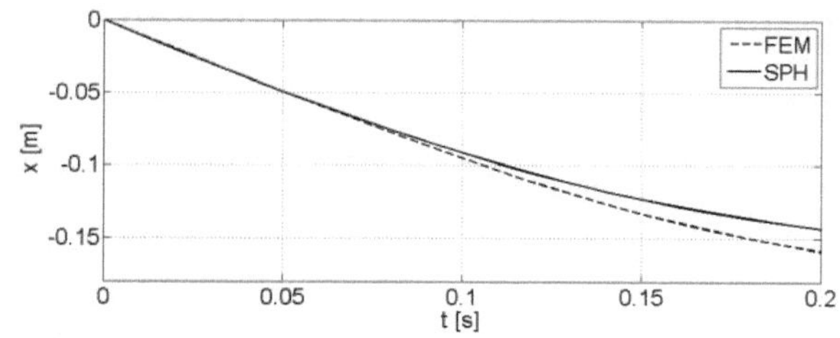

Fig. 3 Water front propagation

Fig. 3 shows the top ball particle displacement during the water ball collapse when crashing the rigid ball.

V. DISCUSSION

The validation tasks presented in the paper prove the SPHCOFEM solver as a tool for modeling fluid flow coupled to the boundary.

The shock tube results show the use of the algorithm in the simplest problems' investigation in 1D. One can see the discontinuity in both pressure and density (see Fig. 1) during the simulation, which are caused by the shock wave not being smoothed ideally.

The broken dam contributes to the own algorithm analysis itself. The results are successfully compared to the experimental data (Fig. 2) and it can be seen that there is not always a necessity to use a too complex model having a high number of particles. The optimization approach is proved to be useful, when a higher number of particles is taken into account. The X-SPH variant improves the results (Fig. 2) even for a lower number of particles, but the time consumption is considerably higher.

The spatial problem of the crashing ball also proved the performance of the algorithm, but in the 3D problems, NNS is usually to be taken into account due to the high calculation time consumption.

VI. CONCLUSIONS

The computational system SPHCOFEM investigating the problems of fluid flow and its interaction with deformable boundaries was developed. The core of the package in C++ is the Lagrangean method for fluid flow simulation called smoothed particle hydrodynamics (SPH) that is coupled via the continuous contact model to the boundary modeled by the finite element method (FEM).

The algorithm presented is implemented as in-house software for investigation of fluid coupling to a deformable boundary in 1D, 2D or 3D. There is a graphical user interface (GUI) for reading the structured input text file describing the particle coordinates and initial conditions as well as the boundary definitions in the MATLAB environment.

The study shows both the advantages and disadvantages of using SPH coupled to FE and its exploitation for investigating real problems.

ACKNOWLEDGMENT

The work was co-financed by the internal research project SGS-2013-026 "Development of biomechanical models for medicine".

CONFLICT OF INTEREST

The author declares that he has no conflict of interest.

REFERENCES

1. Violeau D (2012) Fluid Mechanics and the SPH Method: Theory and Application. Oxford University Press
2. Gockenbach M S (2006) Understanding and Implementing the Finite Element Method. Society for Industrial and Applied Mathematics, Philadelphia
3. Hynčík L (2002) Smoothed particle hydrodynamics theory and practice. Engineering Mechanics 9:1–15
4. Gingold R A, Monaghan J J (1977) Smoothed particle hydrodynamics: theory and application to non-spherical stars. Monthly Notices of the Royal Astronomical Society 181:375–389
5. Lobovský L, Křen, J (2002) Smoothed particle hydrodynamics modelling of fluids and solids. Applied and Computational Mechanics 1:521–530
6. Ellero M, Serrano M, Español P (2007) Incompressible smoothed particle hydrodynamics. Journal of Computational Physics 226:1731–1752
7. Shahriari S, Hassan I G, Kadem L (2013) Modeling unsteady flow characteristics using smoothed particle hydrodynamics. Applied Mathematical Modelling 37:1431–1450
8. Hynčík L (2014) SPHCOFEM User's Manual. University of West Bohemia
9. Henneken E A C (1995) Smoothed Particle Hydrodynamics. Sterrewacht, Leiden
10. Bonet J, Lok T S L (1999) Variational and momentum preservation aspects of Smoothed Particle Hydrodynamic formulations. Computed Methods Apl. Mech. Engrg 180:97–115
11. Hynčík L (2002) Biomechanical model of abdominal organs and tissues for crash tests purposes. University of West Bohemia, Plzeň

Author: Luděk Hynčík
Institute: University of West Bohemia
Street: Univerzitní 8
City: Plzeň
Country: Czech Republic
Email: hyncik@ntc.zcu.cz

Blood Group Determination Using Vivado System Generator in Zynq SoC

Athmasri B. Krishnan and K.P. Peeyush

Amrita School of Engineering, Amrita Vishwa Vidyapeetham, Coimbatore, India

Abstract— **Blood grouping is an essential procedure in blood transfusions, donations and other emergencies. An error in grouping or a cross match can have adverse effect on the patient. Conventional way of blood grouping is performed manually on the basis of agglutination of blood in reaction with chemical reagents. The above method can lead to human errors while handling a large number of samples. Various systems have been developed to automate blood typing, but their excess testing time made it difficult during emergencies. This work emphasizes the FPGA implementation of an image processing algorithm to determine the blood group of a person. To do so, it uses plate test and image processing techniques using Vivado System Generator. Edge detection algorithm is performed in the image captured from the plate test. The number of edges determines whether the sample has agglutination or not. By comparing the number of edges in the image, blood group is classified. The proposed algorithm is implemented on a standard FPGA Development Board and its performance is evaluated by calculating accuracy of classification.**

Keywords— **Plate test, Agglutination, Simulink, System Generator, FPGA**

I. INTRODUCTION

Blood is the most important element in the human body which helps in the transportation of substances like nutrients, oxygen, metabolic wastes etc. It also helps in body regulation functions and system protection. Blood differs among humans in terms of its type or blood group. A blood group system is defined by the International Society of Blood Transfusion (ISBT) as the system where "one or more antigens controlled at a single gene locus or by two or more very closely linked homologous genes with little or no observable recombination between them"[1].

Blood type is determined by the presence or absence of certain protein molecules called antigens and antibodies in the blood. ABO is the most important blood group having A, B, AB and O as the blood groups. These groups will give the details of antigens present on the surface of Red Blood Cells (RBCs). Rh system is also important which determines the presence of "Rhesus factor" on the surface of RBCs.

Plate test or slide test is the common method to determine the blood group of a person. In plate test, antigen containing blood sample is exposed to commercially available chemical reagents having antibodies. These reagents are Anti-A, Anti-B and Anti-D. The result of plate test is interpreted by the occurrence or the non-occurrence of agglutination, which occurs due to the reaction of antigen and antibodies. The combination of agglutination in the reagents will determine the blood group of the sample. This process is done manually by technicians. Only a few numbers of samples can be tested by these methods [2]. While handling a large number of samples and during emergencies, the above method causes human errors. These errors can cause severe effects in patients during blood transfusion.

Different methods have been used to determine blood group automatically using spectrophotometric methods [3], mechatronic devices [4] having both mechanical and electronic parts, but by these methods, complexity increases and time taken for obtaining the result is more. Image processing techniques, especially segmentation can also be used to classify the blood groups [5].

This paper presents an image processing algorithm which is implemented on Zynq 7000 SoC to determine the blood group. The continuous growth in parallelism and functionality of Field Programmable Gate Arrays (FPGAs) in the last few years has contributed to an increasing interest in their use as an implementation platform for various image processing algorithms.

For simplified approach of hardware implementation, the Vivado System Generator is used. System Generator is a part of the Vivado system edition design suite. It is a high-level tool for designing high performance DSP systems using Xilinx All Programmable devices. It helps in automatic code generation of VHDL (VHSIC Hardware Description Language) and Verilog from Simulink. It quickly imports Vivado High-Level Synthesis IP for modeling with Simulink. It automatically generates IP and exports to the Vivado IP Catalog for easy design reuse and model sharing. System Generator supports JTAG communication between a hardware platform and Simulink [6]. This helps in easy implementation of the algorithm in hardware.

This paper is prepared as follows: Part II describes the materials and methods used, part III presents the results and discussions after the implementation of the proposed algorithm in hardware and part IV concludes the paper.

J. Goh and C.T. Lim (eds.), *7th World Congress on Bioengineering 2015*,
IFMBE Proceedings 52, DOI: 10.1007/978-3-319-19452-3_44

II. MATERIALS AND METHOD

Plate test is a method in which mixing of blood sample (antigens) with reagents (antibodies) is done to determine the blood group. Based on the reaction between antigen and antibodies, agglutination occurs in the plate test. Fig. 1 shows agglutination with reagents Anti-A and Anti-D and Fig. 2 shows agglutination with Anti-B and Anti-D

Blood samples of 30 individuals having 20-25 years age and with different blood groups were collected and plate test was performed. Image of plate test result was captured using a camera of 5 Mega pixel resolution. The proposed algorithm was applied to the images captured.

A Simulink model containing image processing techniques is designed in Matlab (version 2013a) with the help of Simulink library browser. One of the blocks provided in its library is System Generator, which is configured for Zynq SoC. The image pre-processing and the post-processing uses Simulink blocks from computer vision and DSP toolbox. Edge detection algorithm uses Xilinx reference blocks. The process is carried out in a computer system having Windows XP as the operating system, Intel Core^{TM} i5 processor @ 3.10GHz, system type is 64bit operating system and 4GB of RAM.

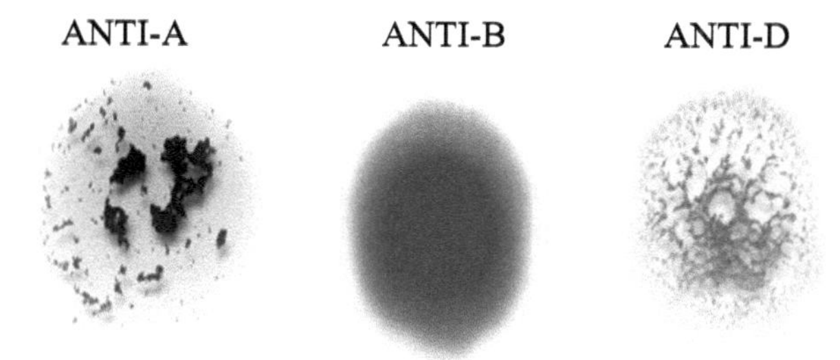

Fig. 1 Plate test for A+ blood group

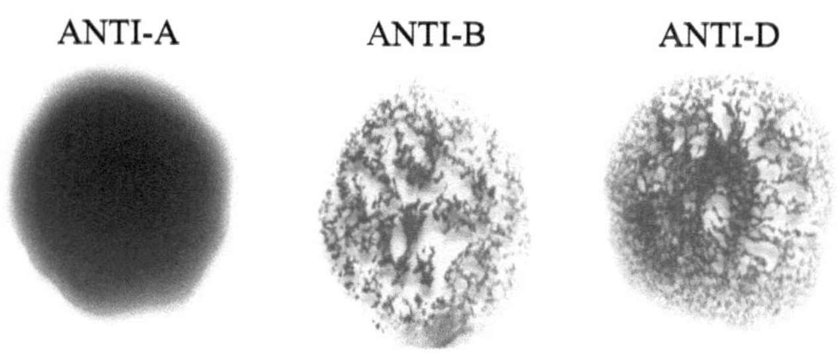

Fig. 2 Plate test for B+ blood group

The model designed is simulated with suitable simulation time and the code generated by System Generator is downloaded to Zynq board. Fig. 3 shows the general design flow of System Generator.

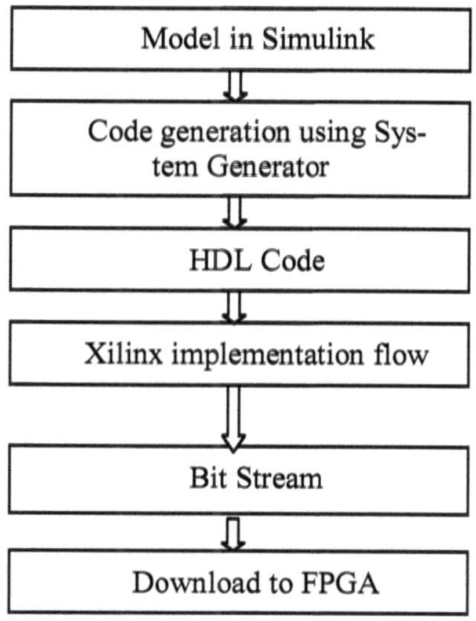

Fig. 3 Design flow of System Generator

A. Image Pre-processing

'Image from File' block reads the input image captured from plate test. From the color image, green plane alone is taken for processing [2]. The System Generator is not able to process the image in 2-D and for easy implementation, 1D is preferred. Thus 2-D to 1-D conversion is applied here.

B. Edge Detection using System Generator

The Xilinx Gateway In and Gateway Out are the boundaries for FPGA or the input and output to the Xilinx portion of Simulink design. So the proposed algorithm should be processed using Xilinx blockset, which is placed in between Gateway In and Gateway Out. Edge detection algorithm is applied to the image using Xilinx blockset.

Edges are a set of connected edge pixels, in which intensity of the image changes abruptly. When agglutination occurs in the plate test, there will be a variation in the intensity of the image. So the number of edges in the image will indicate the occurrence or non-occurrence of agglutination in the plate test [5].

Sobel edge detection algorithm is applied for the image taken. It is the most common method for edge detection of an image. Sobel operator calculates the gradient at each point of the image intensity and it gives the largest possible increase from light to dark and the rate of change in the above direction. It is a first order edge detection operator [7].

$$G_x = \begin{bmatrix} -1 & 0 & 1 \\ -2 & 0 & 2 \\ -1 & 0 & 1 \end{bmatrix} \quad G_y = \begin{bmatrix} 1 & 2 & 1 \\ 0 & 0 & 0 \\ -1 & -2 & -1 \end{bmatrix} \tag{1}$$

G_x and G_y are the convolution kernels for sobel operator which is sensitive to changes in x and y directions respectively. Gradient magnitude is given by:

$$G = \sqrt{G_X^2 + G_Y^2} \tag{2}$$

But in some cases, due to the computational burden for implementing squares and square roots, approximation of the gradient magnitude is given by its absolute values [7]. i.e

$$G \approx |Gx| + |Gy| \tag{3}$$

In Xilinx blocksets, 5x5 filter mask gives the coefficients for Sobel edge detection. It contains 5 n-tap MAC FIR filters, counters, addressable shift register, coefficient ROM, MAC engine and capture register [6]. These filters are clocked 5 times faster than input rate. The throughput of the design is 213 MHz/5 = 42.6 million pixels/second [6] .The values will change depending on the size of the image taken.

C. Image Post-processing

The Post-processing part has the data type converter block, which converts the data type of input image to required output data type. In the System Generator part i.e using Xilinx blocksets, only 1-D signal can be processed. To view the output, 1-D to 2-D conversion is done. Classification of blood group based on the edges detected is included in the post-processing subsystem.

Fig. 4 shows the whole model, containing both image Pre-processing and Post-processing along with the algorithm in Xilinx blocks.

Two video viewers are used to compare the edge detected image by Simulink simulation and to the edge detected image which is implemented on Zynq SoC. The System Generator token in the model is the block which generates the HDL code and it helps in cramming a large number of pixels to smaller spaces. This helps in implementing various image processing techniques for an image of larger size.

III. RESULTS AND DISCUSSION

The implementation of the proposed algorithm in Zynq SoC requires the conversion of entire model of edge detection (Xilinx blocks) to JTAG hardware co-simulation using Vivado System Generator (Fig. 4). In the System Generator token, Zynq xc7z020 and VHDL are selected as part (FPGA part to be used) and language respectively.

After generating the bit file, by providing specific parameters of clocking tab in System Generator, JTAG hardware co-simulation block is created.

From the edge detection, the count of edges is calculated in each region of the 3 reagents. If the count is more with respect to a fixed value, then it indicates the occurrence of agglutination in that particular region [5]. So depending on the count of edges and the sequence of agglutination occurred in the image, blood group is determined.

Fig. 5 shows the input image and the edge detected output image for the blood group A+. The Anti-A and Anti-D region is having more number of edges compared to Anti-B. Thresholding the image gives the real edges which comes in foreground and all other particles in the image goes to background. The output image shown below is the 'output image from FPGA' in Fig. 4.

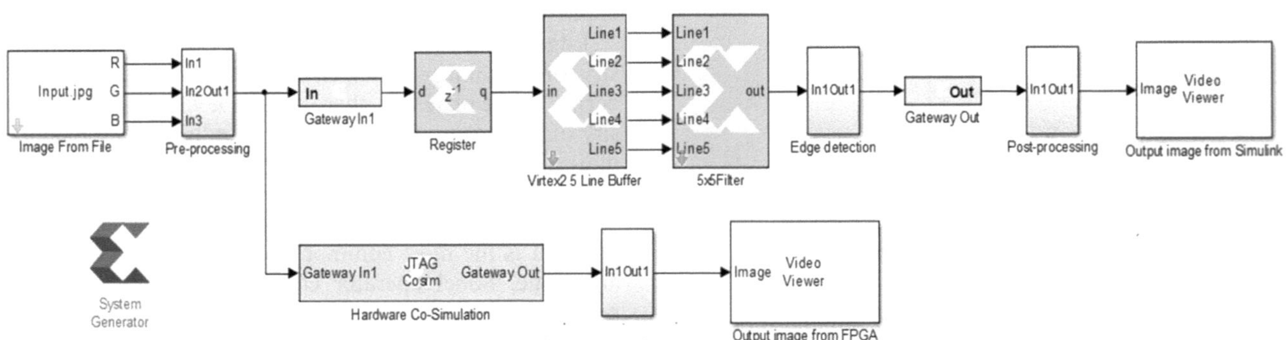

Fig. 4 Designed Simulink Model

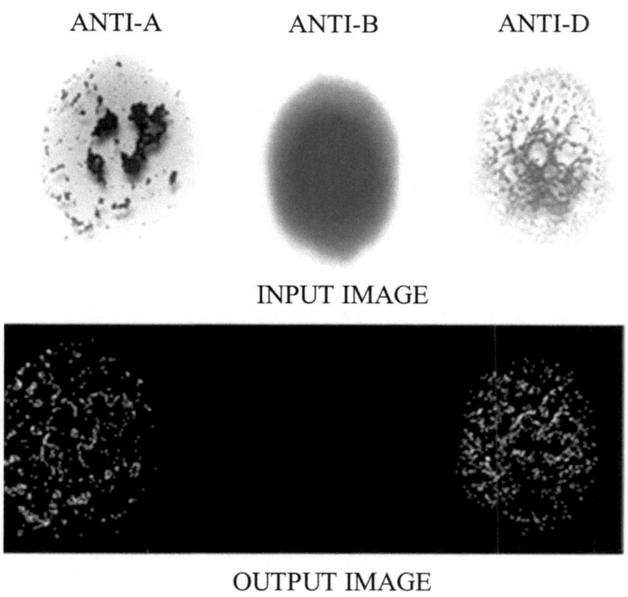

Fig. 5 For blood group A+

The proposed algorithm is applied to the 30 images captured. The hardware system achieved an accuracy of 96.6% in classifying the blood group for the above mentioned images when compared with the results obtained from the conventional techniques used in laboratories. The final hardware implementation requires only 239mW for the total on-chip power and it takes only 30 seconds for getting the results. Table 1 shows the utilization summary of hardware resources on the chip.

Table 1 Device utilization

Resource	Utilization	Available
Flip flops	837	106400
LUTs	479	53200
Memory LUT	45	17400
I/O	2	200
BRAM	6	140
DSP48	5	220
BUFG	4	32
MMCM	1	4

For the proposed hardware, the number of flip flops, memory LUT, LUTs and I/O utilizes only 1% of the resources. DSP48 utilizes only 2% and BRAM utilization is 4%.

IV. Conclusion

The hardware implementation of an image processing algorithm, which is used to determine the blood group of an individual, was presented. The proposed method uses Vivado System Generator for various image processing techniques. The implementation on Zynq 7000 consumes less power with minimum resource utilization. For future works, image acquisition can be done using onboard camera of the Zynq SoC and the proposed algorithm can be used to fully automate a portable device to determine the blood group.

Acknowledgment

The authors would like to thank Dr. N. Madhu Mohan of Biomedical Engineering Department at Amrita Vishwa Vidyapeetham, Coimbatore, for his valuable comments and suggestions to improve the quality of work done. The help rendered by Mr. G. Prakash of CoreEL Technologies, Banglore, for hardware implementation is also gratefully acknowledged.

References

1. ISBT International Society of Blood Transfusion, ‘Blood Group Terminology’ at http://www.isbtweb.org
2. A. Ferraz, V. Carvalho, F. Soares, and C. P. Leao, (2011), “ Characterization of blood samples using image processing techniques,” *Sensors Actuators A. Phys.*, vol. 172, pp. 308–314.
3. S. Pimenta, J. M. Nobrega, F. M. Duarte, G. Minas, and F. O. Soares, (2012) “Automatic Electronic System to Human Blood Typing,” *Procedia Eng.*, vol. 47, pp. 647–650.
4. K. Bezerra, A. Ferraz, V. Carvalho, F. Soares, T. Romanian, and R. Precision, (2012), “Advanced design of a mechatronic system for human blood typing,” *4th Int. Conf. Innov. ,recent Trends Mechantronics,Mechanical Eng. New High-Tech Prod. Dev* ,Bucharest, Romania, (2012), no. 41, pp. 1–7.
5. T. Zarifi and M. Malek, (2013) “FPGA implementation of image processing technique for blood samples characterization,” *Comput. Electr. Eng.*, vol. 40, no. 5, pp. 1750–1757.
6. Xilinx, (2010) “Xilinx System Generator for DSP Reference Guide,” *Interface*, vol. 638, pp. 1–504.
7. R. Gonzalez, R. E. Woods, (2008), Digital Image Processing, Pearson Education, New Jersey.

Author: K.P. Peeyush
Institute: Amrita School of Engineering
Street: Ettimadai
City: Coimbatore
Country: India
Email: peeyushkp@gmail.com

The Design of Interactive Physical Game for Cognitive Ability Detecting for Elderly with Mild Cognitive Impairment

Y.T. Chen[1], C.J. Hou[1], M.W. Huang[2], J.H. Dong[1], J.Y. Zhou[3], and I.C. Hung[1]

[1] Department of Electrical Engineering, Southern Taiwan University of Science and Technology, Tainan, Taiwan
[2] Deputy Superintendent, Bali Psychiatric Center, Ministry of Health and Welfare, Bali, Taiwan
[3] Asia University, Taichung, Taiwan

Abstract—**The problem of population aging is getting worse in many countries. Taiwan will become an aged society in 2017. Alzheimer's disease is the most common form of dementia in the Elderly population. The pathogenesis of Alzheimer's disease at the early stage usually starts slowly and gets worse over time. The preclinical stage of the disease has also been termed mild cognitive impairment (MCI). Detailed neuropsychological testing reveal mild cognitive difficulties and questionnaires are popularly used in clinics. Those procedures are hard to use by the families and health caregiver. A way of ease using for cognitive ability detecting at home should be considered. Therefore, the interesting interactive physical game with body motion sensing device has been developing in this study. The body motion game not only promote the physical and mental health, but also supervise the behavior of player. The behavior information recorded by the body motion sensor device can be used to evaluate the cognitive status of the elderly who play the game. Three major works in this study have been performing: (a) the diverse easy-playing interactive physical game with body motion sensing device has been designing, the parameters for featuring the behavior of player are recorded, (b) the face recognition using computer vision techniques was implemented for identifying the player, (c) the management of the user authorities for accessing player's historical records and personal information. The performance of the proposed system has evaluated by seven psychiatrists, the statistical results showed that the player's cognition statuses can be revealed by the behavior features and game records. This system were used on the spot of some healthcare institute and residential communities. The statistical results showed that the system has good performance for detecting the cognitive abilities of elderly.**

Keywords—**Dementia, Cognition Ability, Body-motion Sensor**

I. INTRODUCTION

Population aging in developing countries become a very serious problem. Because of rising life expectancy and declining birth rates, the proportion of aged population rise acutely. The issues about population aging are getting important. Taiwan will become an aged society in 2017. The proportion of aged population will be higher than 20% by 2025 and Taiwan will be a super-aged society.

In Taiwan, the proportion of population with dementia in the aged population occupied 8.09% in 2013. Symptoms of dementia vary depending on the cause and the area of the brain that is affected. The progressive decline in cognitive function due to damage or disease in the brain beyond what might be expected from normal aging. Alzheimer's disease is the most common form of dementia in the elderly population. The pathogenesis of Alzheimer's disease at the early stage usually starts slowly and gets worse over time. The preclinical stage of the disease has also been termed mild cognitive impairment (MCI). The memory loss is usually the earliest and most noticeable symptom and always is associated with the decline of other cognitive abilities. The symptoms are not only the decline of memory, the affected cognitive abilities include attention, judgement, language and problem solving [1]. Trouble controlling moods and behaviors, personality changes, and having delusion are very possible occurred. To take care of the patient with dementia is a heavy burden to the family and healthgiver.

Detailed neuropsychological tests reveal mild cognitive difficulties and questionnaires are popularly used in clinics. The professional psychologist always apply the mini-mental status examination (MMSE) [2], cognitive ability screening instrument (CASI) [3], and clinical dementia rating (CDR). These tests are very important reference for the diagnosis of dementia. These procedures of using questionnaires and tests are hard to practice by the families and health caregiver in usually daily lives. A way of easy use for cognitive ability detecting at home should be considered.

There are novel techniques developed to detect the symptomatic appearance of dementia. B. Dubois [4] devised a short bedside cognitive and behavioral battery to assess frontal lobe functions. Che [5] extracted the featured qEEG variables from resting eyes-closed continuous EEGs. The results showed statistically significant difference between early dementia group and control subject group. Waragai [6] evaluated the diagnostic value of brain perfusion SPECT using eZIS in patients with various neurodegenerative diseases at a very early stage, within one year from onset. Huang [7] developed several systems including visual, audio, and memory tests. Those systems were designed on

J. Goh and C.T. Lim (eds.), *7th World Congress on Bioengineering 2015*,
IFMBE Proceedings 52, DOI: 10.1007/978-3-319-19452-3_45

Tablet PC for elderly to self evaluate their health status. However, the above-mentioned systems were not easy use in family for homecare of elderly daily lives.

Developing a game system with the body-motion sensor and applying the function of cognitive ability detection in the system may be a useful way for elderly to detect the MCI in the preclinical stage. It can be like an execise or entertainment to use in the daily life. Instead of the boring process of questionnaires and tests, the game can be performed every day to trace the variations of cognitive ability. The Dual-Task Tai Chi designed by Kayama [8] and the KiMentia recognition system proposed by Breton[9] could be the representative systems which combined the functions of exercise/mutimedia and cognitive ability detecting.

Therefore, the interesting interactive physical game with body motion sensing device was developed in this study. The body motion game not only promote the physical and mental health, but also supervise the behavior of player. The behavior information recorded by the body motion sensor device can be used to evaluate the cognitive status of the elderly who play the game.

II. MATERIALS AND MATHEDS

Three major works in this study have been developed: (A) The design of interactive physical games. Five diverse easy-playing interactive physical games with Kinect, the body motion sensing device, has been designed. The parameters for featuring the behavior of the player were recorded. (B) The player identification implemented the face recognition techniques for identifying who the player is. This procedure can ease the operation steps of the system for elderly and facilitate the access of personal information. (C) The management of the user authorities was designed for accessing player's historical records and personal information (Fig. 1).

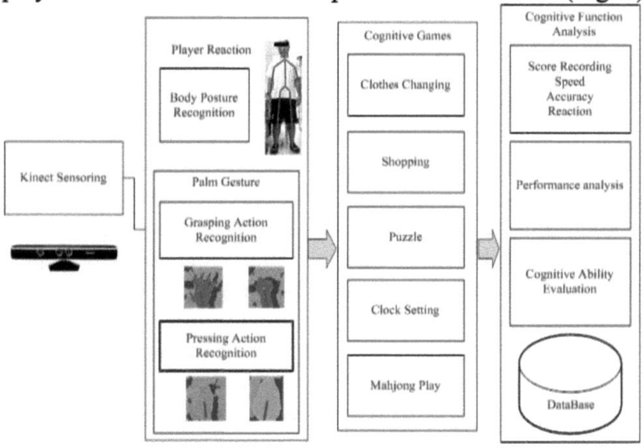

Fig. 1 System block diagram

A. *Interactive physical game design*

This study adopted Kinect, the motion-sensing device, to develop an interactive motion-sensing game for cognitive ability detecting. The player controlled and interacted with the game through a natural way using hand gestures. These games were designed referring to the specific cognitive functions. The performance in the playing history of the player would be recorded and analyzed for cognitive ability evaluation.

Cognition function can be divided into domains of ability including attention, memory, orientation, calculation, construction ability, and language and other higher cognitive functions [10]. Five games were designed in the proposed system to evaluate the cognitive abilities. The parameters, including the accuracy ratio, the time taken to complete the game, and the number of right answer, for every game need to be recorded for the subsequent cognitive ability evaluation. These five games are briefly explained as follows and the main operation screens are shown in Fig. 2.

a) Cloths changing:

The hat, coat, and pants of three kinds of clothes including a uniform, casual dress, and athletic suit can be selected individually from the figure items on the screen. The taken time, the accuracy are recorded.

b) Shopping:

The subject is requested to answer the hypothetical questions about shopping to subtract 7 dollars from 100 dollars, and to keep subtracting 7 dollars from the result. The taken time and the accuracy ratio are recorded.

c) Puzzle:

The subject is required to put pieces of a picture together to compose the picture. Various pictures with different pieces can be selected. The recorded parameters include selected pictures, the taken time, and the accuracy ratio.

d) Clock Setting:

The subject is required to recognize what time the clock is and then select the correct time from the lists in the bottom of screen. The figure of the clock might be pointer-type or numeric-type. The recorded parameters include the mode of clock, taken time, and the accuracy ratio.

e) Mahjong Play:

The player selects the tile of mahjong game from four tiles listed at the below of the screen to complete the order of rules. The accuracy ratio and taken time are saved.

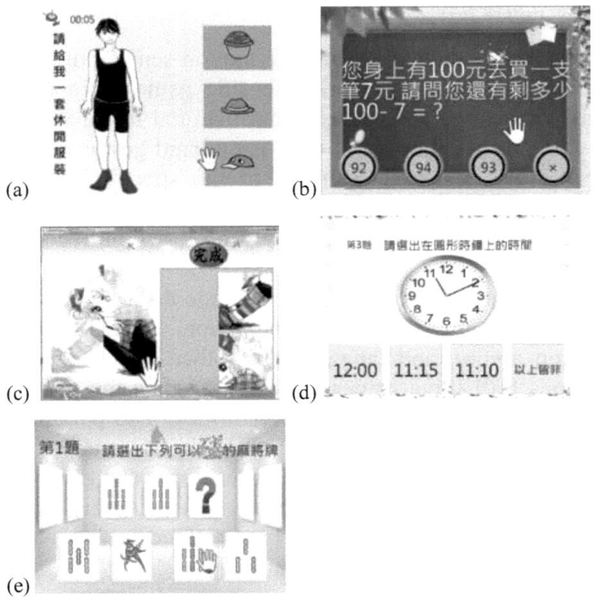

Fig. 2 The five games of the proposed system (a) Cloths changing, (b)Shopping, (c) Puzzle, (d) Clock Setting, and (e) Mahjong Play.

B. Player identification

The Kinect could estimate the body posture and transfer the matched skeleton model to the system. It is very easy to segment the region of face from the synchronous video image. This user identification procedure in the proposed system applied the Scale-invariant feature transform (SIFT) proposed by Lowe [11] to identify the player.

The SIFT algorithm can be applied to the topics of object recognition, video tracking, image matching, etc. SIFT consists of four key stages including 1) *scale-invariant feature detection*: the image is trasformed into a large collection of feature vectors which is invariant to image traslation, scaling, and rotation, 2) *feature matching and indexing*: the consists of storing SIFT keys is indexed and then identify the matching keys from the new image, 3) *cluster identification*: Hough Transform is used to cluster reliable model hypotheses to search for keys that agree upon a particular model pose, and 4) *model verification*: the identified cluster is then subject to a verification procedure in which the solution of linear least squares is evaluated for the parameters of the affine transformation relating the model to the image.

C. Data management

The performance and playing historical records of player would be recorded. The authority management of this proposed system included the administrator (highest), the clinical staff, and personal user and his/her family. The clinical staff can trace and access the information and historical game-play records of players. These records are used to evaluate cognitive ability of the player.

D. Experiments and Statistical analysis

Seven psychiatrists participated in evaluating the correlation between the games and cognitive function which includes attention, memory, orientation, calculation, construction ability, and language and other higher cognitive functions. Totally 168 questionnaire items were designed for surveying these correlations. The correlation scale ranked by psychiatrists was low, moderate or high. The cronbach's alpha [12] is used as an estimate of the reliability or internal consistency of the games. The rule for describing internal consistency using cronbach's alpha, α, is: $\alpha \geqq 0.9$ represented 'excellent', $0.7 \leqq \alpha < 0.9$ represented 'good', $0.6 \leqq \alpha < 0.7$ represented 'acceptable', $0.5 \leqq \alpha < 0.6$ represented 'poor', and $\alpha < 0.5$ represented 'unacceptable'. The α value is represented as

$$\alpha = \frac{n}{n-1}\left(1 - \frac{\sum s_i^2}{s_H^2}\right)$$

where s_i^2 is the variance of the *i*th items, s_H^2 is the variance of the sum of all items, and n is the number of items. Fleiss' kappa [13] is used for accessing the reliability of agreement of the games among seven psychiatrists. The degree of agreement interpreted by the kappa value, κ, includes poor ($\kappa < 0$), slight (κ = 0.01~0.20), fair (κ = 0.21~0.40), moderate (κ = 0.41~0.60), substantial (κ = 0.61~0.80), and almost perfect agreement (κ = 0.81~1.00).

Twenty subjects participated in the game. They accepted an evaluation of MMSE before playing the game. Eleven subjects showed normal in cognitive function and nine subjects had mild cognitive impairment. Each subject had completed the test of playing five games. The parameters such as the accuracy ratio, the time taken to complete the game, and the number of right answer of playing each game were recorded in data management and used for statistically analyzing the correlation between the game and cognitive function. The relationship between the parameters of the game (independent variables) and the MMSE score (dependent variable) was investigated by a stepwise regression analysis method. The multiple regression equation is assumed

$$y = \beta_0 + \beta_1 x_1 + \beta_2 x_2 + \cdots + \beta_k x_k$$

where y is the dependent variable, x_k is the kth independent variables, and β_k is the coefficient of the kth independent variables. The statistical analysis was performed using the Statistical Package for the Social Sciences (SPSS) statistical software for Windows. The study had the approval of Institutional Review Board in Taiwan.

III. Results and discussion

The results of cronbach's alpha analysis showed that α value was 0.961. It revealed the reliability of the games was excellent. The Fleiss' kappa value was 0.50, which meant the reliability of agreement of the games among seven psychiatrists was moderate. The result of the regression analysis obtained the following equation ($R^2 = 0.87$ and F=34.52 (p<0.05) for checking the goodness of fit of regression equation model)

$$y = 10.183 + 0.166x_1 - 0.15x_2 + 0.47x_3$$

where three game parameters, 'x_1, x_2 and x_3' were the number of right answer, the time taken to complete the game and the accuracy ratio. The results showed that the cognitive ability detection in our games was highly correlated with the MMSE.

IV. Conclusions

Five body-sensing game using Kinect for cognitive ability detecting were developed in this study. The performance of the proposed system has evaluated by seven psychiatrists, the statistical results showed that the player's cognition statuses can be revealed by the behavior features and game records. This system were used on the spot of some healthcare institute and residential communities. The statistical results showed that the system has good performance for detecting the cognitive abilities of elderly.

Acknowledgment

The authors acknowledge financial supports for this research from the Ministry of Science and Technology in Taiwan (Grant No. MOST 103-2221-E-218 -010).

References

1. Ambrose CT (2013) Alzheimer's Disease. Am Sci 101(3): 200
2. Folstein MF, Folstein SE, McHugh PR (1975) "Mini-mental state". A practical method for grading the cognitive state of patients for the clinician. J Psychiat Res 12(3): 189–98
3. Teng EL, Hasegawa K, Homma A et al. (19994) The cognitive abilities screening instrument (CASI): a practical test for cross-cultural epidemiological studies of dementia. Int Psychogeriatr 6(1):45-58
4. Dubois B, Slachevsky A, Litvan I et al (2000) The FAB A frontal assessment battery at bedside. Neurology 55:1621-1626
5. Che H, Jung YJ, Im CH et al (2007) Extraction of qEEG Variables to Diagnose Early Dementia. Conf Proc IEEE Eng Med Biol Soc, 2007, pp 23-26
6. Waragai M, Yamada T, Matsuda H (2007) Evaluation of brain perfusion SPECT using an easy Z-score imaging system (eZIS) as an adjunct to early-diagnosis of neurodegenerative diseases. J Neurol Sci 260:57-64
7. Huang YP, Hsin TJ, Liu SI (2013) Design and Implementation of Cognitive Examination Systems on Tablet PC. CME Proc. Int Conf on Complex Med. Eng., Beijing, China, 2013, pp 25-28
8. Kayama H, Nishiguchi S, Yamada M et al (2013) Effect of a Kinect-based exercise game on improving executive cognitive performance in community-dwelling elderly. 7th Int Conf on Pervasive Computing Technologies for Healthcare, pp 362-365
9. Breton ZS de U, Zapirain BG, Zorrilla AM (2012) KiMentia:Kinect based tool to help cognitive stimulation for individuals with dementia. 14th Int Conf on e-Health Networking, Applications and Services (Healthcom), pp 325-328
10. Sheehan B (2012) Assessment scales in dementia. Ther Adv Neurol Disord 5(6): 349–358
11. Lowe DG (2004) Distinctive image features from scale-invariant keypoints. Int J Comput Vision 60(2):1150–1157
12. Cronbach LJ (1951) Coefficient alpha and the internal structure of tests. Psychometrika 16(3):297-334
13. Fleiss JL (1981) Statistical Methods for Rates and Proportions, 2nd edition, John Wiley & Sons.

Author: Chun Ju Hou
Institute: Department of Electrical engineering, Southern Taiwan University of Science and Technology
Street: No. 1, Nan-Tai Street, Yungkang Dist.
City: Tainan City
Country: Taiwan
Email: cjhou@mail.stust.edu.tw

A Quantitative Model for Sepsis Stratification

Jing Xia[1], Min Zhu[1], Shengyu Zhang[1], Molei Yan[2], Guolong Cai[2], Jing Yan[2], and Gangmin Ning[1]

[1] Department of Biomedical Engineering, Zhejiang University, Hangzhou, China
[2] Zhejiang Hospital, Hangzhou, China

Abstract— **Sepsis is a kind of systemic inflammatory response syndrome caused by infection and it endangers the life of patients seriously due to its rapid development progression and high mortality rate. In clinic it is highly demanded to quantitatively stratify the severity of sepsis for individual management. This work aimed to build a quantitative model for sepsis patients which can stratify the disease severity in three levels. For this purpose, clinical data were collected and preprocessed, i.e. screening, normalization and data replenishing. Afterwards, sepsis sensitive parameters were tested and selected, which were utilized as the input of the stratification model. For the model, the algorithm of Support Vector Machine was applied. Eventually, the model was tested in total of 522 clinical cases and an accuracy of 67.5% in stratification was achieved. The performance of the established model is superior to the conventional APACHE scoring method. Preliminary results exhibited that the established model is potential to help improve the patients' management by quickly stratifying the sepsis severity.**

Keywords— **sepsis, stratification model, support vector machine, APACHE scoring**

I. Introduction

Sepsis is a kind of systemic inflammatory response syndrome caused by infection[1]. It has become one of the worldwide healthcare problems due to its rapid development progression and high mortality rate, which is reported as 28.7-49.7%[2,3].

Sepsis is a highly dynamic illness and to quickly assess individuals' sepsis severity is critical for reducing the mortality. Thus, it is of great significance to develop a quantitative approach to identify high risk patients[3,4] and make a quick stratification of sepsis severity. Presently, several severity scoring systems have been used in clinic, such as Acute Physiology and Chronic Health Evaluation II (APACHEII). However, these methods are complicated and impractical[4].

This work aims to build a physiological parameters based quantitative model for sepsis patients that can stratify the severity in three levels.

II. Materials and Methods

The severity stratification process includes three steps. Firstly, statistical analysis of clinical parameters is conducted to screen severity level related features. Secondly, data normalization and replenishing is introduced. Finally, two-layer stratification model is built[5].

A. Data source

Totally 522 sepsis patients were recruited between June 2012 and January 2013. All patients were informed consent to participate in the study.

The patients enrolled had sepsis syndrome with at least one organ dysfunction or hypoperfusion and were divided into three severity levels. The definitions of the three levels are as follows: level one is sepsis plus sepsis-induced organ dysfunction or tissue hypoperfusion; level two is sepsis-induced persisting hypotension despite adequate fluid resuscitation; level three is multi organ dysfunction. The cohort sizes of three levels are 131, 240 and 151, respectively.

Clinical data contain personal information (age, gender, height, weight, etc.), history of disease (diabetes, hypertension, chronic lung disease, blood disease, cancer, cerebrovascular accident, etc.), physical symptoms and examinations (fever, chills, temperature, heart rate, respiratory rate, mean arterial pressure, etc.) and laboratory tests.

B. Data preprocess

First, clinical parameters were statistically tested and screened. Kolmogorov-Smirnov method was carried out to test the normality of continuous data. For attributes in normal distribution, independent Student's t test was applied to analyze the differences among the groups, otherwise, Wilcoxon rank sum test was used. Besides, the inter-group differences of categorical data were analyzed using Chi-square test. For all statistical tests, a p-value of <0.05 was considered significant.

Selected attributes were then scaled through the min-max normalization so that they fell in the identical range [0, 1]. In order to retain the most information of data, attributes

J. Goh and C.T. Lim (eds.), *7th World Congress on Bioengineering 2015*,
IFMBE Proceedings 52, DOI: 10.1007/978-3-319-19452-3_46

with missing data were replenished using the mean imputation method.

C. *Model development*

Support Vector Machine (SVM) was used to construct the model. As one of machine learning algorithms, SVM is based on statistical learning theory and uses the principle of structural risk minimization[6,7]. Based on LIBSVM[8], parameters are adjusted to build the most satisfied models.

Multi-layer stratification model is applied in the study as shown in Fig. 1. In the first layer, the samples are distinguished into cohorts of severity level one and none one. In the second layer, the none one subgroup is identified as severity level two or three. Consequently, as output, patients are stratified into severity level one, two and three.

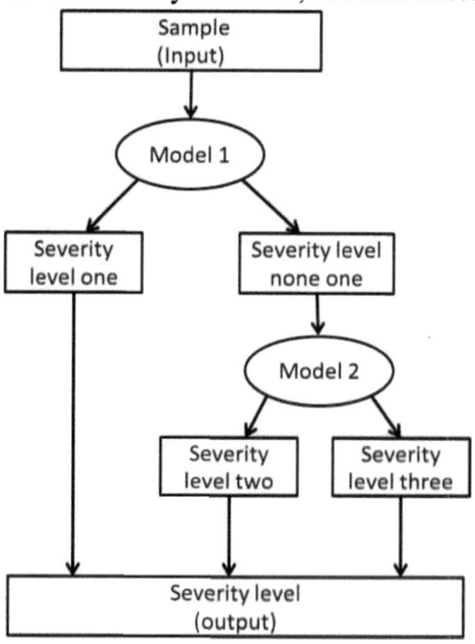

Fig.1 Schematic diagram of the two-layer stratification model for sepsis severity

D. *Model Evaluation*

The validity of the model is tested by the accuracy in predicting patients with different severity levels. Also, the results are compared with the APACHE scoring system[9].

III. RESULTS

A. *Dataset Characteristics*

Table 1 gives the characteristics of the study cohort. The average age of the patients was 63.5 years. The percentages of male and female were 64.0% is and 36.0%, respectively. For the patients, hypertension was the most common disease (63.4%) followed by cerebrovascular accident (15.9%) and diabetes (15.5%).

Table 1 Characteristics of the study cohort

	N(%)
Age(mean±sd)	63.5±19.0
Gender	
Male	334(64.0%)
Female	188(36.0%)
History of disease	
Diabetes	81(15.5%)
Hypertension	190(36.4%)
Chronic lung	72(13.8%)
Blood disease	9(1.7%)
Cancer	41(7.9%)
Cerebrovascular accident	83(15.9%)

B. *Selected Attributes*

For the model 1, 32 attributes were selected, including personal information (age, height), history of disease (cerebrovascular accident), physical symptoms and examinations (fever, dyspnea, obnubilation, cough, hemoptysis, cyanosis, temperature, heart rate, respiratory rate, mean arterial pressure, urine volume) and laboratory tests (triglycerides, arterial PH, direct bilirubin, intravenous PH, albumin, PaO_2, BUN, BE, Cr, SaO_2, Na^+, FiO_2, Cl^-, $PaCO_2$, Plt, PCO_2, PCT, blood lactate). These attributes showed significant difference between severity level one and else group.

For the model 2, 26 attributes were chosen, including personal information (age), history of disease (connective tissue disease), physical symptoms and examinations (chills, anorexia, obnubilation, clammy skin, cough, wheeze, bloating, mean arterial pressure, urine volume) and laboratory tests (direct bilirubin, intravenous PH, ALT, SaO_2, AST, PaO_2, albumin, BE, BUN, blood lactate, CTNI, WBC, Plt, PCT). These attributes showed significance between severity level two and level three group.

C. *Model Prediction*

In all, 371 samples were chosen as training set, and the rest 151 samples as testing set. In the testing set, the sample sizes of severity level one, two and three were 35, 76 and 40, respectively.

The prediction results of model 1 are listed in Table 2. The accuracy, sensitivity and specificity of model 1 were 67.5%, 60.0% and 89.7%, respectively.

The prediction results of model 2 are given in Table 3. The accuracy, sensitivity and specificity of the model 2 were 79.3%, 81.6% and 75.0%, respectively.

Generally, the accuracy of the overall model was 67.5%. And the accuracy of severity level one, two and three was 60.0%, 71.1% and 67.5%, respectively.

As for misclassified samples, there are six different types of misclassification, which are summarized in Table 4.

Table 2 Prediction result of model1

	True (level one)	False (level none one)
Predicted true (level one)	21	12
Predicted false (level none one)	14	104

Table 3 Prediction results of model 2

	True (level two)	False (level three)
Predicted true (level two)	62	10
Predicted false (level three)	14	30

Table 4 Misclassifications of the stratification model

Type of misclassification	Sample size (proportion)
Level 1(real)→Level 2(predicted)	9(6.0%)
Level 1(real)→Level 3(predicted)	5(3.3%)
Level 2(real)→Level 1(predicted)	8(5.3%)
Level 2(real)→Level 3(predicted)	14(9.3%)
Level 3(real)→Level 1(predicted)	4(2.6%)
Level 3(real)→Level 2(predicted)	9(6.0%)

D. Comparison with the APACHE model

Considering the APACHE scores, 19.5 and 24.5 were set as cutoff for classifying the patients into severity level one, two and three. Namely, patients with APACHE score lower than 19.5 points, between 19.5 and 24.5 points, higher than 24.5 points were identified as severity level 1, 2 and 3, respectively.

Only 68 samples were involved in the APACHE scoring for comparison as it requires strict completeness of data.

The accuracy of APACHE classification in severity level one, two and three was 62%,61% and 33%, respectively. In contrast, the corresponding accuracy resulted from the two-layer stratification model was 71%,68% and 78%, respectively.

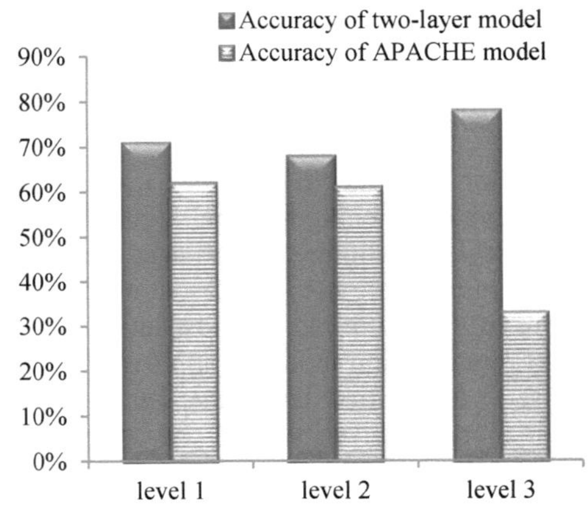

Fig.2 Comparison of the accuracy of APACHE model and two-layer stratification model

IV. Discussion

The prediction results of model 1 and model 2 showed that the first layer model is less sensitive and more specific, while the sensitivity and specificity of the second layer model are almost equal.

From Table 4, it can be seen that the largest proportion of misclassification was the type that level two cases were falsely predicted as level three, followed by level one being falsely predicted as level two and level three being falsely predicted as level two. It indicates that misclassifications are more likely to occur in neighbor levels. It can be interpreted as that these samples have similar clinical symptoms which make it difficult to distinguish them.

As shown in Fig. 2, the two-layer stratification model has a better discrimination than the APACHE model. The APACHE scoring system was designed for various kinds of severe diseases in clinic, while the stratification model was established upon sensitive parameters to sepsis severity. In this aspect, it is not surprised that the stratification model is superior to the APACHE model.

V. Conclusions

The present work proposed an SVM based two-layer stratification model for sepsis. The patients are identified as severity one, two and three with a general accuracy of 67.5%. Compared to the conventional method of APACHE, it has the advantages of better discrimination, more objective and lower requirement for data completeness. In conclusion, the established model has the potency to provide a computational tool for sepsis stratification and help improve the patients' management.

ACKNOWLEDGMENT

This work is supported by the National Nature Science Foundation of China (Grant 81271662), the Department of Science and Technology of Zhejiang Province (Grant 2011R50018), and the Ministry of Health of China (Grant 201202011).

CONFLICT OF INTEREST

The authors declare that they have no conflict of interest.

REFERENCES

1. Levy M M, Fink M P, Marshall J C, et al. 2001 SCCM/ESICM/ACCP/ATS/SIS International sepsis definitions conference. Intensive care medicine. 2003; 29: 530-538.
2. Dellinger R P, Levy M M, Rhodes A, et al. Surviving Sepsis Campaign: International Guidelines for Management of Severe Sepsis and Septic Shock, 2012. Intensive care medicine. 2013; 39: 165-228.
3. Vorwerk C, Loryman B, Coats T J, et al. Prediction of mortality in adult emergency department patients with sepsis. Emergency Medicine Journal. 2009; 26.4: 254-258.
4. Chen C, Chong C, Liu Y, et al. Risk stratification of severe sepsis patients in the emergency department. Emergency medicine journal. 2006; 23.4: 281-285.
5. Fu X, Ren Y, Yang G, et al. A computational model for heart failure stratification. proceedings of the Computing in Cardiology, 2011.IEEE:2011;385-388.
6. Cortes C, Vapnik V.Support-vector networks. Machine Learning. 1995; 20.3: 273-297.
7. Burges C J C. A tutorial on support vector machines for pattern recognition. Data Mining and Knowledge Discovery. 1998; 2.2: 121-167.
8. Chang CC, Lin CJ. LIBSVM: a library for support vector machines. ACM Transactions on Intelligent Systems and Technology. 2011; 2.3: 1-27.
9. Knaus W A, Draper E A, Wagner D P, et al. APACHE II: a severity of disease classification system. Critical care medicine. 1985; 13.10: 818-829.

Author: GangminNing
Institute: Department of Biomedical Engineering, Zhejiang University
Street: Zheda Road 38, Yuquan Campus, Hangzhou,P.R.China
City: Hangzhou
Country: China
Email: gmning@zju.edu.cn

Medical Device to Tackle Neurogenic Bladder

Akriti Chadda, Manish Thigale, and Preethika Britto

Biomedical Engineering Division, School of BioSciences and Technology, VIT University, Vellore, India

Abstract— **Almost one in twenty adults experience urinary incontinence problems according to major test surveys across all the populations irrespective of geography, cultures or backgrounds (Source: rightdiagnosis.com on 9th March 2015). While most of us take the blessing of voluntary voiding of our urinary bladder as a given, there are patients around the world who are not sensitive to the filling up of their bladder or its involuntary leakage. Thus, incontinence has a major impact on the quality of all patients leading to stress and embarrassment apart from other biological problems such as Urinary Tract Infections. Thus, to tackle this problem we have devised an innovative device that not only helps the patient concerned to track how much his/her bladder has filled up, but also helps in voluntary emptying of the same. The device, hence, has three major components, namely: 1) The sensor, 2) The Display and 3) The Micturition Control.**

The sensor circuit will help to convert the volume expansion of the urinary bladder into a resistance value. This value will keep on changing as the sensor bends and will be used as an input to the microcontroller that will control the display panel. The display panel works such that the changes seen are directly proportional to the volume expansion of the bladder. The sensor and the display are connected wirelessly.

As such, the patient can get a clear indication of the filling up of the bladder. The micturition control part of the device consists of a micro-pump around the sphincter that gets activated when the patient presses a switch. The pump deflates to let the patient urinate and inflates again once the bladder is empty.

This device thus hopes to make the urinary incontinence patients more independent and ready to face the world without any feelings of shame.

Keywords— **Neurogenic Bladder, Incontinence, Innovation, Biomedical, Paraplegics.**

I. Introduction

Paraplegia refers to the complete (-plegia) loss of voluntary motor function in the pelvic limbs. Paraplegia generally results from any spinal cord lesions which may have been caused to the second thoracic spinal cord segment. The limbs may be affected equally; however, the asymmetric lesions cause even greater clinical involvement on the ipsi-lateral side.[1]

Urinary incontinence often occurs concomitant with paralysis. [1] Because of the complex nature of the neural control of the lower urinary tract, paraplegic patients generally complain about bladder problems. The main neural circuits controlling the two functions of the bladder—that is, voiding and storage—are trans-spinal so that intact cord connections between the pons and the sacral segments are necessary to sustain physiological control. Furthermore, input from the higher centers is critical in the assessment of appropriate timing of voiding. There are many types of cortical diseases that can affect the centers involved with this. In addition to the spinal pathways and input from higher centres, the peripheral innervation to the bladder is through the pelvic plexus, sacral, and pudendal nerves. [2]

Hence, it is pertinent to note that a need of a medical device is highly necessary which can not only help paraplegic patients with the voiding of the urinary bladder, but also notify them about the filling up of the same.

The currently available medical devices in the market that tackle incontinence such as pessaries, catheters, artificial sphincters and urinary inserts solve one part of the problem – the urinary micturition control. However, besides the problems of Urinary Tract Infections (UTIs) and involuntary micturition, one of the most adverse side-effects of a neurogenic bladder, and hence incontinence is the inability of the patients to know when they need to void their bladder. This leads to continuous embarrassment and make the patients more vulnerable to depression than they already are.

The main aim of this medical device is to innovation in terms of not only micturition control, but also measuring the filling up of the urinary bladder and communicating the same to the patient. In this paper, the proof of concept of this medical device and its complete working has been described.

II. Method

A. Design Approach for the device

The proposed medical device to tackle neurogenic bladder must be able to perform three necessary functions: a) Sense the filling up of the Urinary Bladder, b) Convey the filling up of the bladder to the patient and c) Help the patient control the micturition as and when needed.

J. Goh and C.T. Lim (eds.), *7th World Congress on Bioengineering 2015*,
IFMBE Proceedings 52, DOI: 10.1007/978-3-319-19452-3_47

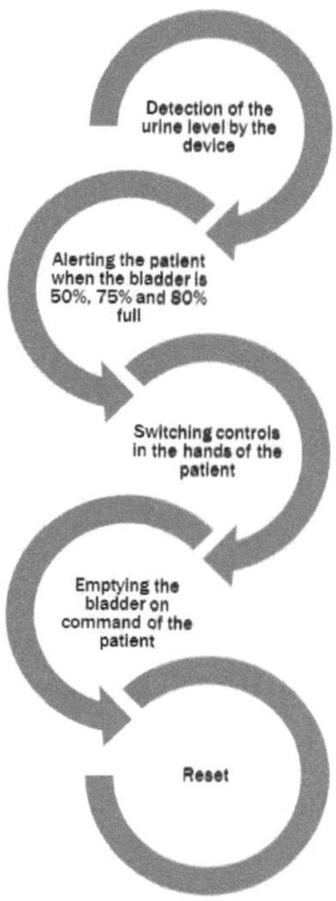

Fig. 1Work Flow of the proposed device

Thus it suffices to say that the medical device has been divided into three broad areas in order to achieve the targeted functionalities. The device can thus be studied under three main titles, namely the a) The Sensor, b) The Display and c) The Micturition control.

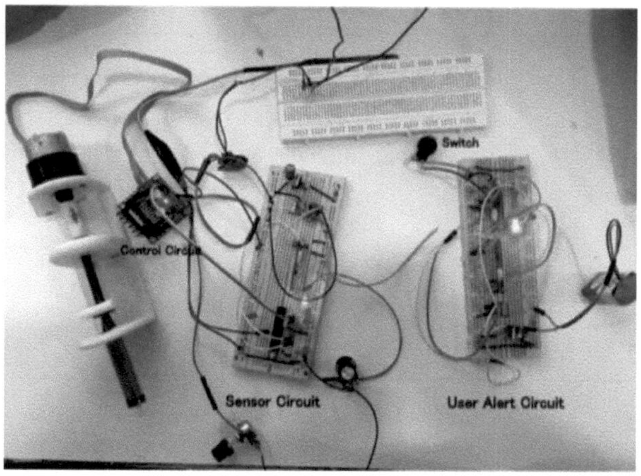

Fig. 2 Working Circuit

B. The Sensor Circuit

The sensor used in our concept is a simple flex sensor which is 2.2" in length. The working of a flex sensor depends on its curvature; as the sensor is flexed, the resistance across the sensor increases.

When the sensor is placed on the urinary bladder, the volume expansion of the bladder (due to increasing amount of the urine stored in it) leads to a change in the curvature of the sensor. As a result of the same, the resistance of the sensor changes.

This change in resistance is converted into voltage signals by using an external voltage divider circuit which provides output in voltage as a linear function of the resistance. These voltage signals are thus processed by the micro-controller and converted into various pressure levels.

C. The Control and Display Circuit

This circuit provides the feedback of the sensor to the user by displaying the 10 levels of the volume expansion of the urinary bladder on 10 3mm red LEDS. This means that as the volume expansion of the bladder increases, the number of LEDs glowing on the LED panel also increases, with 10 being the maximum number of LEDs that can light up.

When the patient demands, he/she can press a simple switch that will activate a motor to help in micturition.

D. The Actuator Circuit

The actuator circuit consists of an H-bridge Motor driver L-298N circuit to control speed and direction of motor.

L298N is a high voltage, high current dual full-bridge driver designed to accept standard TTL logic levels and drive inductive loads such as solenoids, relays, DC and stepping motors. Two enable inputs are provided to disable or enable the device independently of the input signals. The emitters of the lower transistors of each bridge are connected together and the corresponding external terminal can be used for the connection of an external sensing resistor. An additional supply input is provided so that the logic works at a lower voltage.

The actuator circuit can activate either by automatic mechanism or by using manual input from user.

In this proposed device, the actuator circuit turns on when the patient presses a switch and turns off automatically after a delay of 10 seconds, i.e., after the voiding of the urinary bladder has completed.

E. The Microcontroller and The Tranceiver Circuit

In this circuit, the Atmega8 microcontroller has been used to accumulate sensor data using the analog reading functionality of the controller. The controller has an analog to digital converter which reads analog data and converts the analog data to 10 bit digital data and saves it into the RAM for processing. This saved digital data is then processed and the range of sensor is then limited to 0 to 1024 voltage levels. These levels are now converted into levels by taking a range of multiple voltage levels for each level for the purpose of studying the volume expansion of the urinary bladder.

The range is taken linearly as the sensor used here is linear.

Now this final value is transmitted by using 433 MHz RF transmitter to the receiver circuit. The Atmega8 used on the receiver side is used for receiving data and this received data is then displayed by actuating the LED attached to 10 pins of the controller

At the same a time controller is also monitoring the status of buttons. When a button is pressed, it sends a command to the receiver circuit to actuate the motor using a 433 MHz transmitter.

III. RESULTS

In order to analyze the working of the circuit, it is important to know the sensitivity of the sensor. This is mainly because the working of the entire system, whether it is the filling up of the urinary bladder or the indication of the same to the patients, depends of how the sensor detects the filling up of the bladder. The sensitivity lies at 0.157 Kilo-Ohms per Degree bend of the flex sensor, as determined by the manufacturing of the used sensor.

Table 1 Sensitivity of the Flex Sensor

Number of LEDs	0	15	30	45	60	75	90
Resistance (Kilo-Ohms)	31.27	33.89	35.14	36.54	37.61	42.18	45.40

Thus we can easily see the sensor is extremely sensitive and can be used effectively to sense the filling up of the urinary bladder and consequently notify the patient.

Moreover, as the volume expansion of the urinary bladder increases, the number of LEDs that glow on the panel will also increase.

As the volume expansion of the urinary bladder increases, the number of LEDs glowing on the panel also increases. When the sensor reaches it maximum bend, i.e. 90 degrees, all the LEDs on the panel start glowing as can be seen in Fig.3.

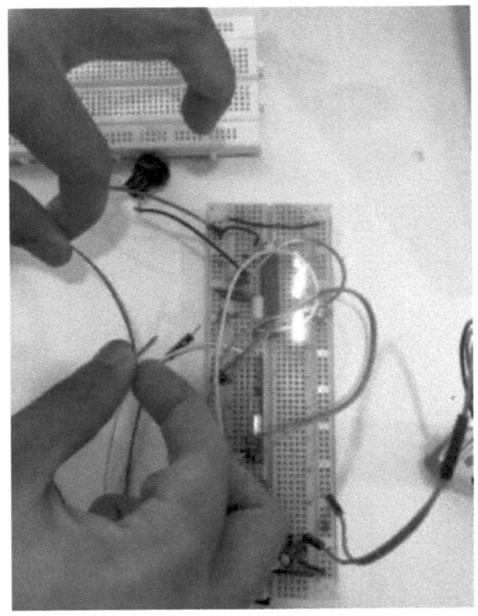

Fig. 3 Full Bend of the Sensor

As already stated the saved digital data is processed and the range of sensor is then limited to 0 to 1024 voltage levels. These levels are now converted into steps by taking a range of multiple voltage levels for each step for the purpose of studying the volume expansion of the urinary bladder.

Table 2 Voltage Levels and Glowing of LEDs

Number of LEDs glowing	Voltage Levels	Actual Voltage Delivered (in Volts)
1	0-102.4	0-0.5
2	102.5-204.8	0.5-1
3	204.9-307.2	1-1.5
4	307.3-409.6	1.5-2
5	409.7-512	2-2.5
6	512.1-614.4	2.5-3
7	614.5-716.8	3-3.5
8	716.9-819.2	3.5-4
9	819.3-921.6	4-4.5
10	921.7-1024	4.5-5

IV. CONCLUSION

The described device will detect the fullness of the urinary bladder of a paraplegic patient. Not only the sensing, it will also notify the patient about the same using a series of LEDs in an LED panel.

The device will allow patients to restore their independence and also allow them to complete their normal everyday tasks without the worry of their incontinence issues.

Overall the device will provide an affordable, practical, and user friendly solution to urinary incontinence to patients of all ages and abilities, as well as both sexes, and individuals with every type of incontinence.

CONFLICT OF INTEREST

The authors declare that they have no conflict of interest.

REFERENCES

1. Kornegay JN1 (1991) Paraparesis (paraplegia), tetraparesis (tetraplegia), urinary/fecal incontinence. Spinal cord diseases. Probl Vet Med.(3):363-77.
2. C J Fowler, K J O'Malley (2003) J Neurol Neurosurg Psychiatry 2003;74:iv27-iv31 Investigation and management of neurogenic bladder dysfunction DOI:10.1136/jnnp.74.suppl_4.iv27
3. T.Gurpinar , D.P. Griffith (1996) "The prosthetic bladder"
4. SIAA (2001) "Handbook developed by the Spinal injuries association of America."

Author: Akriti Chadda
Institute: VIT University
Street:
City: Vellore
Country: India
Email: akritichadda93@gmail.com

Diagnostic Tool for PCOS Classification

B. Padmapriya[1] and T. Kesavamurthy[2]

[1] PSG College of Technology, Department of Biomedical Engineering, Coimbatore, India
[2] PSG College of Technology, Department of Electronics and Communication Engineering, Coimbatore, India

Abstract— **Poly Cystic Ovarian Syndrome (PCOS) is a common disease of the endocrine gland and is otherwise called as Stein-Leventhal syndrome. Generally about 5% to 10% of women at the reproductive age are affected by this disease. The real cause of the disease is not exactly known, but the onset of the disease is characterized by the excessive secretion of insulin resistance androgen. There are many different methods to diagnose this condition. The most effective method is the pelvic ultrasound, which confirms the presence of multiple small cysts in the periphery of the ovaries. The ultrasound scan image gives us the visualization of the follicles. Actually there are about three types of ovaries in women. They are classified based on the number and the size of the follicles as normal ovary, cystic ovary and polycystic ovary. If the numbers of follicles are 12 or more than 12 and the diameter is more than 2-9 mm, it is being classified as polycystic ovary. In the conventional method, the follicles are counted manually by a medical expert and verified by the second person. Therefore, it is operator biased. Also, there may be a possibility of overlapping of follicles during the ultrasonographic examination process which lead to the wrong diagnosis. This led to the development of automatic detection and counting of follicle in the ovarian ultrasound image. This method makes use of the image processing techniques to pre-process and segment the region of interest. The algorithm works in such a way that it automatically detects and counts the number of follicles based on the size of the follicles in the image. Finally, the ovary is classified as PCOS present/PCOS absent. This PCOS diagnostic tool would save time a physician who has to spend time in manual tracing of follicles.**

Keywords— **PCOS, Ultrasound, Diagnostic tool, Image processing, Classification**

I. INTRODUCTION

There are three types of ovaries and is classified as normal ovary, cystic ovary and polycystic ovary. Poly Cystic Ovarian Syndrome or PCOS is a complex hormonal condition affecting up to around 1 in every 10 women at their reproductive age. The diagnosis of PCOS is made if a woman has infrequent or absence of ovulation. PCOS is characterized by clinical/biochemical signs of androgen excess. The other methods to confirm the presence of multiple follicles is the ultrasonography, which is widely used in the diagnosis of diseases due to its low cost, portability, and low risk to patients [1]. The scanned image of the ovary is used for the study of its morphology, abnormalities, ovulation monitoring for its growth and function. This scan helps the physician determine the formation, maturity of the egg and ovulation. [2]. The number of follicles are counted manually by the medical expert and is verified by a second person. There are possibilities of overlapping of follicles which may lead to wrong counting and in turn result in wrong diagnosis[4]. Having considered these limitations of the existing method, a new method of diagnostic tool for automatic counting and classification of follicles have been developed.

II. MATERIALS AND METHODS

A. Materials

The type of probe and frequency of the ultrasound wave emitted determines the quality of the image. The higher the ultrasonic frequency the lesser will be the depth of penetration due to attenuation. The highest frequency, 6 MHz is used to image pelvic organs with a satisfactory resolution [3].

Probes with arrays of piezoelectric elements arranged either in a curvilinear or convex sector shape is developed to provide a better fit on the abdomen. It also offers a wider view field than with the linear-array configuration.

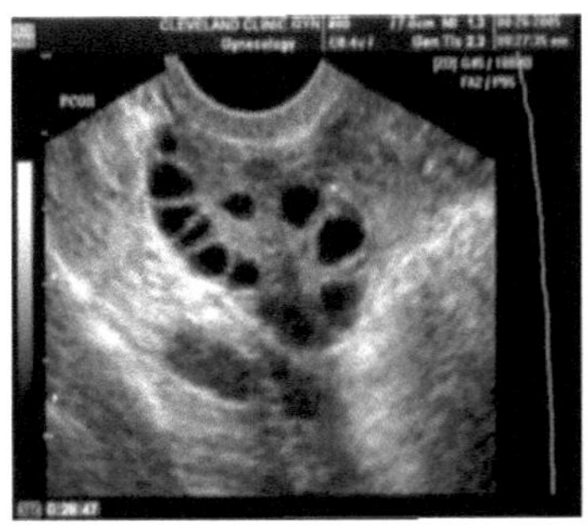

Fig. 1 Poly cystic ovarian ultrasound image

J. Goh and C.T. Lim (eds.), *7th World Congress on Bioengineering 2015*, IFMBE Proceedings 52, DOI: 10.1007/978-3-319-19452-3_48

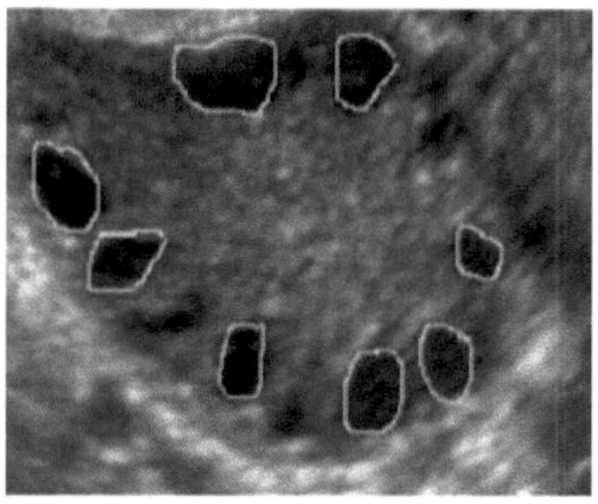

Fig. 2 Manually traced Poly Cystic Ovaries

The Fig. 1 is the PCOS ultrasound image and Fig.2 is the manually traced PCOS ultrasound image by a medical expert which is normally considered as the ground truth.

B. Methodology

The automatic classification of ovaries using the proposed method will not only overcome the shortcomings of manual tracing method but also improves the accuracy and reduces the time to process the image. This method will reduce the inter-observer variability. The method makes use of MATLAB 7 software to process the images. The databases of ultrasound images consisting of normal and polycystic ovarian ultrasound images were collected. The procedure of automatic classification of polycystic ovarian ultrasonographic images was performed on DICOM (Digital Imaging and Communications in Medicine) and JPEG (Joint Photographic Experts Group) images.

The ultrasound ovarian image obtained from the ultrasound imaging modality is given as input to the proposed automatic ovary classification system. The input image is preprocessed so that it can be used efficiently for further processing steps. Preprocessing steps include histogram equalization and filtering. The preprocessed image is then subjected to the threshold to segment the image. Morphological operations are carried out on the segmented image to enhance the boundaries of the image. Edges of the ultrasound images are detected using canny operator. As a result of the above-mentioned steps, the processed image consists of the outer boundary of the follicles in the ovary. The preprocessed image is superimposed on the original image in order to check whether all the ovaries are taken into consideration for counting or not.

Then finally the follicles are checked for the prescribed value of diameter. If the numbers of follicles are 12 or more than 12 and the diameter is more than 2-9 mm, it is being classified as polycystic ovary.

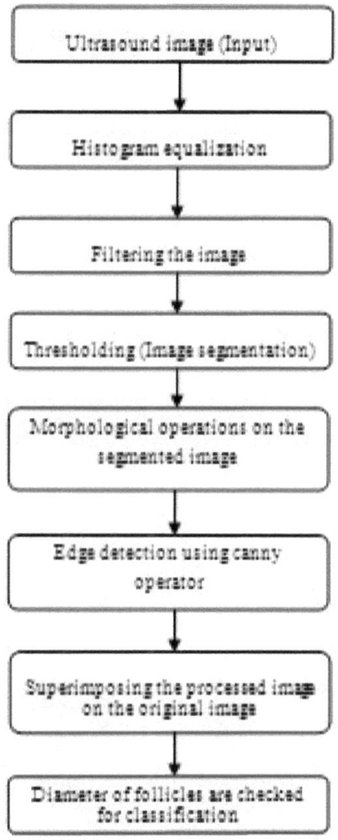

Fig. 3 Flow diagram of the classification of PCOS

The flow diagram of the proposed methodology for the classification of PCOS is shown in Fig. 3.

C. Description of the methodology

Histogram equalization is a method to adjust image intensities to enhance the global contrast. In other words, the intensity values of the image are spread over the entire range. The resultant image after histogram equalization contains uniform intensity distribution [5].

Filtering removes noise and improves the visual quality of the image. Sometimes, due to the presence of noise the edge detecting methods may give false results. A moving average filter smoothes the data by replacing each data point with the average of the neighboring data points defined within the span.

Segmentation subdivides image into its constituent objects or regions. The follicles are to be segmented from the rest of the ovarian image, which contains the dense stroma and blood vessels. The segmentation method partition the image into its regions that are similar according to a set of predefined threshold value T and then extracting the objects. Thus the follicles are extracted.

The morphological operations such as opening and dilation are carried out on the segmented image to get the edges of the follicles [6].

The image obtained after morphological operation is subjected to canny edge detection operation to extract the region of interest.

Thus obtained image is subjected to the classification by feeding the criteria of number of follicles and the size of the follicles to the algorithm. Thus the classification is done.

III. Results and Discussion

The proposed method is implemented on the PCOS ultrasound images and classification of ovaries is done. The Fig.3 shows the results of the process of classification.

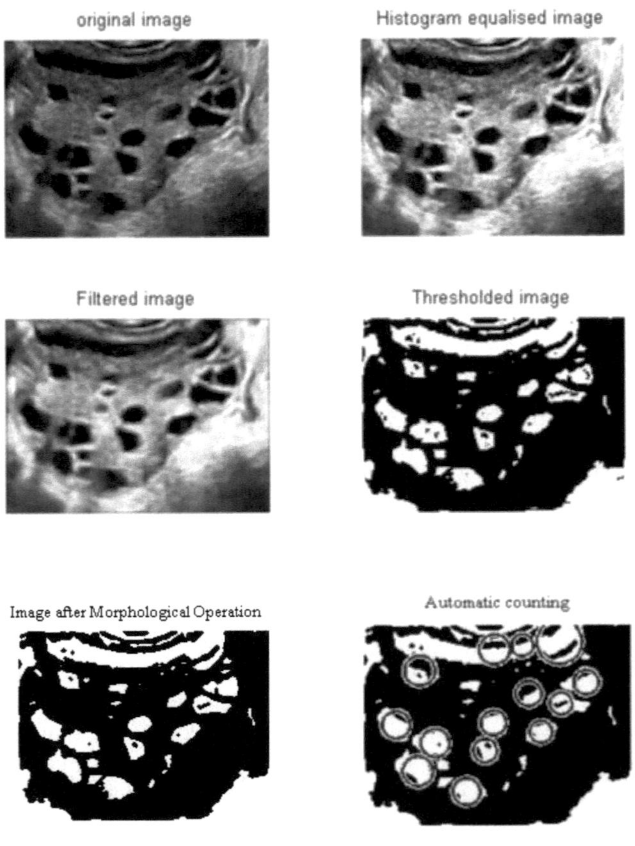

Fig.4 Results of the proposed method

Table.1 Classification Results of Ground truth Vs. Proposed method

Sl. No	Number of follicles (manual)	Classification Result	Number of follicles (proposed)	Classification Result
1.	9	Absent	15	Present
2.	11	Absent	14	Present
3.	13	Present	16	Present
4.	7	Absent	8	Absent
5.	11	Absent	17	Present
6.	12	Present	13	Present
7.	4	Absent	7	Absent

The results in Table.1 show the efficiency in the detection and classification of the proposed diagnostic tool for PCOS. The proposed method is efficient in detecting and counting the follicles as compared to the manual method in all the ultrasound images of ovaries. It is because many follicles overlap or cannot be seen with the naked eyes. Therefore, the number of follicles detected by the proposed algorithm is high in number. The correlation coefficient is 0.9.

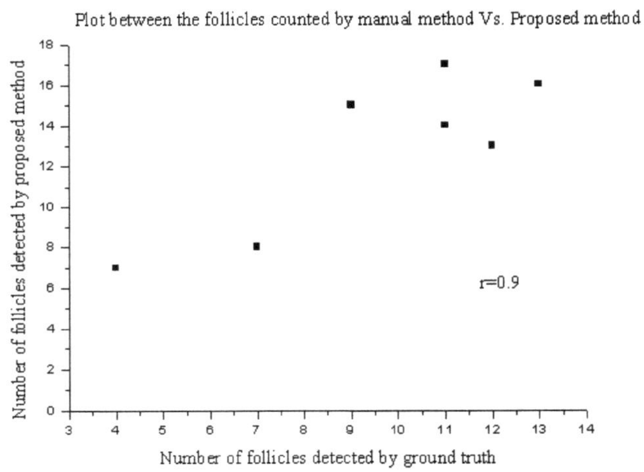

Fig.5 Plot between manual Vs. Proposed method

The average recognition rate of the manual method is 78%[7]. The average recognition rate of the proposed method is 88%.

IV. Conclusion

The proposed diagnostic tool for PCOS is very efficient in detecting and counting follicles from ultrasound images using image processing techniques.

Conflict of Interest

"The authors declare that they have no conflict of interest".

References

1. Simon Carter & Andrea Tubaro (2001) Standardisation of Bladder Ultrasonography 1-36. Available from: <http://wiki.ics.org/ file/view/Bladder+ultrasonography.pdf>. [7th August 2002].
2. William Herring (2011) Available from: http://www.Learning Radiology: Recognizing the Basics (With STUDENT CONSULT Online Access), 2e Paperback. [28 April 2011].
3. Holmes, JH (1967) Ultrasonic studies of the bladder. The Journal of Urology 97(4): 654-663.
4. Mehrotra, P. Chatterjee, J. Chakraborty, C. Ghoshdastidar, B & Ghoshdastidar, S (2011) Automated screening of Polycystic Ovary Syndrome using machine learning techniques, Conference Proceedings of 2011Annual IEEE India Conference, Hyderabad, India, pp.1-5.
5. Raju, A, Dwarakish, GS & Venkat Reddy, D (2013) A Comparative Analysis of Histogram Equalization based Techniques for Contrast Enhancement and Brightness Preserving, International Journal of Signal Processing, Image Processing and Pattern Recognition, 6(5):353-366.
6. Rafael R Gonzalez & Richard E Woods 2002, Digital Image Processing, Second Edition, Prentice-Hall, Inc. New Jersey.
7. Potocnik, B & Zazula, D (2001) Suppressing the System Error in the Measurement Model of the Prediction-Based Object Recognition Algorithm: Ovarian Follicle Detection case, Proceedings of the 2nd International Symposium on Image and Signal Processing and Analysis, IEEE- ISPA 2001, pp.196-201.

The address of the corresponding author:

Author: B. Padmapriya
Institute: PSG College of Technology
Street: Peelamedu
City: Coimbatore
Country: India
Email: priyadhileep@yahoo.co.in

An Integrated Processing System for Temporal Neuron Analysis

Han-Wei Dan, Chia-Yi Tseng, Cheng-An J. Lin, and Yuh-Show Tsai

Chung Yuan Christian University, Department of Biomedical Engineering, Taoyuan City, Taiwan

Abstract— **Microtubule's structure in neurons with neurodegenerative disease is very different from normal neurons. For example, one of the major characteristic of Alzheimer's disease is abnormal changes of Tau protein in brain, producing intracellular neurofibrillary tangles and colocalized with microtubule. Microtubules are instructors of neuronal morphogenesis, and the fragmentation of microtubule in neurites is an indicator of abnormal neurons. Analysis of time-lapse microscope images can provide quantitative information of neurons' status under different development or disease stages; however, there are few researches building methods for extracting microtubule features in neurons. An automated system for processing and analyzing temporal microtubule images may be helpful for understanding morphological dynamics in neurons. To have a united criterion for the localization of single particle in continuous images, image registration is important. In this paper, the ImageJ plugin, StackReg, was used for registration, and followed by the brightness analysis in neurite. Part of dendrite in tubulin images of primary hippocampal neurons treated with NMDA is cropped and the intensities along the skeletonized dendrite are recorded. The dendrite with NMDA at 5 minutes started to have a huge variation on the intensity distribution. The global intensity decreased while the range of changing broadened. In conclusion, the intensity distributions were different for neurons with varied health conditions, and could be shown from quantitative data of microscope image. In the future, we tend to develop an automated image processing system for microtubule images, and build a standard for identifying healthy and abnormal neurons based on these results.**

Keywords— **Neuron image processing, neurite, microtubule, quantitative analysis**

I. INTRODUCTION

Neurodegenerative diseases are caused by incomplete function of nerve system, which usually accompany with abnormal proteins in neurons. For instance, the major characteristics of Alzheimer's disease (AD) are mutation of beta amyloid (Aβ) and redistribution of Tau protein, which may induce extracellular amyloid plaques and intracellular neurofibrillary tangles [1]. Tau is major protein associated with microtubule. So in AD neurons, Tau and microtubule are colocalized with neurofibrillary tangles, making huge diversification on protein morphologies between AD and healthy neurons [2]. Since microtubule plays an important role in cytoskeleton, architectures of neuronal network can be affected by morphogenesis of microtubule. In other words, fragmentation or disconnection of microtubule in neurites is an indicator of abnormal neurons [3]. Most of the current studies are focus on analysis the whole neuron, and barely any method for extracting microtubule features in neurons.

Analysis of microscope images during a time interval can provide essential information for neuronal status under different stages of disease. However, biologists suffer in manually approach, which is extremely labor-intensive and time-consuming; furthermore, the accuracy cannot be verified since all procedures rely on experts and results may be subjective. To solve these problems and increase the proceeding speed, an automated analysis system with united criterion is necessary. Combination of micrographs and digital image processing can easily obtain quantified information, and use for unbiased statistical analysis.

In this paper, to achieve above goals, we used an automatic method for aligning continuous images, and output quantitative results based on semi-automatic image processing, for understanding the differences of single neuron under normal circumstance and after treating with NMDA.

II. MATERIALS AND METHODS

A. Tools and Materials

a) Images

The neuron images are acquired from hippocampal neurons as following steps. Primary hippocampal neurons transfected with cDNA encoding enhanced green fluorescence tubulin conjugated protein are treated with or without N-Methyl-D-aspartic acid or N-Methyl-D-aspartate (NMDA), an agonist of NMDA receptor. That is to mimic the sub-lethal glutamate excitotoxicity of ischemic stroke. The smooth organization of microtubules is disturbed during the NMDA treatment in comparison to vehicle (DMSO) condition.

b) Operating System and Development Environment

The hardware (PC) for development is with 2.00 GHz Pentium® Dual- Core CPU, and 1.20 GHz 3.00GB RAM. The operating system is 64-bit Windows 7. All the functions are built with libraries of 64-bit Java JRE 7 and ImageJ 1.48(s). Integrated development environment is Eclipse LUNA.

J. Goh and C.T. Lim (eds.), *7th World Congress on Bioengineering 2015*,
IFMBE Proceedings 52, DOI: 10.1007/978-3-319-19452-3_49

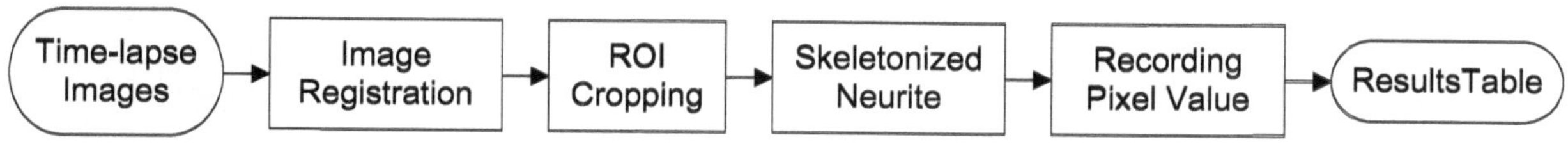

Fig. 1 Flow chart of image data acquisition

B. Flow Diagram

The flow for extraction the quantitative data of neuron was as Fig. 1. After acquiring the time-lapse images, automatic method, StackReg, was used for image registration. The targeted neurite was cropped and skeletonized as a region of interest (ROI) for recording the intensity (pixel value) at each point. The pixel values of every slice in image stack were outputted for further analysis.

C. Image registration

Continuous live images acquired during a time interval may have motion artifacts resulted from mobile objects like living cells or unsteady platform of hardware devices, and affect the comparison based on serial feature and influence statistical results. Image registration is the process of aligning two or more images geometrically, can be applied to correct the position difference between slices.

StackReg [4] refers to the abbreviation of stack registration, which proposes to align or to match slices in an image stack. Each slice is the template for the alignment of next slice, so this method presents an adaptive mapping pattern that may be suitable for continuous images within a long time, in other words, the first slice has huge difference between the last slice. StackReg is based on pyramid approach to minimize the mean square intensity difference between a reference and a test set. In order to suit the least-squares criterion for registration, a nonlinear optimization method, the Marquardt-Levenberg [5] algorithm is applied.

D. Brightness Analysis

Neuron image at time 0 (first slice) was used to crop ROI for recording the intensity, as yellow circle shown in Fig. 2.

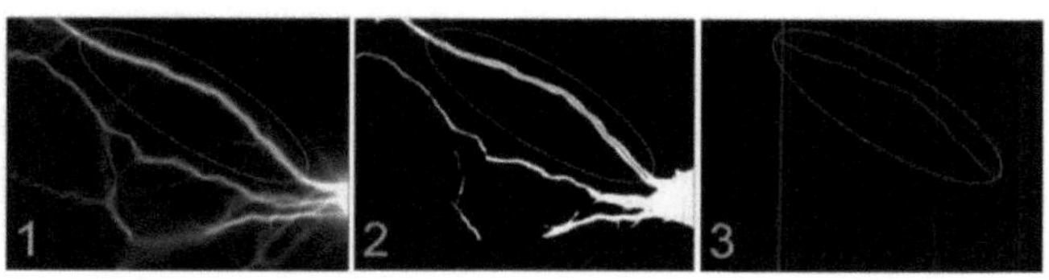

Fig. 2 ROI cropping and targeted dendrite for brightness analysis

The most complete and longest dendrite was the target for analysis. To get the area, first, used Otsu's method [6] to make image into binary image, as Fig. 2-2; then skeletonized binary part and kept the skeleton in yellow region like Fig. 2-3. The red line in Fig. 2-1 was the targeted dendrite for brightness analysis. In this paper, statistical value including mean (μ), standard deviation (σ) and coefficient of variation (CV) were calculated to evaluate the trend and change rate of intensity.

III. RESULTS

A. Image registration

Fig. 3 was part of the result of StackReg for neuron stack. Left column were images before registration and right column were the result (labels on the top-left were acquired time). As the area of image acquisition move left, StackReg modified images to a proper position. This method could align well by one click for program execution, and was very suitable for integrating into an automated system. Also,

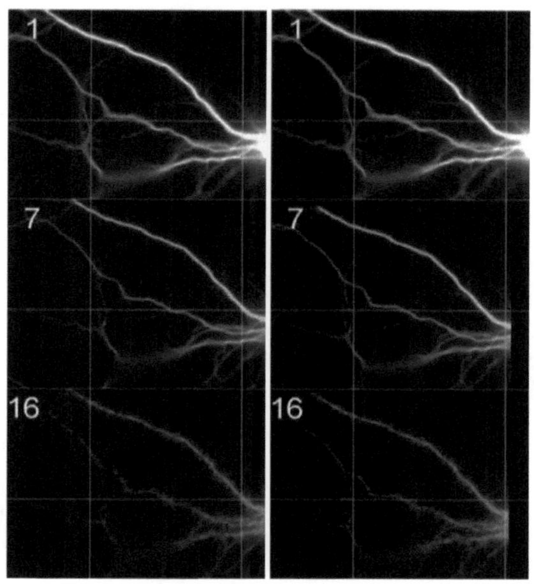

Fig. 3 Result of neuron image registration
(Left column: before registration, right column: after registration)

StackReg could adaptively change the alignment criterion, because the method for it to do registration was based on the differences between neighbor slices. For the growing or moving cells, adaptive method will be more appropriate to avoid the big morphological differences between template and the slices too far from it.

B. Brightness Analysis

Primary hippocampal neurons started to be treated with NMDA at 5th minute. In Fig. 4, the mean value of intensity decreased while the standard deviation increased slightly over time. The change rate was clearer by showing as the coefficient of variance (CV) in Fig. 5. CV is the ration of standard deviation to the mean, and can be used to present the distribution of data. Intensity distribution had huge vibration at the 10th minute, which was treated with NMDA in 5 minutes. It meant that there was uneven distribution of intensity along targeted dendrite. The big differences between the first and last slice also showed that the different morphologies of healthy and abnormal (even dead) neurons could be told by quantitative data from image processing.

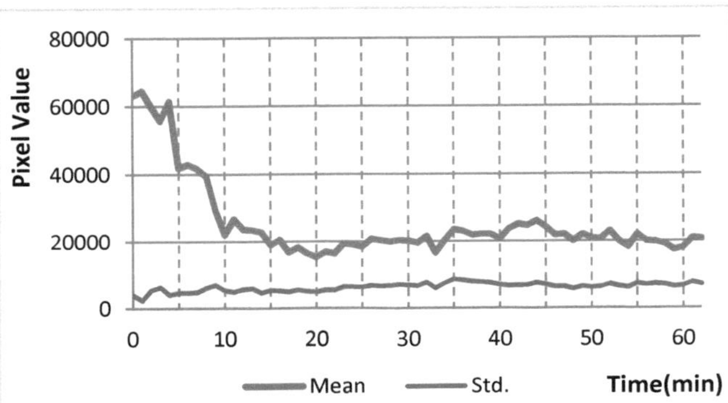

Fig. 4 Mean and standard deviation of intensity (pixel value) along the targeted dendrite at different time (minute)

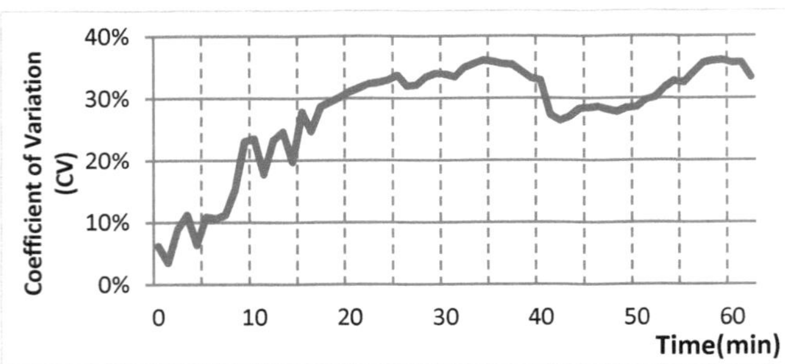

Fig. 5 CV of intensity along the targeted dendrite at different time (minute)

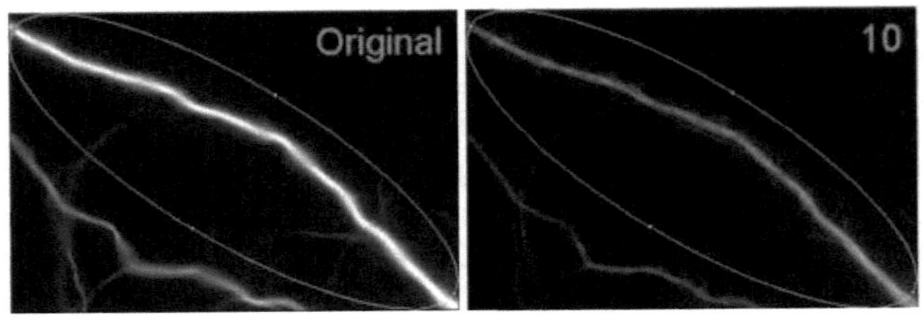

Fig. 6 Raw images of neuron at different time (minute)

IV. Conclusion and Discussion

The image registration method, StackReg, could already be implemented to the image processing system without any parameters setting. However, it should be tested for robustness in different types of cellular images for practical application. The idea of adaptive template for image alignment could also apply to be the mask for observation of neuron growth (or changes) between time-lapse images.

Based on the result of brightness analysis, there were big differences in mean, standard deviation, and CV value of intensity between healthy and abnormal neurons. CV started to sharply increase in the image at 10th minute, and there was microtubule fragmentation in dendrite as shown in Fig. 6. The quantitative information can truly represent the status of neurons. Collecting enough images and quantitative data export from this system, a classification rule or model for identifying normal neuron or unhealthy neuron may even be found after further studying.

In the future, the system for time-lapse neuron image processing and analysis will be developed. Different from the other tools for processing neuron images, this study will focus on the changes of intensity and morphologies in microtubules of time-lapse neuronal micrographs, to build an automated and user-friendly temporal neuron analysis system for biologist.

References

1. Zempel H, Thies E, Mandelkow E, Mandelkow E-M (2010) Abeta oligomers cause localized Ca(2+) elevation, missorting of endogenous Tau into dendrites, Tau phosphorylation, and destruction of microtubules and spines. J Neurosci 30 (36):11938-11950. doi:10.1523/jneurosci.2357-10.2010
2. Wang JZ, Liu F (2008) Microtubule-associated protein tau in development, degeneration and protection of neurons. Prog Neurobiol 85 (2):148-175. doi:10.1016/j.pneurobio.2008.03.002
3. Poulain FE, Sobel A (2010) The microtubule network and neuronal morphogenesis: Dynamic and coordinated orchestration through multiple players. Molecular and Cellular Neuroscience 43 (1):15-32. doi:http://dx.doi.org/10.1016/j.mcn.2009.07.012
4. StackReg at http://bigwww.epfl.ch/thevenaz/stackreg/

5. Thevenaz P, Ruttimann UE, Unser M (1998) A pyramid approach to subpixel registration based on intensity. Image Processing, IEEE Transactions on 7:27-41. doi:10.1109/83.650848
6. Otsu N (1979) A Threshold Selection Method from Gray-Level Histograms. Systems, Man and Cybernetics, IEEE Transactions on 9 (1):62-66. doi:10.1109/TSMC.1979.4310076

Author: Yuh-Show Tsai
Institute: Chung Yuan Christian University, Department of Biomedical Engineering
Street: 200 Chung Pei Road
City: Taoyuan City
Country: Taiwan, R.O.C.
Email: yuhshow@gmail.com

Estimation Shell Elasticity of Lipid-Coated Microbubbles Based on the Linearized Marmottant Model

Ming-Huang Chen and Jenho Tsao*

Graduate Institute of Biomedical Electronics and Bioinformatics, National Taiwan University, Taipei, Taiwan

Abstract— **Lipid-coated microbubbles have been proven to be very effective ultrasound contrast agents. Knowledge of resonant frequencies of contrast microbubbles is important for the optimization of ultrasound contrast imaging and therapeutic techniques. The Marmottant model is considered as a realistic model for microbubbles with nonlinear elastic shell. Based on a linearized Marmottant model, an analytic solution of the resonant frequency can be derived. An equivalent shell elasticity is defined for this analytic solution and enable the resultant resonant frequency equation to predict the resonant frequency of nonlinear elastic microbubbles. A least-square error estimator is proposed for estimating the equivalent shell elasticity. It is shown that based on the equivalent shell elasticity, the linearized model can predict the nonlinear pressure-dependent property of nonlinear elastic microbubbles.**

Keywords—**Ultrasound contrast agents, Lipid-coated microbubbles, Resonant frequency, Marmottant model.**

I. Introduction

Lipid-coated microbubbles have been proven to be very effective ultrasound contrast agents. Knowledge of resonant frequencies of contrast microbubbles is important for the optimization of ultrasound contrast imaging and therapeutic techniques. It has been shown that the lipid shell can fold, buckle and shed as the lipid shell is compressed. To model these changes in the shell microstructure, Marmottant proposed a nonlinear elastic model for the surface tension $\sigma(R_0)$ of lipid shell [1], where R_0 is the bubble radius. The Marmottant model is considered as a realistic model of bubble shell which can simulate the 'compression-only', pressure-dependent resonance frequency, thresholding and skewing behaviors of lipid-coated microbubbles [2-4].

The Marmottant model is parameterized by the shell elasticity χ and initial (or resting) surface tension $\sigma(R_0)$, beside the others. By simulation using the Marmottant model, it was found that the resonant frequency $f_r(R_0)$ of lipid-coated microbubbles depends not only on its shell parameters: the shell elasticity and initial surface tension, but also on the excitation pressure P_{ex} [2,5]. To understand the properties of the resonant frequency of lipid-coated microbubbles, an analytical expression for the resonant frequency is preferred. However due to the nonlinearity of the Marmottant model, there is no analytical solution for the resonant frequency.

This is not the case for microbubbles with linear elastic shell, which can have a close-form solution for its resonant frequency. The close-form solution is to be called resonant frequency equation. In this study, a *linearized* Marmottant model [6] is employed to study the resonant behavior of lipid-coated microbubbles. This study is trying to know whether the resonant frequency equation of this linearized model is able to predict the resonant frequency of microbubble with nonlinear shell elasticity.

For the linear elastic model, the resonant frequency of microbubble is a function of an equivalent shell elasticity as $f_r^l(R_0;\chi_{eq})$, to be defined latter. The parameter, equivalent shell elasticity $\chi_{\rm eq}$, can be estimated by minimizing the average squared error between the measured resonant frequency $f_r(R_0)$ and the predicted resonant frequency $f_r^l(R_0;\chi_{eq})$, that is.,

$$\min_{\chi_{\rm eq}} \sum_{\rm R} \left[f_r(R_0) - f_r^l(R_0;\chi_{eq})\right]^2.$$

The solution is a least-square error (LSE) estimate of the equivalent shell elasticity $\hat{\chi}_{\rm eq}$.

By the equivalent shell elasticity, it is shown that the resonant frequency equation can predict the pressure-dependent behavior of microbubbles with nonlinear shell elasticity. The details about the shell model will be given in section II. Simulation studies including the prediction of resonant frequencies are given in section III. Lastly, some conclusions will be given in section IV.

II. Theories

A. Marmottant model

The generalized bubble dynamics equation for a phospholipid-coated microbubble [1-3] is

$$\rho\left(R\ddot{R}+\frac{3}{2}\dot{R}^2\right)=\left(P_0+\frac{2\sigma(R_0)}{R_0}\right)\left(\frac{R_0}{R}\right)^{3\gamma}\left(1-\frac{3\gamma\dot{R}}{c}\right)-\frac{2\sigma(R)}{R}-4\mu\frac{\dot{R}}{R}-4\kappa_s\frac{\dot{R}}{R^2}-P_0-P_A(t) \tag{1}$$

* Corresponding author.

J. Goh and C.T. Lim (eds.), *7th World Congress on Bioengineering 2015*, IFMBE Proceedings 52, DOI: 10.1007/978-3-319-19452-3_50

where $R, \dot{R}$ and $\ddot{R}$ are the radius, velocity, and acceleration of the bubble wall with time variable suppressed. R_0 is the initial bubble radius and $P_0 = 100$ kPa is the ambient pressure. The properties of the surrounding water are described by the viscosity $\mu = 10^{-3}$Pa s , the density $\rho = 10^3$ kg/m^3, and the sound speed $c = 1500$ m/s. The driving pressure is $P_A(t) = P_{ex}A(t)\sin(\omega t)$, where P_{ex} is the peak excitation pressure, $A(t)$ is the amplitude tapper and ω is the driving frequency. γ is the polytropic exponent of ideal gas, which is set to be 1.07.

The effect of the phospholipid shell is taken into account through shell viscosity κ_s and effective surface tension $\sigma(R)$ [1], which is assumed to depend on the radius R and the concentration of phospholipid molecules on the surface of bubble. In Marmottant model [1,2], the surface tension $\sigma(R)$ is a three segments function for modeling the buckling and rupture behavior of bubbles coated with compressible shell. The surface tension in the three regimes is given by:

$$\sigma(R) = \begin{cases} 0 & \text{if } R \le R_b \quad \text{(buckled)} \\ \chi\left(\frac{R^2}{R_b^2}\right) & \text{if } R_b < R < R_r \quad \text{(elastic)} \\ \sigma_W & \text{if } R \ge R_r \quad \text{(ruptured)} \end{cases} \tag{2}$$

where χ is the elasticity of the shell and σ_W is the surface tension of the gas-water interface. The shell buckles for radii below the buckling radius $R_b = R_0/\sqrt{1+\sigma(R_0)/\chi}$ and is in the ruptured state for radii larger than $R_r = R_b\sqrt{1+\sigma_W/\chi}$. The initial state is defined by the initial surface tension $\sigma(R_0)$.

B. Linearized Marmottant model

By simulations and experiments, it was shown that resonance frequency of phospholipid-coated microbubble is excitation-pressure dependent [2,5]. For investigating the resonance behavior of UCA microbubbles, analytical expression for the pressure dependent property of microbubble is preferred. However, due to the nonlinearity of surface tension in Marmottant model, resonance frequency of bubble can only be found by solving the generalized Rayleigh-Plesset equation numerically. In the development of the microbubble spectroscopy for characterizing the behavior of ultrasound contrast agents [6], the Marmottant model is linearized to allow the resonance frequency of bubble have an analytical expression. The linearized surface tension is

$$\sigma_L(R) = \sigma_W + 2\chi\left(\frac{R}{R_0} - 1\right) \tag{3}$$

which is to be referred as linearized Marmottant model in latter use. By using the linearized Marmottant model in the generalized Rayleigh-Plesset equation, the resonance frequency can be solved analytically to be

$$f_r = f_0\sqrt{1 - \frac{\delta^2}{2}}, \tag{4}$$

where f_0 is the eigenfrequency and δ is the damping coefficient of the microbubble, and they are given below:

$$f_0 = \frac{1}{2\pi}\sqrt{\frac{1}{\rho R_0^2}\left[3\gamma P_0 + \frac{2(3\gamma-1)\sigma_W}{R_0} + \frac{4\chi}{R_0}\right]}, \tag{5}$$

$$\delta = \frac{R_0}{c} + \frac{4\mu}{\rho R_0^2\omega_0} + \frac{4\kappa_s}{\rho R_0^3\omega_0}. \tag{6}$$

It is found that the shell elasticity χ depends greatly on the driving pressure and is responsible for the decrease of resonance frequency for increasing excitation pressure P_{ex}[5,7]. This means that the shell elasticity in EQ.5 is a variable depends on the excitation pressure, which will be named as *equivalent shell elasticity* and denoted as $\chi_{eq}(P_{ex})$ or simply χ_{eq}. Then, the resonant frequency becomes

$$f_r^l(R_0; \chi_{eq}) = \frac{1}{2\pi}\sqrt{\frac{(2-\delta^2)}{2\rho R_0^2}\left[3\gamma P_0 + \frac{2(3\gamma-1)\sigma_W}{R_0} + \frac{4\chi_{eq}(P_{ex})}{R_0}\right]} \tag{7}$$

and the resonance frequency becomes a P_{ex} dependent variable. For convenience, the above equation will be referred to as the resonance frequency equation of the linearized Marmottant model.

III. Simulation Studies

A. Simulation method

The properties and performances of this LSE estimate $\hat{\chi}_{eq}$ is investigated by simulation studies with the relevance parameters, bubble radius R, shell elasticity χ, initial surface tension $\sigma(R_0)$ and excitation pressure P_{ex}, being varied over their possible values. Using the Marmottant model, we simulated the response of lipid-coated microbubbles by solving the generalized Rayleigh-Plesset equation for the radius-time curve $R(t)$.

The driving pressure $P_A(t)$ is a Hanning tapered 13μs long pulse. The excitation pressure P_{ex} and driving frequency are varied for different excitation conditions. The shell elasticity χ and the shell viscosity κ_s were set to 2.5 N/m and 6 · 10−9 kg /s to simulate the BR14 UCA (Bracco Research SA, Geneva, Switzerland), which is used for *in vitro* experiments [2]. The procedures to find the resonant frequency a bubble is same as what are done in [2]. The resonant frequencies are found at the maximum response of the simulated resonant curves and denoted as $f_r(R_0)$. The initial surface tension in the Marmottant model was set to 0, 0.01 or 0.02 N/m to test its effects on the pressure-dependence behavior. There eleven excitation pressures simulated: 1, 5, 10,15,..., 50 kPa. Five bubbles with different resting sizes, R_0 = 1.6, 2.0, 2.4, 2.8 and 3.2 μ m, are simulated.

B. estimation of $\hat{\chi}_{eq}$

The resonant frequencies of the simulation cases are presented in Fig.3 and will be discussed latter. The least-square estimates of the equivalent shell elasticity $\hat{\chi}_{eq}$ for the simulated resonant frequencies are presented in Fig. 1. The RMS (Root mean squared) error of the least-square estimates are shown in Fig. 2. The simulation results are summarized below:

1) As shown in Fig. 1, it is clear that the equivalent shell elasticity is pressure-dependent. Actually, this dependence results from the pressure-dependent property of resonant frequency $f_r(R_0)$.
2) $\hat{\chi}_{eq}$ is P_{ex} and $\sigma(R_0)$ dependent
3) Different combination of P_{ex} and $\sigma(R_0)$ can yield a common $\hat{\chi}_{eq}$.
4) $\hat{\chi}_{eq} \leq \chi_{max}$; $\chi_{max} = 2.5$N/m
5) When excitation pressure is large enough (40 kPa in Fig. 1), $\hat{\chi}_{eq}$ approaches a constant,
6) When $\sigma(R_0) = 0.00$ N/m, $\hat{\chi}_{eq}$ keeps constant about 0.2 N/m and the frequency matching error in Fig.2 is minimal about 0.05MHz and
7) The frequency matching error increases for the excitation pressures varying between 0 and 50 kPa (Note: this is the range for the pressure-dependent behavior of the simulated bubbles).

C. Resonant frequencies and Prediction of $f_r(R_0)$

The resonant frequencies in Fig.3 are the results of the simulations based on the Marmottant model for (a) $\sigma(R_0) = 0.02$N/m, (b) $\sigma(R_0) = 0.01$N/m and (c) $\sigma(R_0) = 0.0$N/m. The resonant frequencies are plotted over excitation pressure. They show that the pressure dependent property of resonant frequency happens for bubbles with $\sigma(R_0) > 0$N/m only. For bubbles with $\sigma(R_0) = 0$N/m, their resonant frequencies are pressure independent.

Since the resonance frequency equation is analytic, it has an application for resonance frequency prediction. Fig.4 shows the predicted resonant frequencies of the test bubbles using the equivalent shell elasticity shown in Fig. 1. By comparing Fig. 3 and Fig. 4, it is shown that the resonance frequency equation can predict the resonant frequencies of the test bubbles. The prediction is quite good for $\sigma(R_0) = 0.01$N/m and $\sigma(R_0) = 0.0$N/m.

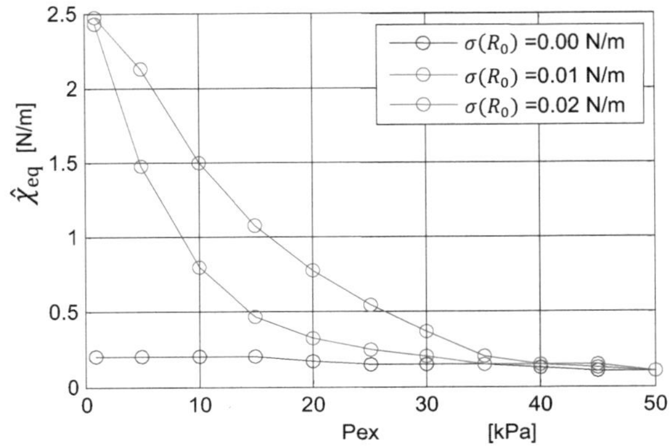

Fig. 1 Estimates of the equivalent shell elasticity for different $\sigma(R_0)$

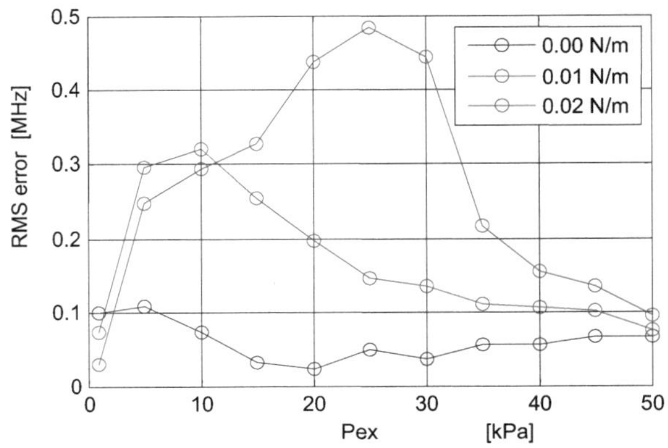

Fig. 2 Frequency matching errors for different $\sigma(R_0)$

IV. CONCLUSIONS

The elasticity of nonlinear elastic shell can be estimated based on linear elastic shell model. The result is an equivalent shell elasticity. The analytic form of the resonant frequency of linear elastic bubble can be used for predicting the resonant frequency of nonlinear elastic bubble. The equivalent shell elasticity is a pressure dependent parameter, which enables the resonant frequency equation to predict the pressure dependent property of nonlinear elastic microbubbles.

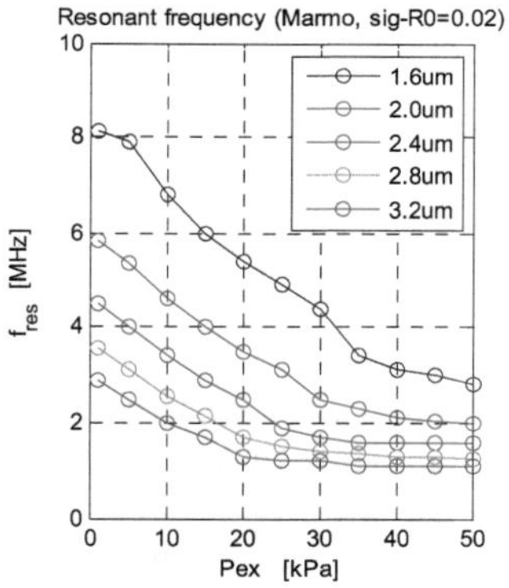

(a)

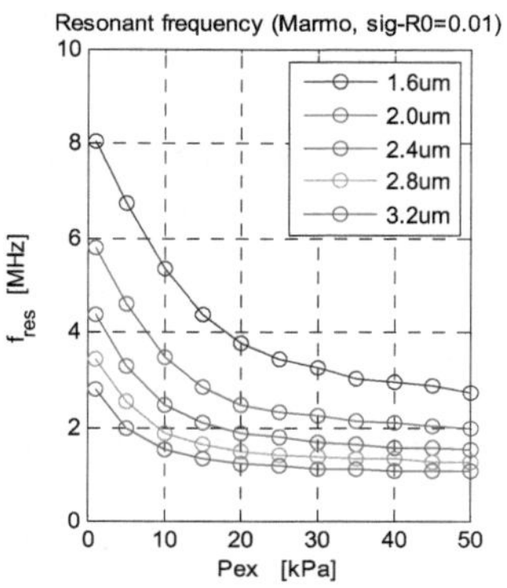

(b)

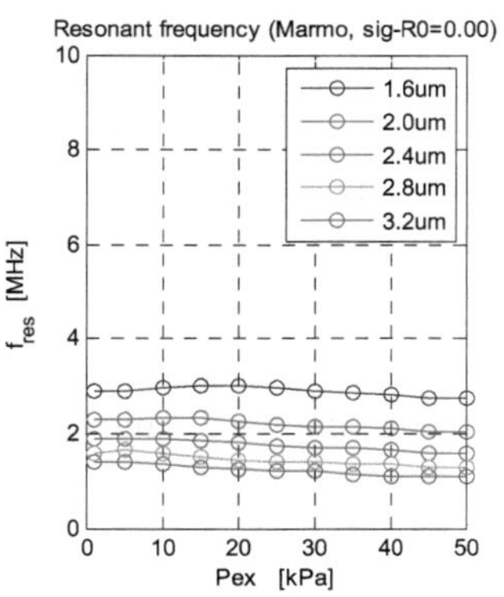

(c)

Figure 3 The resonant frequencies of the simulations based on the Marmottant model for
(a) $\sigma(R_0) = 0.02\text{N/m}$,
(b) $\sigma(R_0) = 0.01\text{N/m}$ and
(c) $\sigma(R_0) = 0.0\text{N/m}$

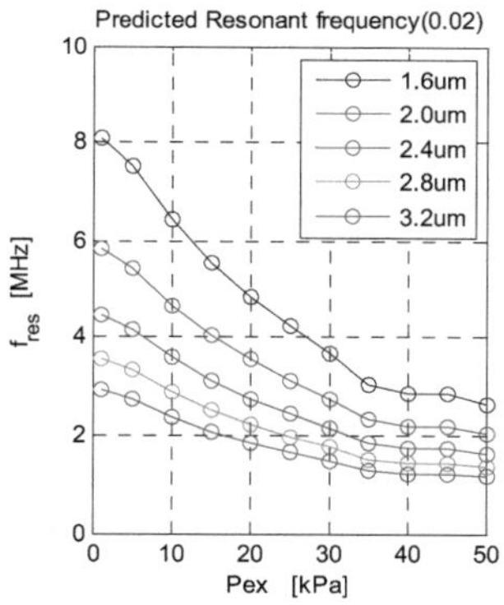

(a)

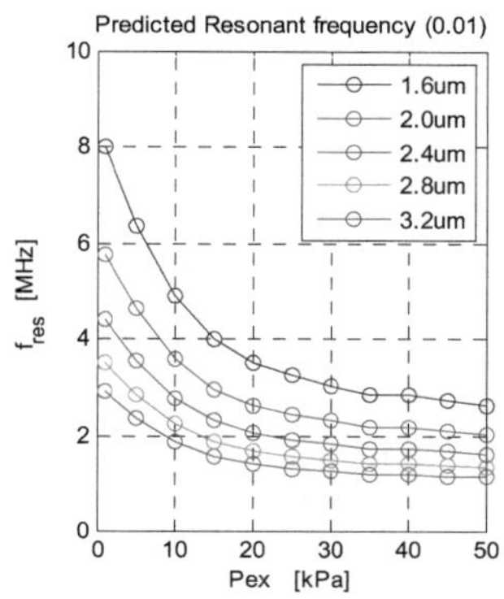

(b)

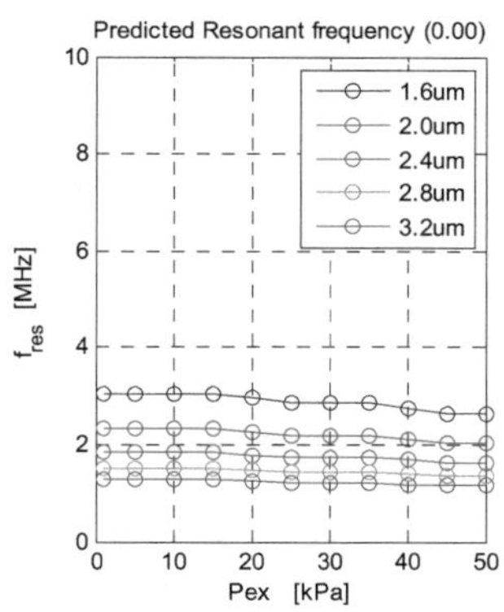

(c)

Figure 4 The predicted resonant frequencies of the tested bubbles based on the $\hat{\chi}_{eq}$ given in Fig. 1. For (a) $\sigma(R_0) = 0.02\text{N/m}$,
(b) $\sigma(R_0) = 0.01\text{N/m}$ and
(c) $\sigma(R_0) = 0.0\text{N/m}$

ACKNOWLEDGMENT

This work is supported by the Ministry of Science and Technology, Taiwan, ROC.(MOST 103-2221-E-002-015)

REFERENCES

of appearance in the text. Examples of citations for Journal articles [1], books [2], the Digital Object Identifier

1. Marmottant P, Van der Meer S, Emmer M, Versluis M, de Jong N, Hilgenfeldt S, Lohse D., "A model for large amplitude oscillations of coated bubbles accounting for buckling and rupture," J Acoust Soc Am ;118:3499–3505, 2005.
2. Overvelde, M., Garbin, V., Sijl, J., Dollet, B., De Jong, N., Lohse, D., & Versluis, M. (2010). Nonlinear shell behavior of phospholipid-coated microbubbles. *Ultrasound in medicine & biology,36*(12), 2080-2092.
3. Sijl J, Dollet B, Overvelde M, Garbin V, Rozendal T, De Jong N, Lohse D, Versluis M., "Subharmonic behavior of phospholipidcoated ultrasound contrast agent microbubbles," J Acoust Soc Am 2010;128:3239–3252.
4. Sijl J, Overvelde M, Dollet B, Garbin V, de Jong N, Lohse D. ''Compression-only'' behavior: A second-order nonlinear response of ultrasound contrast agent microbubbles. J Acoust Soc Am 2011;129:1729–1739.
5. Gong Y., Cabodi M., Porter T.M.," Acoustic investigation of pressure-dependent resonance and shell elasticity of lipid-coated monodisperse microbubbles" Appl Phys Lett;104:074103, 2014.
6. S. M. van der Meer, B. Dollet, M. M. Voormolen, C. T. Chin, A. Bouakaz, N. de Jong, M. Versluis, and D. Lohse, "Microbubble spectroscopy of ultrasound contrast agents," J. Acoust. Soc. Am. 121, pp. 648–656, 2007.
7. van Rooij, T., Luan, Y., Renaud, G., van der Steen, A. F., Versluis, M., de Jong, N., & Kooiman, K. (2015). Non-linear Response and Viscoelastic Properties of Lipid-Coated Microbubbles: DSPC versus DPPC. Ultrasound in medicine & biology.

Author: Jenho Tsao
Institute: Graduate Institute of Biomedical Electronics and Bioinformatics, National Taiwan University
Street: No 1, Sec. 4, Roosevelt road
City: Taipei
Country: Taiwan, ROC
Email: jhtsao@ntu.edu.tw

Development of a Modular FPGA Based Digital Beamformer for PC Based Ultrasound Imaging System

P. Busono, Y. Suryana, A. Barkah, R. Febryarto, T. Handoyo, Riyanto, and A. Fitryanto

Center for Information and Communication Technology, BPPT, Jakarta

Abstract—**A digital transmit and receive beamformer for PC based ultrasound imaging was developed. This system consists of an analog front end, FPGA control board, ultrasound probe, and computer. Analog front end consists of programmable pulser and receiver circuit. The pulser circuit uses high voltage pulse generator to produce monocycle electrical pulses with the adjustable center frequency and amplitude up to 50 V. The receiver circuit uses a AD9272 chip, which is an 8-channel integrated analog front end module. Each of the 8 signal paths consists of a low noise amplifier (LNA), a digitally programmable variable gain amplifier (VGA), anti aliasing filter (AAF), and a 12-bit, 40 Mega Samples Per Second (MSPS) analog-to-digital converter (ADC). The probe is a 64 element linear probe which can produce pulses with center frequency of 3.5 MHz. The Xillinx Spartan 6 FPGA based beamformer consists of 8 channel transmit beamformer controller block and 8 channel receive beamformer blocks. The receive beamformer is composed with delay stage block, apodization blok, summation block, Hilber transform block, and envelope detection block. The implementation result shows that the system was able to process the RF signals.**

Keywords— **Ultrasound, digital beamformer, FPGA**

I. INTRODUCTION

Modern medical ultrasound scanners are used for imaging the internal structures of the body which are displayed in gray-scale B-mode images. Currently, it has become the most widely used medical imaging modality since it is safe, portable, inexpensive, non invasive, easy to use and display image in real time. Even as a mature technology, advances are still being made, ranging from improvement in transducer design, imaging hardware and new signal processing algorithm [1]

An ultrasoud image was produced by firing the internal target with focused ultrasound beam transmitted from ultrasound probe and using the same probe acquiring the echoes reflected from the internal target under investigation. Digital signal processing and scan conversion algorithms are used to process the RF data and produce the 2-D ultrasound image, respectively [1]

Digital beamformer has a vital role in ultrasound imaging system. It has two main functions: steering the transmitted ultrasound beam and defining a focal point within the target from which location of the returning echo is derived. The beamforming process needs a high delay time resolution to avoid the deteriorating effects of the delay quantization lobes on the image dynamic range and signal to noise ratio. However, to achieve a high timing resolution using over-sampling tehnique a huge data volume has to be acquired and processed in real time. It need high capacity storage. To avoide such problem, sampling process was just conducted above the Nyquist rate and interpolation was performed to achieve the required delay resolution.

This paper describes the progress that has been made in the design and construction of a prototype real-time FGPA-based digital beamformer for 64-channel 3.5 MHz linear arrays probe. The design of digital beamforming and how it was implemented on FPGA is described. Prior to the real application, simulated RF data were used to test the performance of the receive beamformer. The result shows that system be able to process RF signals and display the 2D image. Images from small targets were used to assess both axial and lateral resolutions of the system.

II. METHODOLOGY

The imaging hardware which consists of an 8 channel analog front end circuit were first designed and contructed to provide real experimetal data. Digital beamformer was designed and implemented on low cost Xillinx Spartan 6 development board of Opal Kelly [15]. The echo signals were acquired using 64-element 3.5 MHz ultrasound probe. A scan conversion algorithm was developed and implemented using Matlab for displaying the 2D ultrasound image.

A. Analog Front End

The analog front-end circuit consists of 5 boards: 8 channel programmable transmit beamformer board, 8 channel pulser board, 8 channel TX-RX board, 8 to 64 multiplexer and 64 to 8 demultiplexer HV switch board, 8 channel receiver board. The transmit beamformer, pulser, and TX-RX boards contains LM96570 Octal Digital Beamformer, LM96550 Octal High Voltage Pulser, LM96530 Octal Transmit/Receive Switch Chips of National

J. Goh and C.T. Lim (eds.), *7th World Congress on Bioengineering 2015*, IFMBE Proceedings 52, DOI: 10.1007/978-3-319-19452-3_51

Semiconductor, respectively. The receiver board incorporates the 8 channel AD9272 from Analog Device.

The programmable transmit beamformer board serves as the signal generator by outputting programmable digital pulse patterns. The LM96570 beamformer provides an 8-output channels designed to drive the positive and negative pulse control inputs of LM96550. Each channel launches an individually programmable pulse pattern with a maximum delay of 102.4μs in adjustable in increments of 2 ns. The length of a fired pulse pattern was set to 8 pulses and can be extended to 64 pulses. The pulse patterns and delay settings can be programmed into and read out from the individual channel controls via a four-wire serial interface [11].

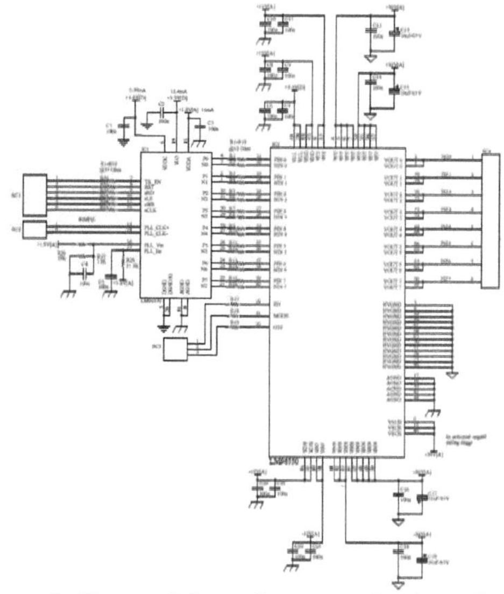

Figure 1. Transmit beamformer and pulser circuit

The programmed pulse patterns are sent out by activating the transmit signal TX_EN. Upon a rising edge of the transmit signal TX_EN, the delay counter of each channel starts counting according to the programmable delay profile. When the counter reaches the 17-bit programmed delay value, the programmed pulse pattern is sent out continuously at the programmed frequency until it reaches the length of the pulse pattern. These output pulse patterns then drive the inputs of the high voltage pulser, LM96550.

The LM96550 pulser board contains eight high-voltage pulser with integrated diodes generating ±50V bipolar pulses with peak currents of up to 2A and pulse rates of up to 15 MHz. The LM96550 has an input buffer at its logic interface. It is powered by VLL (2.5 to 5 Volt) and controlled by EN. When EN is set to HI, and the channel inputs PIN n or NIN n is HI will result in a positive or negative pulse at the channel's output pin [12]. The pulser outputs drive both the inputs of the LM96530 T/R switch board. Fig. 1 shows the transmit beamformer and pulser circuitry.

The LM96530 TX/RX board is triggered by 8, 5-volt, rising edge signals. The signals pass through 8 1-of-8 HV demultiplexers (Supertex HV2301, Supertex inc.) so that 8 adjacent elements are chosen. Triggered by these rising-edge signals, the pulser circuit generates a 50 volt impulse of 10 ns duration for each chosen element. Immediately following the transmitter/receiver (T/R) switch, 8 8-to-1 multiplexers are used to select the corresponding activated element out of 64 elements. The multiplexer and demultiplexers are controlled by the same control signal to maintain the synchronization. In LM96530 chip, a voltage clamping was installed. It is to protect the following circuit. The LM96530 TX-RX switchs as shown in Fig. 2 are directly connected to transducer elements and receiver board. Since, a transducer element has the dual functions of transmitting and receiving ultrasound energy. During the transmitting phase, high voltage pulses are applied to the ceramic elements. A typical transmit/receive (T/R) switch, LM96530, can consist of four high voltage diodes in a bridge configuration. Although the diodes ideally block transmit pulses from the sensitive receiver input, diode characteristics are not ideal, and the resulting leakage transients imposed on the inputs can be problematic [13].

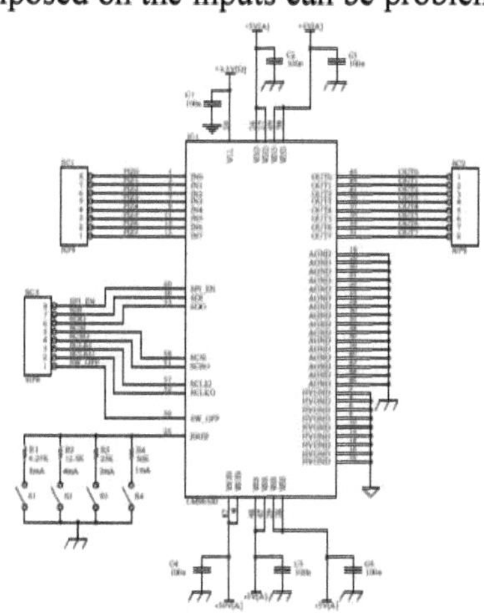

Figure 2. TX/RX switch circuit

The receiver board incorporates 8 channel AD9272 chip from Analog Device. Echo signals from chosen elements enters the AD9272 input channells. Each channel contains both TGC signal path and a CW Doppler signal path. Common to both signal path, the LNA provides user-adjustable input impedance termination. The CW Doppler path includes a transconductance amplifier and a crosspoint switch. The TGC path includes a differential VGA, an antialiasing filter, and a 12-bit, 10 MSPS to 80 MSPS analog-to-digital converter. The LNA has capability amplifying the signals up to 21 dB. VGA has a -42 dB to 0 dB controllable gain stage. Anti aliasing filter (AAF) has programmable second order low pass filter from 8 MHz to 18 MHz and programmable high pass filter which can be set from 775 kHz to 11.5 MHz. The function of antialiasing filter is used to reject dc signals and to prevent noise outside the frequency range of the signals corrupting the signal quality. This anti aliasing filter is placed prior to the ADC. The signal paths are fully differential throughout to maximize signal swing and reduce even-order distortion; however, LNA is designed to be driven from a single-ended signal source. The LNA supports differential output voltages as high as 4.4 V p-p with positive and negative excursions of ±1.1 V from a common-mode voltage of 1.5 V. The LNA differential gain sets the maximum input signal before saturation. The gains are set through the SPI [10].

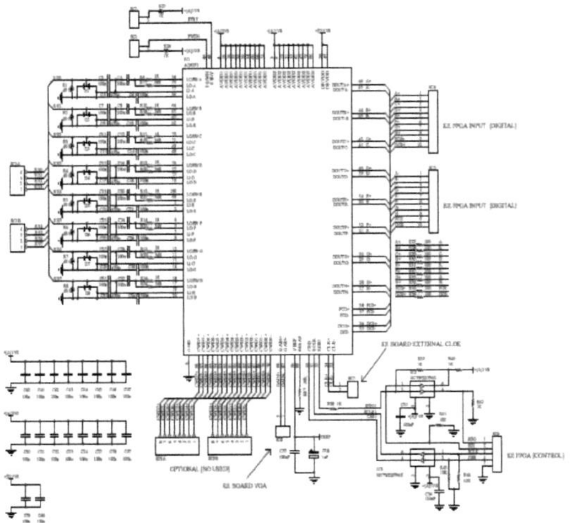

Figure 3. Eight channels receiver circuit

Because ultrasound is a pulse system and time-of-flight is used to determine depth, quick recovery from input overloads is essential. Overload can occur in the preamp and the VGA. Immediately following a transmit pulse, the typical VGA gains are low, and the LNA is subject to overload from T/R switch leakage. With increasing gain, the VGA can become overloaded due to strong echoes that occur near field echoes and acoustically dense materials. Fig. 3 shows the 8 channels receiver circuit.

B. FPGA Based Digital Beamformer

The digital beamformer can be split into two main functional blocks: transmit and receive beamformer. Each block contains complex multiplication, additional, and delay operations. There is also need for control logics which handles the I/O transfers and the internal data transfers. In general receive beamformer can be devided into three stages : delay stage, summation stage, and digital signal processing stage.

Delay stage in receive beamformer consists of coarse and fine delay unit, dynamic focusing and apodization filter. The coarse delay unit is an integer multiple of the clock periode. It is accomplised by using a programmable 4-tap FD FIR filter which generates fine delays less than one clock period.

In the delay unit, the received digital RF data are delayed according to the integer part of the delay information for up to 512 clock cycles. In principle, this corresponds to FIFO operation of variable length with 512 entries of 10 bit, which means that this unit is preferably implemented as a power efficient memory with one read and one write operation per cycle using a pointer based addressing scheme.

The fine delay is implemented using the distributed arithmetic FIR filters, which use full-parallel, fixed coefficient, finite impulse response (FIR) digital filter structures, are used to implement a fine delay from 2 ns to 8 ns. Immediately after those filters, a 8:1 bus multiplexer is used to select the fine delay according to the delay coefficient. For both the coarse delay and fine delay, the delay coefficients are precalculated and stored in the state machine.

In the present design, the dynamic focusing procedure is performed by updating the receive delay for different focus depth. Instead of calculating the delay coefficients in real time, the calculation results are stored in FPGA and are updated in time according to the depth of the echoes. The transition between two states is triggered by a 10-bit counter. In this work, windowing functions (Hamming) is used as apodization functions to reduce the side lobes.

In the summation stage, the final beam data which is also called as beam line is obtained by summing up the output from all the channels. Total 64 beam lines were obtained for 64 elemen probe.

The DSP stages consists of Hilbert transform, envelope detection, compressed dynamic range, and scan converter algorithms to obtain the 2D images. The scan converter code was written in Matlab and implemented in the personal computer. All the algorithms were writen in VHDL and implemented on Opal Kelly FPGA development board, which contains Xilinx Spartan 6 series FPGA chip[15]. The photograph of the electronic board with 64 element probe is shown in Fig.4.

In order to evaluate the performance of the digital beamformer, simulated and real data were used. The simulated data were generated using FIELD II with center frequency of the probe of 3.5 MHz [14]. The real data were obtained from wire target of multipurpose phantom (ATS Laboratories).

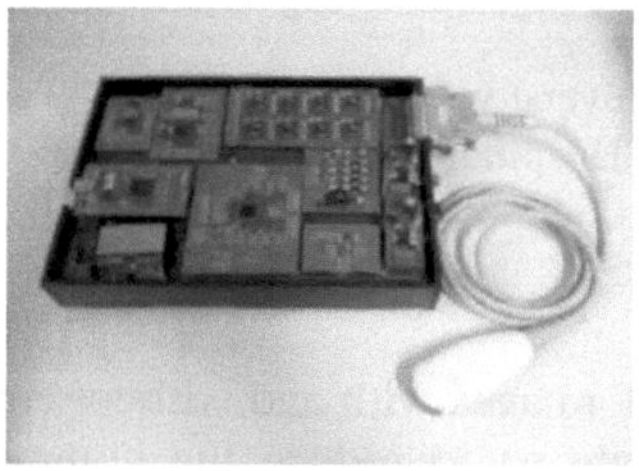

Figure 4. Electronic board for analog front end, FPGA based digital beamformer and 64 elemen probe.

III. RESULTS

A. RF Data

The beamformer had been used to process the simulated RF data. Hilbert transform of simulated RF data of cyst phantom is shown in 6.

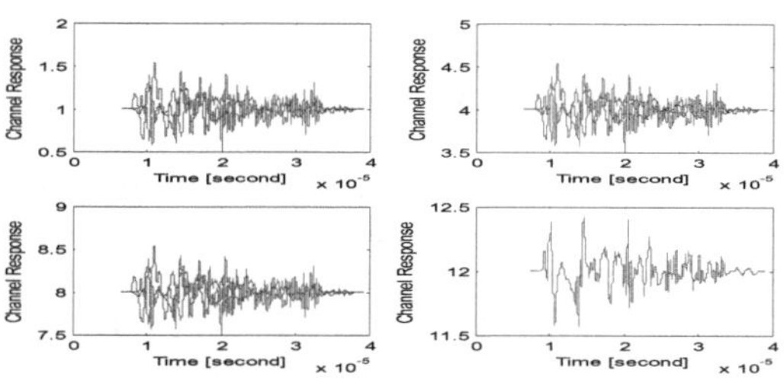

Figure 6. Hilbert transform of simulated RF data.

B. 2D Image

Figure 7 shows the 2D image of wire phantom. The scan conversion software is capable to reconstruct 2D images from experimental RF data.

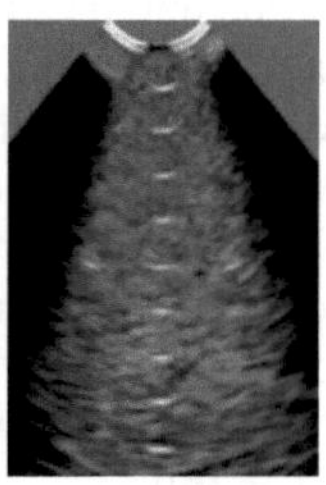

Figure 7. 2D image of wire phantom.

IV. CONCLUSIONS

An electronic system which consists of transmit beamformer, pulser, TX/RX switch, and receiver has been developed for acquiring the experimental data.

The FPGA based digital beamformer had been succesfully developed for processing the RF signals. The digital beamformer was implemented on Xillinx Spartan 6 FPGA chip. Beamformer performance had been succesfully firstly tested with the simulated RF data. A scan conversion algorithm had also been developed for contructing the 2D image. The sytem had been tested for imaging the small wire of multipurpose phantom.

ACKNOWLEDGMENT

The research was funded by the Agency for the Assessment and Application of Technology, Republic of Indonesia, for fiscal year 2011-2014.

REFERENCES

1. W. N.McDicken, Diagnostic Ultrasonics: Principals and Use of Instruments, JohnWiley & Sons, (1976).
2. Huang SW, Li PC (2006) Arbitrary waveform coded excitation using bipolar square wave pulsers in medical ultrasound. IEEE Trans Ultrason Ferroelectr Freq Control 53:106–116.
3. Hassan M.A, Kadah Y.M. (2013) Digital Signal Processing Methodologies for Conventional Digital Medical Ultrasound Imaging System, American Journal of Biomedical Engineering 3(1): 14-30.
4. Steinberg, B.D.(1992) Digital beamforming in ultrasound," IEEE Transactions on Ultrasonics, Ferroelectrics, and Frequency Control, vol. 39, pp. 716–721.
5. Koelling, T. (2010) Xillinx FPGAs in Portable Ultrasound Systems, White Paper: 7 Series, Virtex-6 and Spartan-6 FPGAs.
6. Tomov B.G. and Jensen JA (2001) A new architecture for a singlechip multi-channel beamformer based on a standard FPGA," in Proc. IEEE Ultrason. Symp., pp. 1529–1533.
7. Cao P.J. and Shung K.K (2002) Design of a real time digital beamformer for a 50 MHz annular array transducer," in Proc. IEEE Ultrason. Symp., pp. 1619–1622.
8. Tomov BG and Jensen J.A (2003) Compact FPGA-based beamformer using oversampled 1-bit A/D Converter", IEEE Trans. on Ultrasonics, Ferroelectrics, and Frequency Control.
9. Chang-Hong Hu, Xiao-Chen Xu, Jonathan M. Cannata, Jesse T. Yen, and K. Kirk Shung (2006), Development of a Real-Time, High-Frequency Ultrasound Digital Beamformer for High-Frequency Linear Array Transducers, IEEE Trans. on Ultrasonics, Ferroelectrics, and Frequency Control, Vol. 53, No. 2.
10. AD9272 datasheet http://www.analog.com
11. LM96570 data sheet http://www.ti.com/
12. LM96550 data sheet http://www.ti.com/
13. LM96530 data sheet http://www.ti.com/
14. Jensen, J.A.(2000) Ultrasound Imaging and Modeling, Department of Information Technology, Technical University of Denmark.
15. https://www.opalkelly.com/

Use macro [author address] to enter theaddress of the corresponding author:

Author: Pratondo Busono
Institute: BPPT
Street: PTIK, Gedung Teknologi 3 Lt.3
City: Tangerang Selatan
Country: Indonesia
Email: pratondo.busono@bppt.go.id

The Development of Pancreatic Cancer CAD System for CT and US Images

Jenn-Lung Su[1], Guan-Wen Chen[1], Yu-Chieh Hsu[1], Chia-Hung Pan[1], and Ming-Hui Lin[2]

[1] Department of Biomedical Engineering, Chung Yuan Christian University, Chung Li, Taiwan
[2] Division of Gastroenterology and Hepatology, Tao-Yuan General Hospital, Taoyuan, Taiwan

Abstract— **The pancreatic cancer is extremely fatal. Due to limitations of anatomic location and condition, physicians are hard to make precise diagnoses of patients from traditional ultrasound (US) or CT images. The purpose of this study is to develop a computer-aided diagnosis (CAD) system for pancreatic tumor by the selected features from CT and US images. In this study, the following steps are included: (1) Segment images by applying GVF SNAKE; (2) Select features by applying t-Test; (3) Identify normal tissues, adenocarcinoma tumors, pseudo tumors, cystic tumors, and pseudo cyst by SVM and SOM for CT and US images, respectively. (4) Finally, totally diagnosed 69 US images and 136 CT images were used to evaluate system performance. In order to improve this system, different numbers of features were selected in three different stages for CT and US images. The results show this CAD system has the best performance to identify all images by applying 2 features (Area, NRL_MA) and 4 features (l_Average, g_Entropy, c_Entropy, Area) in US images and contrast injected CT images, respectively. Moreover, the tumor area is the most important morphological feature for tumor classification in US images and the adenocarcinoma tumor has lower value of "Entropy" in contrast injected CT images. In most cases, the performance (sensitivity, specificity, and accuracy are higher than 0.9) of this developed system is good enough for clinical study. However, US CAD system and CT CAD system have better performances on identifying tiny pancreatitis tumors and cystic tumors, respectively. We suggest physicians to diagnose tumors by the aid of US CAD system, and diagnose cysts by CT CAD system; consequently, reduce cost and improve the diagnostic accuracy.**

Keywords— **Pancreatic cancer, Ultrasound images, CT images, CAD system, Image processing.**

I. INTRODUCTION

Pancreas is an important organ in digestive system; however, owing to the physiological location, it is hard for physicians to diagnose. In addition, different kinds of pancreatic tumors, its fatality, severity, treatment effect are all different; what's more, the patients with pancreatic cysts are more predisposed to pancreatic cancers. Nowadays, abdominal ultrasound (US) and computed tomography (CT) are the most two common ways for physicians to make diagnosis.

According to the prior research, the pancreatic tumors in CT images tend to have the distinguishable textural features in both CT images with and without contrast injection, which may allow us to reduce the contrast injection for patients. On the other hand, the morphological features are a lot distinguishable in US images, which tend to be regarded as a criterion for identifying malignant tumors from benign tumors by the surface area and the edge roughness [1]. In general, morphological features are a lot significant than textural features when classifying images.

In this study, we utilized some image processing technologies to reduce the errors of image segmentation; besides, improved the diagnostic performance of the system; furthermore, reduce the probability of injection contrast for patients.

II. MATERIALS AND METHODS

The database had total 205 pathologically proven digital images of pancreas. It included 69 US images (13 normal tissue, 18 Adenocarcinoma tumor, 6 Pancreatitis pseudo-tumor, 23 Cystic tumor, and 9 Pancreatitis pseudo-cyst), 68 CT images (6 normal tissue, 12 Adenocarcinoma tumor, 3 Pancreatitis pseudo-tumor, 16 Cystic tumor, and 31 Pancreatitis pseudo-cyst), and 69 injected contrast medium CT (IcmCT) images (6 normal tissue, 10 Adenocarcinoma tumor, 6 Pancreatitis pseudo-tumor, 13 Cystic tumor, and 33 Pancreatitis pseudo-cyst) were used to develop and evaluate this CAD system, respectively. The image format and resolution of each US image was 24 bits and 640×480 pixels; each CT image was 8 bits and 512×512 pixels.

The proposed image processing flowchart in this study was shown in Figure 1. First of all, we separated our flow chart into CT images and US images, inspecting pancreatic tumors by CT and US CAD systems, respectively. In order to save time for system operation, an experienced physician has manually extracted the region of interest (ROI) sub image in an US or CT image with/without image processed. Each step is described as follow:

(1)Preprocessing: We utilized median filter and histogram equalization to get rid of the noises in CT images and US images, respectively. In consequence, we are able to gain some information that we can't get it from the original images [2].

J. Goh and C.T. Lim (eds.), *7th World Congress on Bioengineering 2015*,
IFMBE Proceedings 52, DOI: 10.1007/978-3-319-19452-3_52

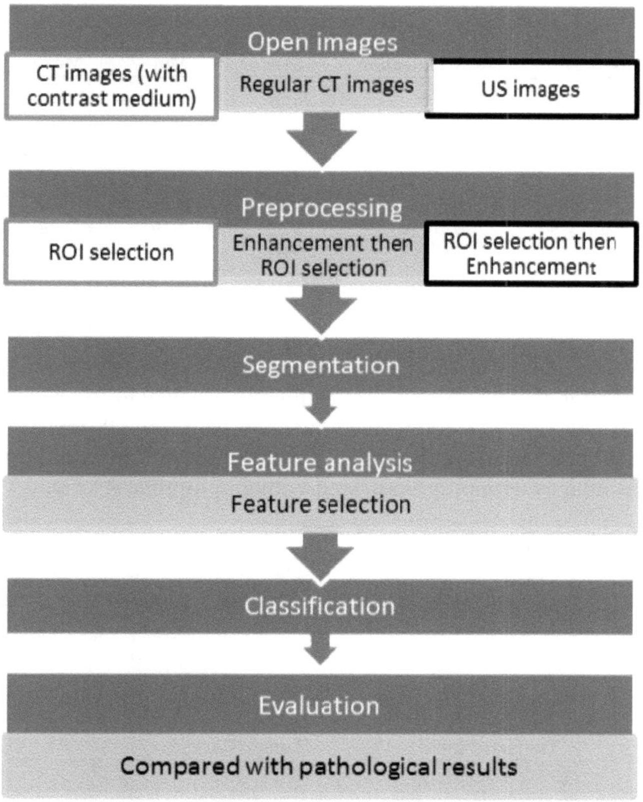

Fig.1 Proposed system development flowchart

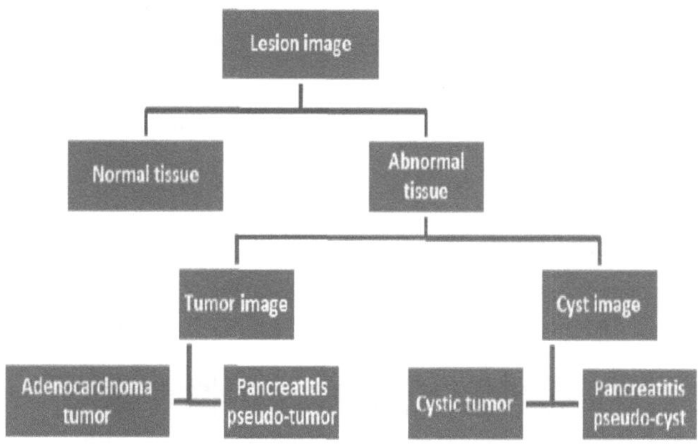

Fig. 2 Proposed image classification flowcharts

(2) Image segmentation: The Gradient Vector Flow Snake method was used to segment the tumors images and also avoids the edge discontinuousness as well as contours noises; in consequence, get the features of tumors [3].

(3) Feature analysis: According to our prior research, figure out the significant features in both CT and US images by applying t-test is an effective approach for identifying different kinds of tumors. In this study, we categorized pancreas tumors into adenocarcinoma tumor; pancreatitis pseudo-tumor, cystic tumor, and pancreatitis pseudo-cyst (see Figure 2). And the statistically effective features ($p < 0.05$) were selected and tested for image classification.

(4) Classification: Support Vector Machines (SVM), Self-Organizing Map (SOM) were used for classifying tumors in CT images and US images, respectively. In addition, there are three stages in our classification such as (1) stage 1: identify the tumors from the normal tissues; (2) stage 2: identify the cysts from the tumors; (3) stage 3: identify the adenocarcinoma tumors from pancreatitis pseudo-tumors, and the cystic tumors from the pancreatitis pseudo-cysts.

(5) Evaluation: Finally, system was test and evaluated by using train set and test set method and comparing with pathological results of patients. When it comes to system evaluation, Kappa value must be more than 0.4 which is reliable. Moreover, we also compared the results of CT images with/without contrast injection, in order to provide the information which may lead to reduce the needs of injecting medical contrast medium into patients.

III. Results and discussion

A. Results for CT CAD system

First of all, we separated CT images into CT images with contrast injection and CT images without contrast injection, and ROI was picked shown as Figure 3 and 4. In this study, we utilized features as l_Average, c_Average, g_Entropy, c_Entropy, Area, Lesion_Entropy, and Lesion_Mean to identify the tumors in the CT images with contrast injection. It turned out that Accuracy is 0.9354, shown in Table 1[4].

On the other hand, we utilized l_Average, g_Entropy, c_Entropy, Area to identify tumors in CT images, it turned out that accuracy is 0.9677, shown in Table 2. And the time cost to classify an image in different Stages is shown in Table 3.

In three different stages, the features of l_Average, g_Entropy, c_Entropy, Area should be taken into account first. Lower l_Average means the edge of tumor is smoother; furthermore, the tumor is more likely to be benign.

On the contrary, the edge of tumor is rougher; the tumor is more likely to be malignant. Besides, Entropy means the variety of the grayscale; physiologically, it shows the structures of our body. To a certain extent, Entropy can tell the tumors from the normal tissues. In general, the normal tissues tend to be with lower entropies, and different kinds of tumors are with various entropies as well. In addition, Area means the surface area of a tumor which is a helpful criterion for telling a tumor, as shown in Table 4.

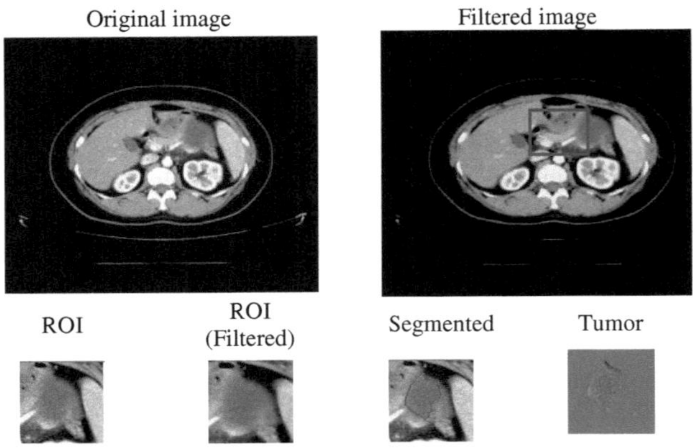

Fig. 3 The ROI of CT images without contrast injection

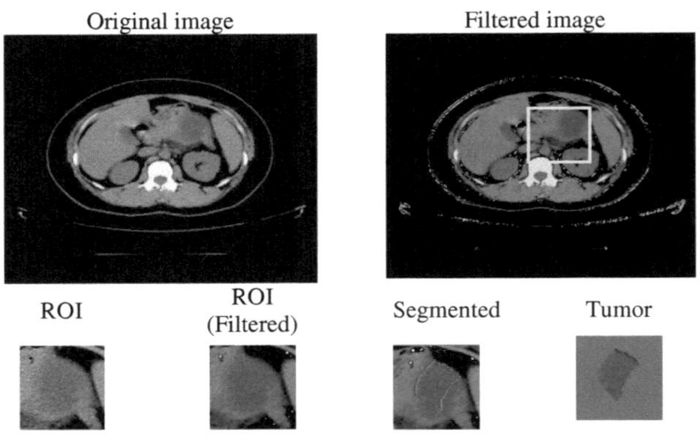

Fig. 4 The ROI of CT images with contrast injection

Table 1 System evaluation on CT images with contrast injection

	Accuracy	Sensitivity	Specificity	Kappa
Stage1	1	1	1	1
Stage2	0.96	0.87	1	0.91
Stage3(Tumors)	1	1	1	1
Stage3(Cysts)	0.95	1	0.94	0.89

Table 2 System evaluation on CT images without contrast injection

	Accuracy	Sensitivity	Specificity	Kappa
Stage1	1	1	1	1
Stage2	0.93	0.71	1	0.79
Stage3(Tumors)	1	1	1	1
Stage3(Cysts)	1	1	1	1

Table 3 Single cycle time of SVM classification

CT images	with contrast injection	without injection
	Single cycle time(sec)	
Stage 1	13	12
Stage 2	12	11
Stage 3	11	10

Table 4 Significant features for CT images

Features	Contrast injection	Max	Min	Mean
I_Average	With	87.94	17.23	28.0
	Without	88.20	18.36	27.96
g_Entropy	With	9.09	6.77	7.26
	Without	9.05	6.69	7.18
c_Entropy	With	9.15	6.75	7.29
	Without	9.12	7.29	7.23
Area	With	23683	811	2511
	Without	23693	1023	2537

B. Results for US CAD system

Secondly, we disposed US images by means of histogram equalization to enhance the contrast of the US images, shown in Fig 5. Accordingly, we separated US images into US images with preprocess and without preprocess as well.

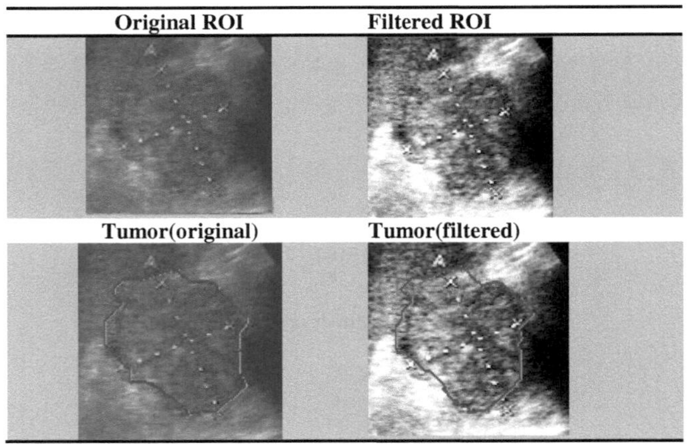

Fig. 5 The result of preprocessed US image

Table 5 Significant features for US images

Stage	Features						
1	Ori.	Area	Circu.	Mean	M.A	R.I	Skew
	Pre.	Area	Circu.	Mean	M.A	R.I	
2	Ori.	Area	Circu.	S.D.	M.A	R.I	
	Pre.	Area	Circu.	S.D.	M.A	R.I	
3 (tumors)	Ori.	Area	Circu.	M.A	R.I		
	Pre.	Area	Circu.	M.A	R.I		
3 (cysts)	Ori.	Area	Mean	S.D.	M.A		
	Pre.	Area	Mean	S.D.	M.A	R.I	

Table 6 System evaluation on US images without preprocess

	Accuracy	Sensitivity	Specificity	Kappa
Stage1	0.957	0.964	0.923	0.862
Stage2	0.946	0.958	0.938	0.891
Stage3(Tumors)	0.875	0.889	0.833	0.684
Stage3(Cysts)	1.000	1.000	1.000	1.000

Table 7 System evaluation on US images with preprocess

	Accuracy	Sensitivity	Specificity	Kappa
Stage1	0.942	0.964	0.846	0.810
Stage2	0.929	0.917	0.938	0.854
Stage3(Tumors)	0.917	0.944	0.833	0.778
Stage3(Cysts)	1.000	1.000	1.000	1.000

After that, we utilized 5, 5, 4, and 4 features in stage1, 2, 3(tumors), 3(cysts), respectively, shown in Table 5. It turned out that the evaluation of US images with and without preprocess are shown in Table 6 and 7. Moreover, it takes 40, 32, and 27 sec in stage 1, 2, and 3, respectively.

In 3 different stages, the features of Area, Circularity, Mean-Area, and Roughness-Index should be taken into account first. According to the clinical experience, a tumor with larger surface area is more likely to be a malignant tumor, vice versa. Besides, a benign tumor tends to be well-rounded, i.e. the circularity should be approximately 1; furthermore, a benign tumor tends to be smooth, i.e. the roughness-index should be as lower as possible. In consequence, our research result definitely matches the clinical experience.

C. *Comparison for different modality*

Compared the result of CT and US CAD system, their preprocessing process and significant features are not similar, due to their different imaging process.

In 3 classification stages; by comparison, the evaluation of CT CAD on stage2 (tumors and cysts) is a little lower than US CAD system, the overall evaluation, however, both CT and US CAD system are acceptable, shown in Table 8.

According to the physician's advice, it is better to use US images for telling the tiny lesion; on the contrary, it is better to use CT images for telling the tumors with the larger scale.

Table 8 Comparison of CT and US CAD system

Stage	Image types		Sens.	Spec.	Acc.	Kappa
1	CT	with	1	1	1	1
		without	1	1	1	1
	US		0.911	0.846	0.899	0.695
2	CT	With	0.87	1	0.96	0.91
		with-out	0.714	1	0.93	0.79
	US		0.91	0.846	0.92	0.85
3 (tumors)	CT	With	1	1	1	1
		without	1	1	1	1
	US		0.88	0.83	0.87	0.68
3 (cysts)	CT	With	1	0.94	0.95	0.80
		without	1	1	1	1
	US		0.95	1	0.96	0.92

The result of our research matches the clinical experience; therefore, we suggest physicians to use US for identifying the solid tumors; on the other hand, to use CT for identifying cysts in order to both reduce the waste of medical resource and make the diagnosis more precisely.

IV. Conclusion

In this study, a CAD system is developed for enhancing images and providing doctors with reliable features as well as information. Textural and morphological features were utilized to analyze pancreas CT images; furthermore, we enhanced the images by preprocess for the sake of reducing the probability of injecting contrast for patients. We also figured out that the morphological features in US images are more significant than the textural features. According to our result, there are both 4 features are significant in CT and US images.

With the CAD systems that we developed, physicians are able to make diagnoses more precisely and avoid misdiagnoses; moreover, to combine the other kinds of medical images in the future to improve the medical quality is possible in the near future.

Acknowledgment

This work was supported by the National Science Council, R.O.C., under Grant NSC102-2221-E-033-005.

References

[1] Jenn-Lung Su, The Development of CAD system for Pancreatic Tumors, NSC Projects Final Report, 2010, NSC99-2221-E-033-040.

[2] Sha-Kang Mou, Digital Image Processing, Pearson Education, Taipei, 2003: 373-385.

[3] Krit Somkantha, Nipon Theera-Umpon, and Sansanee Auephanwiriyakul, "Boundary detection in medical images using edge following algorithm based on intensity gradient and texture gradient features," *IEEE Transactions on Biomedical Engineering*, 2011, 58(3): 567-573.

[4] Yu Guan-Wen Chen, The Development of Computer-Aided Diagnosis System for Pancreatic Tumor Recognition, *Master Thesis*, Department of Biomedical Engineering, CYCU, Chung-Li, 2015.

[5] Yu-Chieh Hsu, The Development of Computer Aided Diagnosis System for Pancreatic Tumor Recognition in Ultrasound Images, *Master Thesis*, Department of Biomedical Engineering, CYCU, Chung-Li, 2015

Author: Jenn-Lung Su, Ph.D.
Dept. of Biomedical Engineering
Chung Yuan Christian University,
Chung Li, Taoyuan City, 32023,
Taiwan
E-mail: jlsu@cycu.edu.tw

Wavelength Optimization of Spectral Near-infrared Optical Tomography Using Available Laser Diodes

Min-Chun Pan[1,2], Liang-Yu Chen[2], Chung-Chen Yan[2], and Min-Cheng Pan[3,*]

[1].Department of Mechanical Engineering, National Central University, Taoyuan City 320, Taiwan
[2] Graduate Institute of Biomedical Engineering, National Central University, Taoyuan City 320, Taiwan
[3] Department of Electronic Engineering, Tung-Nan University, New Taipei City 222, Taiwan

Abstract—**As optimized wavelengths are used, a near-infrared (NIR) tomographic imaging system using multi-wavelengths in continuous wave has the potential to provide accurate information of chromophores. In this paper, we discuss wavelength optimization with commercial laser diodes. Through theoretical analysis, we use the residual norm (R) and the condition number (κ) representing the uniqueness of this matrix problem (Eq. (6)) and the smooth singular-value distribution of each chromophore, respectively. The optimum wavelengths take place as large R and small κ. We found 37 wavelengths in total from the 633nm-980nm range available in the market to discover optimum wavelength sets for a broad range of chromophore combinations. From a complete set of 37 wavelengths, there exist 7770 ($C^{37}{}_3$), 66045 ($C^{37}{}_4$), and 435897 ($C^{37}{}_5$) combinations of three, four, and five wavelength sets, respectively, for accurately estimating chromophores (HbO_2 and HbR) plus the scattering prefactor *A*. With the numerical calculation, top 10 wavelength sets were selected based on high R and low κ. Furthermore, it is found that 650, 730, 905, *705*, and 690 nm are optimal with a same part for three, four, and five wavelength sets, respectively.**

Keywords— **Tomographic imaging, chromophore, wavelength optimization, residual norm, condition number.**

I. INTRODUCTION

As optimized wavelengths are used, a near-infrared (NIR) tomographic imaging system using multi-wavelengths in continuous wave has the potential to provide accurate information of chromophores (oxyhemoglobin (HbO_2), deoxyhemoglobin (HbR), water (H_2O), lipid) with diminishing crosstalk between these parameters. Previously, there mainly were two groups [1, 2] studying and publishing relevant research about the optimum wavelength sets: one is that the wavelengths were chosen from the 650-930nm range, spaced in 6 nm intervals and the other discussed six discrete wavelengths selected from the entire spectral range of 650 to 930 nm with a separation of 4 nm.

However, these analyses are useful but their laser systems must be tuning with a fixed interval of 4 or 6 mm, implying expensive laser systems were operated. Instead, lots of commercial laser diodes are currently available, which are economical, low-cost maintenance, and miniaturized. Therefore, we discuss wavelength optimization with these commercial laser diodes in this paper.

In the following, Sec. 2 describes the theoretical method of wavelength optimization. Subsequently, we present three, four, or five optimized wavelengths for estimating two optical parameters, HbO_2 and HbR, selected through theoretical analysis from 37 commercially available wavelengths in total. Finally, a concluding remark is given.

II. METHOD

The diffusion equation of continuous wave light can be written as

$$\nabla \cdot D(\mathbf{r})\nabla\Phi(\mathbf{r}) - \mu_a(\mathbf{r})\Phi(\mathbf{r}) = -S(\mathbf{r}) \tag{1}$$

where $\Phi(\mathbf{r})$ is the photon density at position $\mathbf{r}$, $S(\mathbf{r})$ is the isotropic source term as well as μ_a and D denote the optical absorption and diffusion coefficients, respectively. Further, D is equal to $1/3(\mu_a + \mu_s^{'})$ where $\mu_s^{'}$ is the reduced scattering coefficient. Given a set of boundary data at a given wavelength, the aim of the inverse problem is to recover the distribution of optical parameters in tissue. By means of the first order Taylor series to expand $\mathbf{\Phi}$, one can get Eq. (2),

$$(\mathbf{\Phi}^M) \approx (\mathbf{\Phi}^C) + \left[\frac{\partial \mathbf{\Phi}^C}{\partial \boldsymbol{\mu}_a}\right](\Delta\boldsymbol{\mu}_a) + \left[\frac{\partial \mathbf{\Phi}^C}{\partial D}\right](\Delta D), \tag{2}$$

where $\mathbf{\Phi}^M$ is the measured data and $\mathbf{\Phi}^C$ is the calculated data from the forward model.

From Eq. (2), the inverse problem can be formulated as

$$\begin{bmatrix} \frac{\partial \mathbf{\Phi}^C}{\partial \boldsymbol{\mu}_a} & \frac{\partial \mathbf{\Phi}^C}{\partial D} \end{bmatrix} \begin{pmatrix} \Delta\boldsymbol{\mu}_a \\ \Delta D \end{pmatrix} = (\mathbf{\Phi}^M - \mathbf{\Phi}^C), \tag{3}$$

* Corresponding author.

J. Goh and C.T. Lim (eds.), *7th World Congress on Bioengineering 2015*,
IFMBE Proceedings 52, DOI: 10.1007/978-3-319-19452-3_53

or simply denoted as $\mathbf{J}\Delta\chi = \Delta\mathbf{\Phi}$, where $\mathbf{J} = \left[\partial\mathbf{\Phi}^C/\partial\boldsymbol{\mu}_a \quad \partial\mathbf{\Phi}^C/\partial D\right]$ is the Jacobian matrix, i.e., the rate of change of model data with respect to optical parameters, and $\Delta\chi$ means $\begin{pmatrix}\Delta\boldsymbol{\mu}_a \\ \Delta D\end{pmatrix}$, the increment of optical parameters. The goal of Eq. (3) is to reach the measured data $\mathbf{\Phi}^M$ from the calculated data $\mathbf{\Phi}^C$ estimated from the forward model, i.e., minimizing the difference between the measured data and the calculated data while estimating optical parameters, μ_a and D, with the increment variation. However, such an inverse problem is nonlinear and ill-posed; to cope with this, Tikhonov regularization is usually implemented. Therefore, the inverse problem is formulated as an optimization of the damped least squares problem,

$$\min_{\Delta\chi}\{Q_{Tk}(\Delta\chi)\} = \min_{\Delta\chi}\left\{\|\mathbf{J}\Delta\chi - \Delta\mathbf{\Phi}\|_2^2 + \lambda^2\|\mathbf{L}\Delta\chi\|_2^2\right\} \quad (4)$$

where $\mathbf{L}$ is the dimensionless regularization matrix and λ is the regularization parameter. Specifically, Eq. (4) is called as zero-order Tikhonov regularization, provided that the identity matrix is adopted as regularization matrix, i.e., $\mathbf{L} = \mathbf{I}$. Subsequently, Eq. (5) can be obtained as

$$(J^TJ + \lambda^2 I)\Delta\chi = J^T\Delta\mathbf{\Phi}. \quad (5)$$

Generally, Eq. (5) is the basic calculation widely used for the inverse problem to estimate the optical parameters (μ_a and D) of a high-scattering matter injected with a light.

As known from Beer's law, the absorption coefficient at a given wavelength can be formulated by a linear summation of the extinction coefficients (ε) [3] and their corresponding chromophore concentrations (C); i.e., it can be expressed as

$$\mu_a(\lambda_n) = \sum_{i=1}^{L}\varepsilon_i(\lambda_n)C_i \quad (6)$$

where i and L are the i-th chromophore and the number of chromophores, respectively, n is the n-th wavelength, and there are N wavelengths in total. Empirically, the reduced scattering parameter is given by

$$\mu_s'(\lambda) = A\lambda^{-s_p} \quad (7)$$

where A is the scattering amplitude and S_p is the scattering power.

Through theoretical analysis [1], here, we use the residual norm (R) and the condition number (κ) representing the uniqueness of this matrix problem (Eq. (6)) and the smooth singular-value distribution of each chromophore, respectively. Furthermore, both parameters are defined as

(1) Uniqueness:

$$R = \left\|\mathbf{1} - \mathbf{E}\left(\mathbf{E}^T\mathbf{E}\right)^{-1}\mathbf{E}^T\mathbf{1}\right\| \quad (8)$$

where $\mathbf{E}$ is the extinction coefficient matrix normalized by the wavelength expressed as Eq. (9),

$$\mathbf{E} = \begin{bmatrix} \frac{\varepsilon_1(\lambda_1)}{\lambda_1^{s_p}} & \cdots\cdot & \frac{\varepsilon_L(\lambda_1)}{\lambda_1^{s_p}} \\ \cdots & \cdots\cdot & \cdots \\ \frac{\varepsilon_1(\lambda_N)}{\lambda_N^{s_p}} & \cdots\cdot & \frac{\varepsilon_L(\lambda_N)}{\lambda_N^{s_p}} \end{bmatrix} \quad (9)$$

(2) Condition number:

$$\kappa = \|\varepsilon(\lambda)\|\|\varepsilon^{-1}(\lambda)\| \quad (10)$$

As has been known, the optimum wavelengths take place as large R and small κ, shown as the top-left corner in Fig.1.

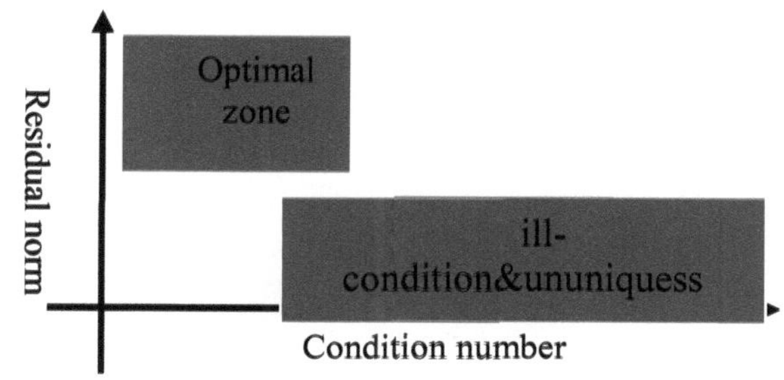

Fig. 1 Schematic of the distribution of residual nom and condition number.

Maximizing the residual norm (R) for a unique solution and minimizing the condition number (κ) for each chromophore to make similar contributions to absorption were used as criteria where R is the norm of wavelength-dependent matrixes and κ is the norm of the matrix of chromophore extinction coefficients. R and κ can be interpreted as the distinguishability parameters representing the uniqueness and stability of solution, respectively.

III. Results and Analysis

As described in introduction, a tunable laser system was previously required; however, there is no analysis and discussion concerning those wavelengths of laser diodes available in the commercial market. We found 37 wavelengths in total from the 633nm-980nm range available in the market to discover optimum wavelength sets for a broad range of chromophore combinations. The scatter plot of the residual and condition number was calculated for two chromophores, hemoglobin and deoxyhemoglobin. From a complete set of 37 wavelengths, there exist 7770 ($C^{37}{}_3$), 66045

($C^{37}{}_4$), and 435897 ($C^{37}{}_5$) combinations of three, four, and five wavelength sets, respectively, for accurately estimating chromophores (HbO_2 and HbR) plus the scattering prefactor A. Therefore, each point in the scatter plot represents a different combination set of 3, 4, or 5 wavelengths from the spectra 633 to 980 nm, as shown in Figs. 2, 4, and 6. Furthermore, the local up-right part is magnified, of which the region highlighted satisfies the criteria of large residual and small condition number, as shown in Figs. 3, 5, and 7.

A. Three Wavelength Sets

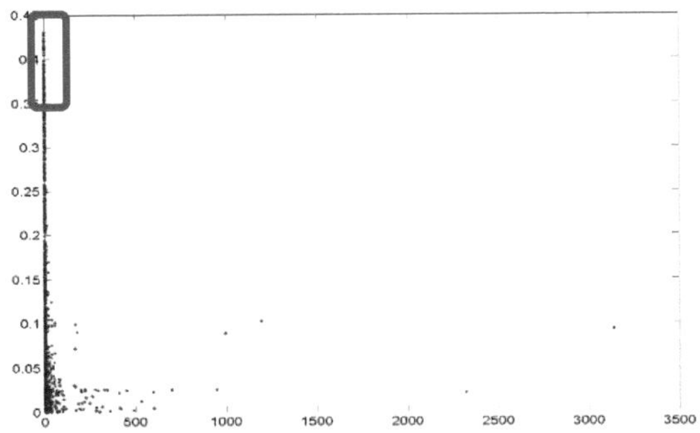

Fig. 2 Scatter plot of the residual and condition number; each point represents a different combination set of three wavelengths.

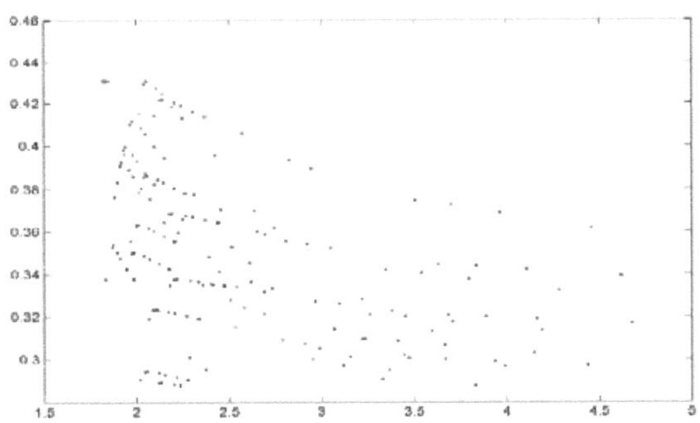

Fig. 3 Local magnification of Fig. 2.

With the help of Fig. (3), top 10 different combination sets of three wavelengths were selected as extreme points with high R and low κ, based on low κ rather than high R, as shown in Table 1.

Table 1 Top 10 different sets of three wavelengths

Top 10	R	κ	wavelengths		
1	2.055085	0.430449	**650**	**730**	**905**
2	2.043565	0.429020	650	730	915
3	2.111004	0.427022	650	730	880
4	2.140691	0.424099	650	730	870
5	2.140082	0.421696	650	705	905
6	2.126096	0.421023	650	705	915
7	2.207547	0.420034	650	705	880
8	2.243244	0.418624	650	705	870
9	2.194587	0.418288	650	730	860
10	2.305959	0.415895	650	705	860

B. Four Wavelength Sets

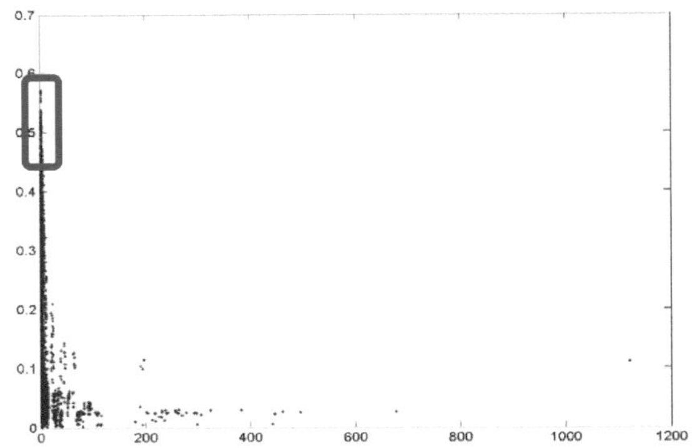

Fig. 4 Scatter plot of the residual and condition number; each point represents a different combination set of four wavelengths.

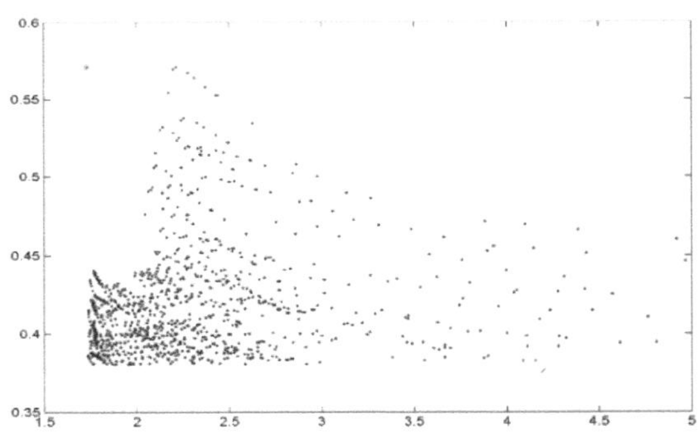

Fig. 5 Local magnification of Fig. 4.

Similarly, top 10 different combination sets of four wavelengths were also selected as extreme points with high R and low κ, based on low κ rather than high R, as shown in Table 2.

Table 2 Top 10 different sets of four wavelengths

Top 10	R	κ	wavelengths			
1	2.214408	0.570611	**650**	**705**	**730**	**905**
2	2.200795	0.569117	650	705	730	915
3	2.279708	0.566967	650	705	730	880
4	2.314012	0.563874	650	705	730	870
5	2.374586	0.557824	650	705	730	860
6	2.174103	0.554065	650	705	730	937
7	2.432230	0.552515	650	705	730	852
8	2.440576	0.552355	650	705	730	850
9	2.257986	0.537846	650	690	730	905
10	2.243220	0.536557	650	690	730	915

C. Five Wavelength Sets

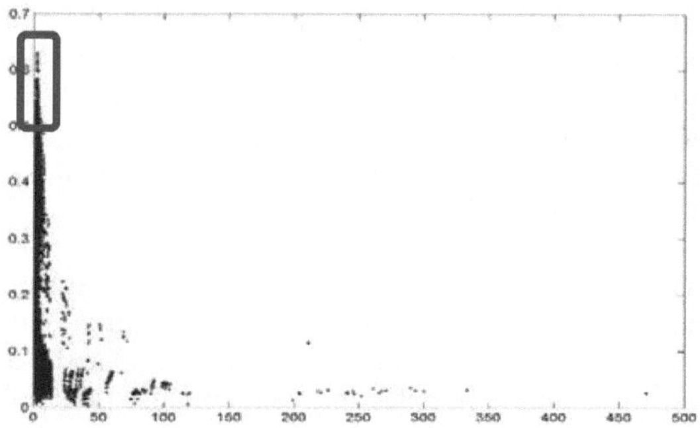

Fig. 6 Scatter plot of the residual and condition number; each point represents a different combination set of five wavelengths.

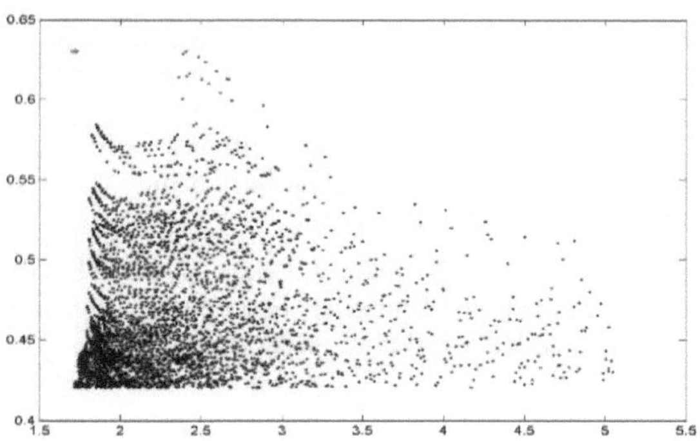

Fig. 7 Local magnification of Fig. 6.

Same as previously, top 10 different combination sets of five wavelengths were selected as extreme points with high R and low κ, based on low κ rather than high R, as shown in Table 3.

Table 3 Top 10 different sets of five wavelengths

Top 10	R	κ	wavelengths				
1	2.404570	0.630178	**650**	690	**705**	**730**	**905**
2	2.388165	0.628760	650	690	705	730	915
3	2.482524	0.626656	650	690	705	730	880
4	2.523151	0.623681	650	690	705	730	870
5	2.592608	0.617971	650	690	705	730	860
6	2.424929	0.616305	650	685	705	730	905
7	2.408115	0.614916	650	685	705	730	915
8	2.363395	0.614114	650	690	705	730	937
9	2.657929	0.613001	650	690	705	730	852
10	2.667320	0.612865	650	690	705	730	850

IV. CONCLUSIONS

In this paper, we have studied the wavelength optimization for reconstructing images of chromophores, HbO_2 and HbR. As a result, it is found that 650, 730, 905, *705*, and 690 nm are optimal with a same part for three, four, and five wavelength sets, respectively.

ACKNOWLEDGMENT

This research was sponsored by the Ministry of Science and Technology through grants NSC 101-2221-E-008-093-MY3 and MOST 103-2221-E-236-010.

CONFLICT OF INTEREST

The authors declare that they have no conflict of interest.

REFERENCES

1. Corlu A, Choe R, Durduran T et al. (2005) Diffuse optical tomography with spectral constraints and wavelength optimization. Appl Opt 44 :2082–2093
2. Eames M, Wang J, Pogue B, Dehghani H (2008) Wavelength band optimization in spectral near-infrared optical tomography improves accuracy while reducing data acquisition and computational burden. J Biomedical Optics 054037:1-9
3. Extinction coefficients at http://omlc.ogi.edu/spectra

Vein Tracking Using 880nm Near Infrared and CMOS Sensor with Maximum Curvature Points Segmentation

Anggunmeka Luhur Prasasti, Richard Karel Willem Mengko, and Widyawardana Adiprawita

Biomedical Engineering (BME), STEI
Institut Teknologi Bandung (ITB)
Bandung, Indonesia

Abstract— **Injection error usually occurs in obese patients and children (especially dehydrated children). More than one injection errors may cause uncomfortable condition for patient and can lead to hematoma. The tools which is used to obtain vein image tends to be expensive because it uses high intensity infrared camera with CCD sensor along with highly sensitive to infrared so that vein image can be seen very clearly. CMOS sensors can be an alternative to CCD because the price is more affordable and readily available. However, vein image from CMOS sensor is not as good as CCD sensor so it needs advanced image processing to get better vein image quality.**

The 880nm near-infrared light, CMOS sensor and 880nm optical filters, help clarify the veins existence up to 3mm below the skin surface. Image acquisition results have low contrast and spreading noise resembles info (vein) that thresholding can not be conducted to separate the info and background. Adaptive smoothing and retinex is used as preprocessing to eliminate noise and to enhance information and background contrast. Maximum curvature points methode is quite good in segmentation for vein location tracking. It can be used in phlebhotomy robot and calculation of many things related to vein changes, for example in detecting vein collapse which is very dangerous and important to be known. Testing on white skin subjects have higher accuracy and precision than the dark and thick skin subjects. This system can track hand vein with 83.92% accuracy and 94.28% precision rate.

Keywords—vein tracking; CMOS sensor; 880nm near-infrared; adaptive smoothing; retinex, maximum curvature points.

I. INTRODUCTION

Blood vessels which can not be seen and sensed by hand, (mostly exist in obese people and dehydrated children) cause injection error on patient. More than one injection errors may cause uncomfortable condition for patient and can lead to hematoma, more over conduct surgery or skin incision to get blood vessels to be injected directly due to the difficulty in finding patient's veins location. Vein can be seen more clearly by using near-infrared camera because blood molecule (Hb) absorbs near-infrared light very well[1]. The use of near-infrared camera can allow medical personnel to view patient's veins more clearly to minimize injection error.

The tools which is used to obtain vein image tends to be expensive because it uses high intensity infrared camera with CCD sensor along with highly sensitive to infrared so that vein image can be seen very clearly. CMOS sensors can be an alternative to CCD because the price is more affordable and very readily available. However, vein image from CMOS sensor is not as good as CCD sensor so it needs advanced image processing to get better vein image quality[2].

Some studies suggest that the near-infrared light source 880nm greatly help clarify the existence of veins up to 3mm below the skin surface[7]. This research uses near-infrared 880nm light source with CMOS sensor and optical filters 880nm to track the dorsal hand vein and antibrachii vein. Image acquisition results have low contrast and spreading noise resembles info (vein) that thresholding can not be conducted to separate the info and background[3]. Adaptive smoothing and retinex is used as pre-processing to eliminate the noise and to enhance contrast between information and background[5]. Maximum curvature points methode is quite good in segmentation, separating the info and background, so that quantification of vein position can be obtained more accurate and precise[6].

Automation vein segmentation results can be developed in phlebhotomy robot. Vein location quantification data can facilitate the calculation of many things related to vein changes, for example in detecting vein collapse which is very dangerous and important to be known.

II. HARDWARE DESIGN

A. *Designing the Near-Infrared Light Source*

80% of sun light consist of infrared that can not be seen by human. Each wavelength of infrared light is absorbed uniquely for each compound, molecule, and every type of material. Blood molecule (Hb) absorbs near-infrared light very well, especially 760nm and 880nm[1],[7]. So, for indoor use, 880nm near-infrared light source is needed.

This system uses 51 880nm LEDs (QED121 fairchild) which are arranged as 17 parallel. Each parallel consists of 3 LEDs in series added a 150 ohm resistors.

J. Goh and C.T. Lim (eds.), *7th World Congress on Bioengineering 2015*,
IFMBE Proceedings 52, DOI: 10.1007/978-3-319-19452-3_54

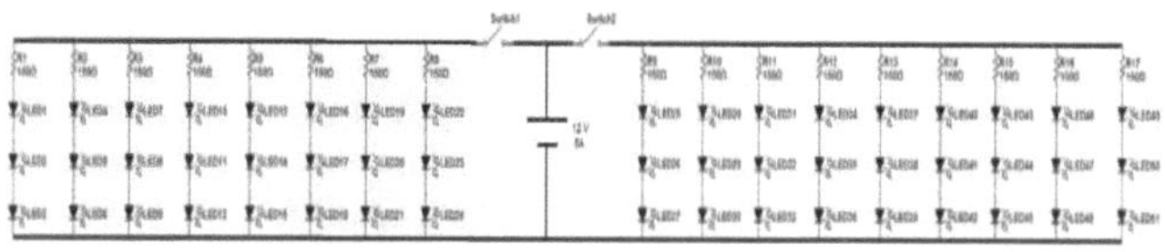

Fig 1. Light source circuit

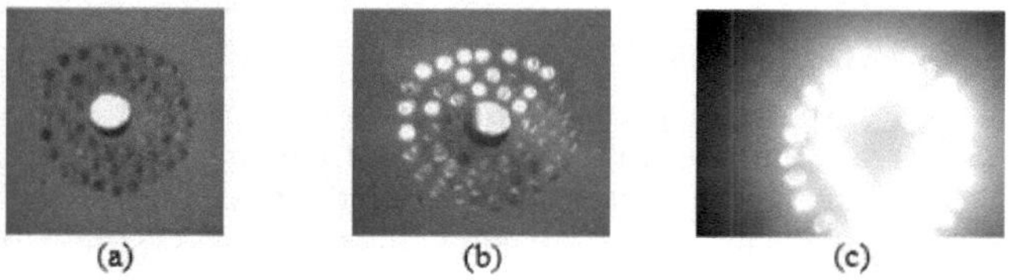

Fig 2. 880nm near-infrared LEDs (a) seen by human eyes, (b) seen by smartphone, (c) seen by near-infrared camera

B. Designing the Near-Infrared Webcam

This system uses A4Tech PK-710G webcam that consist of CMOS sensor, lens, and infrared-cut filter. Infrared-cut filter should be replaced by 880nm optical filter so that only the near-infrared light is passed and the other light spectrum were blocked, including visible light.

C. Using Diffuser

Diffuser is used to spread the light, lessen the harshness of the light, and block shadows. But, it will reduce the light intensity so, many LEDs is needed.

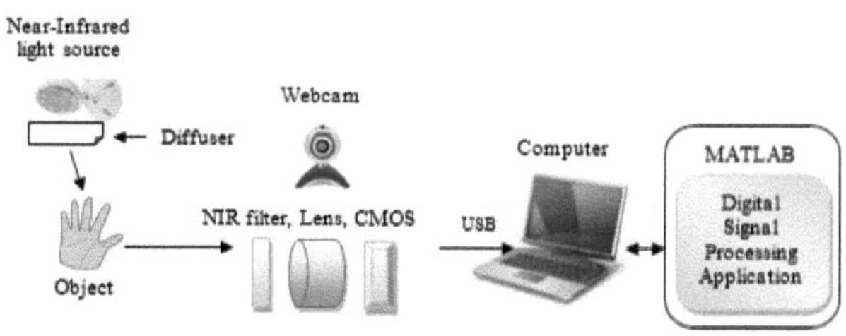

Fig 3. Hardware system

III. Software Design

Software design consist of pre-processing, segmentation, post-processing, and tracking.

A. Pre-Processing

First, changing RGB image acquisition into a grayscale channels to facilitate computing.

$$Gray = 0.4 * R + 0.45 * G + 0.15 * B \qquad (1)$$

That composition is the optimal of RGB coefficient from the study of retina blood vessels[8].

Second, selecting region of interest (ROI) based on the ability of the lens can focus on objects and distance capabilities of LED can illuminates objects. Initial image size from acquisition is 640x480 pixel. From many experiments, the best distance of image acquisition is 16cm and ROI is 200x200 piksel on the center of image.

Third, grayscale normalization to repair image contrast by stretching its histogram through the equations below:

$$N(i,j) = \begin{cases} M_0 + \sqrt{\dfrac{V_0(I(i,j)-m)^2}{V}}, & I(i,j) > m \\ M_0 - \sqrt{\dfrac{V_0(I(i,j)-m)^2}{V}}, & I(i,j) \le m \end{cases} \qquad (2)$$

N (i,j) is the normalization image while I (i,j) is the input image in grayscale format. Vo is the variance, Mo is the average of the normalized image, and m is the average of the input image. In this system, the value of Vo and Mo has been determined that Vo = 150 and Mo = 255 [5]. V is the difference between the largest and smallest values of the input image I (i,j).

$$V = Max(I(i,j)) - Min(I(i,j)) \qquad (3)$$

$$m = \sum_{i=0}^{L-1} r_i p(r_i) \qquad (4)$$

Fourth, Adaptive smoothing filter to eliminate noise. Noise in near-infrared image spread uniformly and has composition that resembles a blood vessel information. So that, common methods to eliminate noise such as Gaussian smoothing filter is not good enough because it will further eliminate the value of the blood vessel information itself. Therefore, it is required a more adaptive smoothing filter without damage the value of the blood vessels info.

$$I_0(x,y) = \frac{1}{N(x,y)} \sum_{i \in D} \sum_{j \in D} I(x+i, y+j) w(x+i, y+j)$$

$$N(x,y) = \sum_{i \in D} \sum_{j \in D} w(x+i, y+j)$$

$$I'_0(x,y) = \max\{I_0(x,y), I(x,y)\}$$

$$w(x,y) = \exp\left(-\frac{|I(x+1,y)-I(x-1,y)|+|I(x,y+1)-I(x,y-1)|}{K}\right) \qquad (5)$$

Where I(x,y) is the original input image, N(x,y) is a normalization factor. D is a filter size (3x3). w(x,y) is the weight of the adaptive smoothing filter. This variable is non-negative monotonically decreasing function, which represents the amount of discontinuity at each pixel. K is a constant that set 10.[5]

Fifth, retinex methode that can calculate the ratio between two different areas in an image to clarify the edge/boundary object area and overcome the differences in lighting sensation that received by the object.

$$R(x,y) = \log I(x,y) - \log [E(x,y) * I(x,y)] \qquad (6)$$

Where I (x,y) is the input image, E (x,y) is the estimate lighting function which is the result of grayscale image normalization, and R (x, y) is the image output.

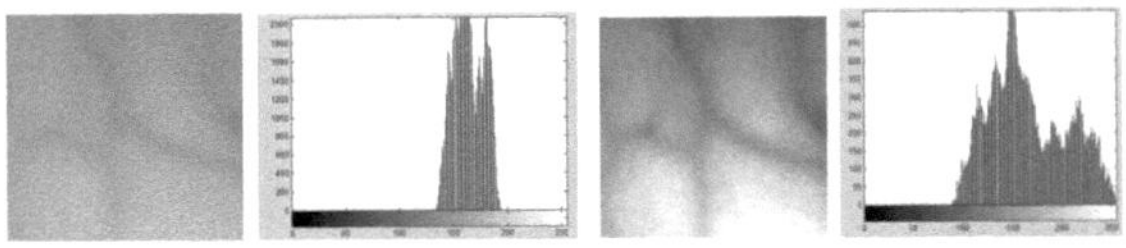

Fig 4. ROI image-hist and pre-processing image-hist

B. Maximum Curvature Points Segmentation

Segmentation is used to seperate object (vein) and background (skin). Thresholding with any level, global thresholding like OTSU methode, and any edge detection system have bad segmentation result for blood vessels near-infrared image. Maximum curvature points[6] has good segmentation result for this system. These are the steps of maximum curvature points with input pre-processing image:

1) Cross Sectional Profile Pf(z)
Horizontal, Vertikal, 45° ke bawah, 45° ke atas

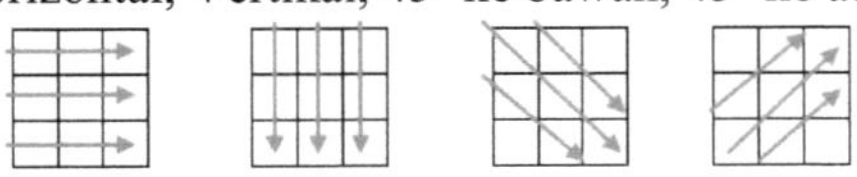

2) Curvature of Profil, calculate positive value (concave) that will be the center of vein

$$\kappa(z) = \frac{d^2P_f(z)/dz^2}{\{1 + (dP_f(z)/dz)^2\}^{\frac{3}{2}}}. \quad (7)$$

3) Scoring, height (intensity) times width (vein width)

$$S_{cr}(z_i') = \kappa(z_i') \times W_r(i). \quad (8)$$

4) Mapping, map the score into the new frame V(x,y)

$$V(x_i', y_i') = V(x_i', y_i') + S_{cr}(z_i'). \quad (9)$$

5) Neighboorhood checking, 2 pixel left and 2 pixel right

$$\begin{aligned} C_{d1}(x,y) &= \min\{\max(V(x+1,y), V(x+2,y)) \\ &+ \max(V(x-1,y), V(x-2,y))\}. \end{aligned} \quad (10)$$

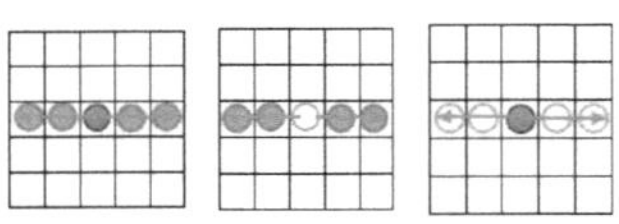

(a) (b) (c)

Condition: (a) connecting line, (b) making gap, (c) eliminating noise.

6) Thresholding, $G(x,y) = Max\ (C_{d1}, C_{d2}, C_{d3}, C_{d4})$ (11)

7) Labeling, binarization 1 if (x,y) ≥ T, otherwise 0

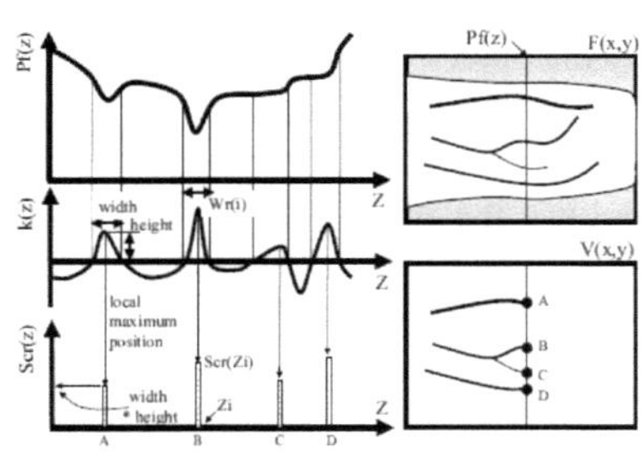

Fig 5. Relationship among profile, curvature, and probability score of veins[6].

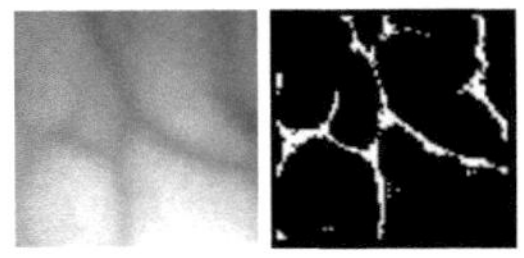

Fig 6. Segmentation result

C. Post-Processing

Post-processing consist of morphology process: Bridge to connect vein line, Erode to eliminate the small area which is not vein, Dilate to return its size with structure element line (3,3), and Thinning to streamline the blood vessels to be more focused and precise in determining the position / track blood vessels

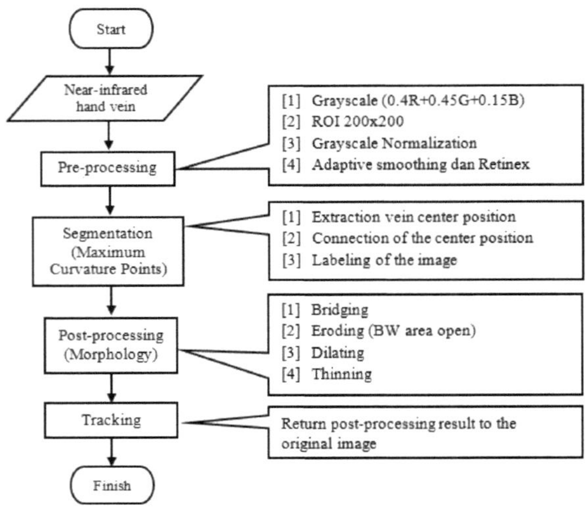

Fig 7. Software process

D. Tracking

Image from post-processing contain vein information, valued 1 (white) and their coordinates have been saved. Those informations are returned to the original image and labeled in red to make it clearer. Blood vessels location can be quantified for various purposes of calculating the veins automatically.

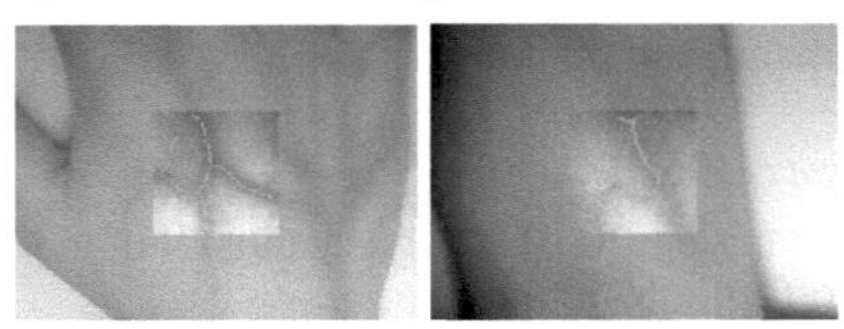

Fig 8. Tracking vein on dorsal (left) and antiibrachi (right)

IV. TESTING

There are 4 category of testing: Normal skin (brown), white skin, dark and thick skin, and children. Testing was accompanied by medical personnel of Rajawali hospital in Bandung who will assess the accuracy and precision of this system visually and manually.

- True positive = vein detected and it is true.

- False positive = vein detected but in fact it doesn't exist.
- True negative = vein not detected and it is true.
- False negative = vein not detected but in fact it exist.

Accuracy = (TP+TN) / {(TP+FN)+(FP+TN)}

Precision defines there is no false positive.

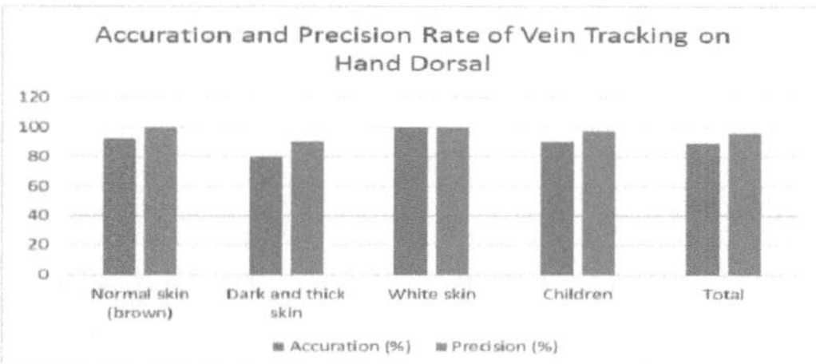

From 21 testing subjects, vein tracking on the hand dorsal has 89.28% accuracy and 96.43% precision. The best vein tracking is on white skin subject and the worst is on dark and thick skin.

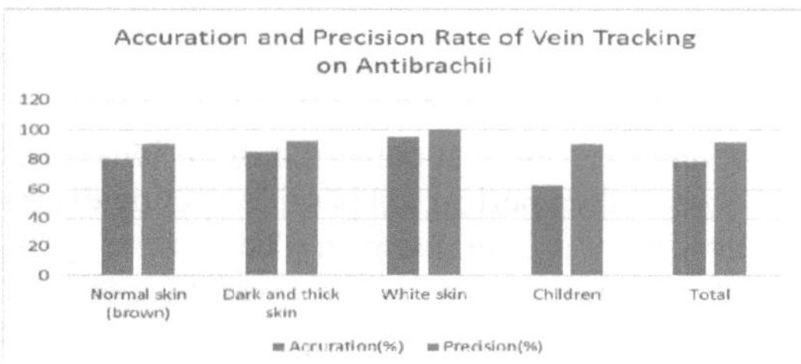

From 21 testing subjects, vein tracking on antiibrachi has 78,57% accuracy and 92,14% precision. The best vein tracking is on white skin subject and the worst is on children.

V. Conclusions

Hand dorsal vein is more easily detected than antibrachii because its position is closer to the skin surface. Subjects with the clean and white skin have higher accuracy and precision than subjects with black and thick skin. White skin has better vein contrast because vein which absorbs near-infrared light become dark and white skin as background reflects more light become lighter. Error detection is caused by the noise which spread uniformly and always moving on the object surface, as well as the presence of thick hairs on the skin surface. The average of whole computation time of this system in blood vessels detection is 1 second, so it is still good enough for real time systems. This system can clarify the hand veins appearance and help the medical personnel in determining the location of injection so it can minimize injection error.

References

The template will number citations consecutively within brackets [1]. The sentence punctuation follows the bracket [2]. Refer simply to the reference number, as in [3]—do not use "Ref. [3]" or "reference [3]" except at the beginning of a sentence: "Reference [3] was the first ..."

Number footnotes separately in superscripts. Place the actual footnote at the bottom of the column in which it was cited. Do not put footnotes in the reference list. Use letters for table footnotes.

Unless there are six authors or more give all authors' names; do not use "et al.". Papers that have not been published, even if they have been submitted for publication, should be cited as "unpublished" [4]. Papers that have been accepted for publication should be cited as "in press" [5]. Capitalize only the first word in a paper title, except for proper nouns and element symbols.

For papers published in translation journals, please give the English citation first, followed by the original foreign-language citation [6].

[1] José Sanchez del Rio Saez, dkk. "Optical system, method and computer program for detecting the presence of a living biological organism." Paten US8766189 B2.

[2] Shi Zhao, Yiding Wang, dan Yunhong Wang. 2007. "Extracting Hand Vein Patterns from Low-Quality Images: A New Biometric Technique Using Low-Cost Devices." IEEE Fourth International Conference on Image and Graphics, pp667-671.

[3] R. Fuksis, M. Greitans, O. Nikisins, M. Pudzs. 2010. "Infrared Imaging System for Analysis of Blood Vessel Structure." ISSN 1392-1215 Electronics and Electrical Engineering 2010. No.1(97), pp45-48.

[4] Goh Kah Ong Michael, Tee Connie, dan Andrew Teoh Beng Jin. 2010. "Design and Implementation of a Contactless Palm Print and Palm Vein Sensor." IEEE 2010 11th Intl Conf Contol, Automation, Robotics, and Vision, pp1268-1273.

[5] Wang, Hua-Bin, dan Liang Tao. 2012. "Novel Algorithm for Enhancement of Hand Vein Image Based on Adaptive Filtering and Retinex Methode." IEEE International Conference on Infomration Science and Technology, pp 857-860.

[6] Naoto Miura, Akio Nagasaka, Takafumi Miyatake. 2005. "Extraction of Finger-Vein Patterns Using Maximum Curvature Points in Image Profile." MVA2005 IAPR Conference on Machine Vision Application, pp 347-350. Japan.

[7] Masaki Watanabe. 2009. "Palm Vein." Encyclopedia of Biometrics, pp 1027-1033: Springer

[8] Agung W. Setiawan, dkk. 2003. "Karakterisasi Struktur Pembuluh Darah Retina untuk Pengkodean Citra Warna Asimetris Menggunakan Kuantisasi Vektor." Disertasi: Intitut Teknologi Bandung.

[9] Zia-ur Rahman. 1995. "Properties of a Center/Surround Retinex: Part 1 –Signal Processing Design." NASA Technical Memorandum 110188.

Oxygen Utilization during Passive Cycling Exercise

T. Saitoh and K. Niizeki

Department of Bio-Systems Engineering, Yamagata University, Yonezawa, Japan

Abstract— **We investigated the dynamics of oxygen utilization during passive cycling exercise. This study included 5 healthy volunteers. The passive cycling exercise test was a 2-minute rest followed by the subjects performing passive cycling exercise for 3 min, followed by a 2-minute rest. The control experiment consisted of a 2-minute rest followed by active cycling exercise at an intensity of 26 W for 3 min, followed by a 2-minute rest. During the tests, ventilation levels, pulmonary gas exchange rates, heart rates, and hemoglobin concentrations of the vastus lateralis muscles were continuously measured. During active cycling exercise, oxygen uptake immediately increased by 393.9 ± 138.7 mL/min, deoxyhemoglobin and total hemoglobin concentrations decreased exponentially by 0.10 ± 0.05 μM and 0.12 ± 0.06 μM, respectively. During passive cycling exercise, oxygen uptake immediately increased by 227.2 ± 42.2 mL/min, deoxyhemoglobin and total hemoglobin concentrations decreased exponentially by 0.12 ± 0.06 μM and 0.12 ± 0.05 μM, respectively. Deoxyhemoglobin dynamics during passive cycling exercise were similar to that during active cycling exercise. The findings of this study indicated that aerobic energy metabolism was utilized for muscle contraction during passive cycling exercise; however, oxygen utilization during passive cycling exercise was lower than that during active cycling exercise.**

Keywords— **passive exercise, oxygen extraction, oxygen uptake.**

I. INTRODUCTION

Pulmonary oxygen uptake reflects systemic oxygen utilization and muscle deoxygenation reflects the balance between muscle oxygen uptake and delivery. Devices to measure pulmonary gas metabolism rates are widely used, while near-infrared spectroscopy, a spectrum measurement method based on absorption of light in the near-infrared radiation wavelength region, is used to noninvasively measure muscle deoxygenation in biological tissues [1-3].

The kinetics of oxygen uptake and active muscle deoxygenation during exercise has been studied [4, 5]. However, there are few studies on oxygen utilization in contracting muscles during passive exercise. Therefore, in this study, we aimed to investigate the dynamics of oxygen utilization during passive cycling exercise.

II. METHODS

A. Subjects

Five healthy men (mean ± SD: age, 22.8 ± 1.1 years; height, 170.2 ± 6.9 cm; body mass, 65.0 ± 7.5 kg) participated in this study. Subjects were informed of the risks associated with the experimental protocol and provided written consent before participation in this study.

B. Protocol

The subjects were seated on the saddle of electronically braked cycle ergometers (AFB6010R; ALINCO, Japan). The passive cycling exercise test consisted of an initial 2-minute rest; the subjects then performed passive cycling exercise for 3 min, followed by a 2-minute rest. The passive cycling exercise used two ergometers, with a pedal of one ergometer connected to a pedal of another ergometer (Fig. 1). During the passive cycling exercise, each subject put his feet on the pedals of ergometer and relaxed both legs, while the researcher pedaled at 60 rpm.

The control experiment consisted of a 2-minute rest followed by active cycling exercise at an intensity of 26 W (lowest load of the ergometer) and rhythm of 60 rpm for 3 min, followed by a 2-minute rest.

Fig. 1 Ergometer in the passive cycling exercise test

J. Goh and C.T. Lim (eds.), *7th World Congress on Bioengineering 2015*, IFMBE Proceedings 52, DOI: 10.1007/978-3-319-19452-3_55

C. Measurements

Ventilation levels and pulmonary gas exchange rates were measured during the tests using a system for pulmonary gas exchange measurement with true breath-by-breath analysis (K4 b2; COSMED, Italy). Heart rates were calculated from R-R interval of a three-lead electrocardiogram. Changes in oxy- and deoxyhemoglobin concentrations of the vastus lateralis muscles were continuously measured at 10 Hz by near-infrared spectroscopy (PocketNIRS Duo; DynaSense, Japan). Electromyograms of the vastus lateralis muscles were measured at 100 Hz using a data acquisition system for life science research (PowerLab; ADInstruments).

D. Data analysis

All data except electromyogram data were interpolated to 1-s intervals, filtered for aberrant data points, and ensemble-averaged to yield a single response for each subject.

Electromyogram data during passive and active cycling exercise were transformed to spectrum components by power spectrum analysis and a frequency of mean power was evaluated for each data. Differences between passive and active cycling exercise were analyzed using a paired *t*-test. Statistical significance was defined as $P < 0.05$.

III. Results

A. Gas exchange rate

The time courses of oxygen uptake during the tests are shown in Fig. 2. During active cycling exercise, oxygen uptake immediately increased by 393.9 ± 138.7 mL/min and then increased linearly. In contrast, after starting passive cycling exercise, oxygen uptake immediately increased by 227.2 ± 42.2 mL/min and then decreased slightly. The shape of time course plots of carbon dioxide output were similar to that of oxygen uptake.

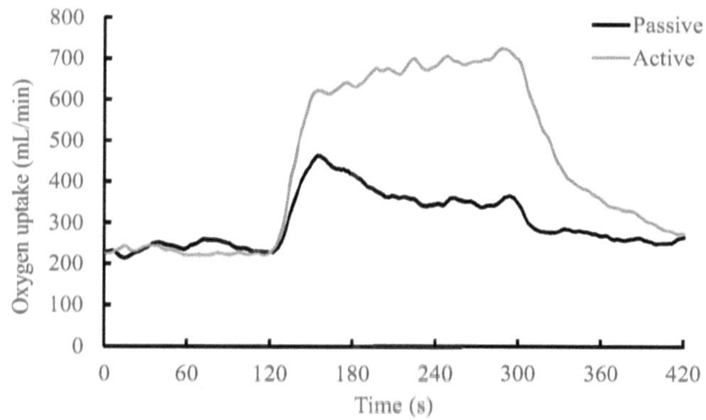

Fig. 2 Oxygen uptake during exercise tests

B. Hemoglobin concentrations

The time courses of deoxyhemoglobin and total hemoglobin concentrations during the tests are shown in Fig. 3 and 4, respectively. During passive cycling exercise, deoxyhemoglobin and total hemoglobin concentrations decreased exponentially by 0.12 ± 0.06 μM and 0.12 ± 0.05 μM, respectively. On the other hand, during active cycling exercise, deoxyhemoglobin and total hemoglobin concentrations decreased exponentially by 0.10 ± 0.05 μM and 0.12 ± 0.06 μM, respectively. Deoxyhemoglobin and total hemoglobin dynamics during passive cycling exercise were similar to that during active cycling exercise.

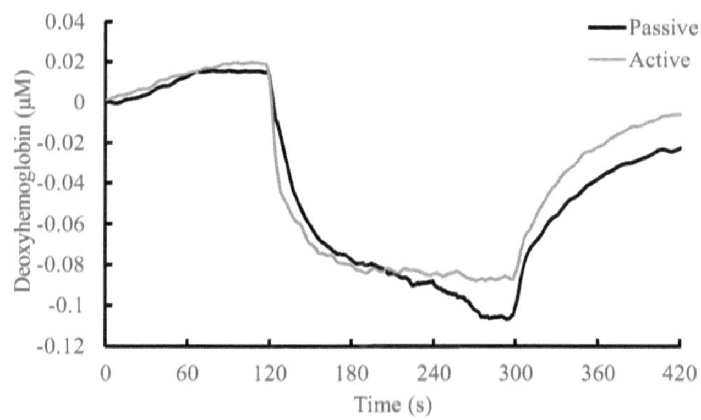

Fig. 3 Deoxyhemoglobin concentration during exercise tests

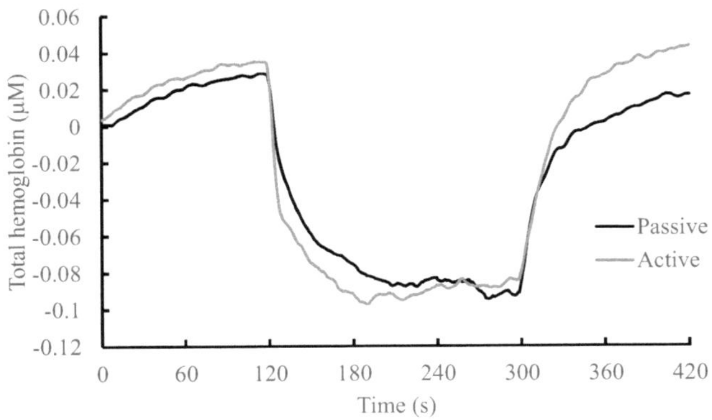

Fig. 4 Total hemoglobin concentration during exercise tests

C. Heart rate

The time courses of heart rate during the tests are shown in Fig. 5. After starting active exercise, the heart rates immediately increased 14.5 ± 5.9 beats/min. In contrast, during passive cycling exercise, the heart rate increased slightly by 6.7 ± 2.6 beats/min, and then immediately returned to the rest level.

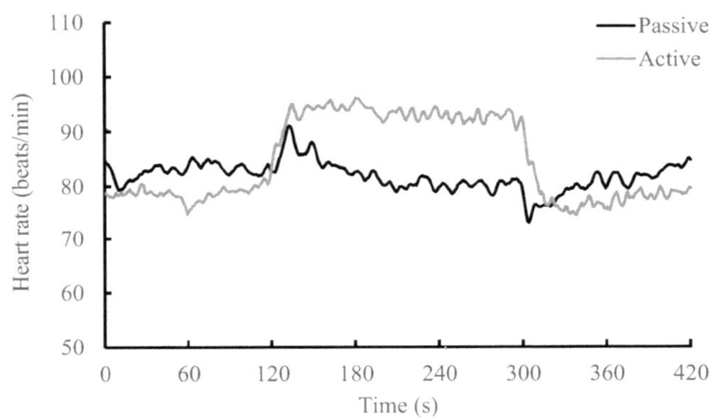

Fig. 5 Heart rate during exercise tests

D. Electromyogram

Electromyogram mean power frequencies during passive and active cycling exercise are shown in Fig. 6. The mean power frequency during passive cycling exercise was significantly lower than during active cycling exercise (passive vs. active: 47.7 ± 6.6 vs. 72.6 ± 3.3 Hz; $P < 0.01$).

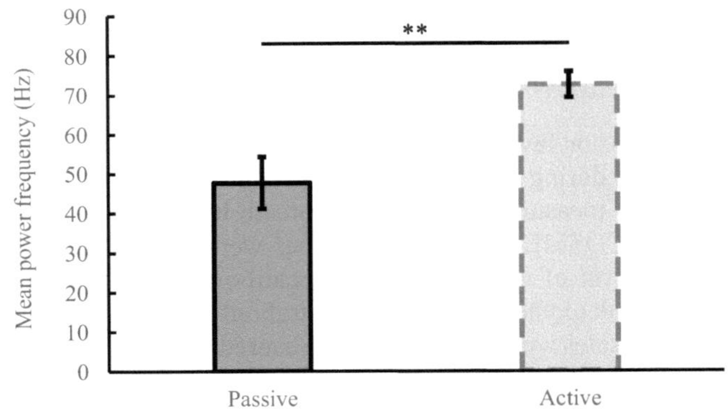

Fig. 6 Electromyogram mean power frequency during passive and active cycling exercise
**: $P < 0.01$

IV. Discussion

This study examined the dynamics of oxygen utilization during passive cycling exercise. In this study, oxygen uptake increased during passive cycling exercise, though it was at a lower level than that during active cycling exercise. Deoxyhemoglobin dynamics during passive cycling exercise was similar to that during active cycling exercise. These findings indicate that, although oxygen utilization during passive cycling exercise was lower than that during active exercise, aerobic energy metabolism was utilized for muscle contraction during passive cycling exercise.

Deoxyhemoglobin and total hemoglobin dynamics during passive cycling exercise were almost similar to that during active cycling exercise, and heart rates during passive cycling exercise were almost steady, but increased during active cycling exercise. Deoxyhemoglobin and total hemoglobin levels reflect oxygen extraction and blood vessel diameter, respectively. Therefore, if heart rate is assumed to reflect the velocity of blood flow, it is possible that difference in oxygen uptake between passive and active cycling exercise was due to changed heart rates; i.e., changed blood flow velocity.

It is known that lower mean power frequency in electromyograms reflects increased utilization of slow muscle fibers, while higher frequency reflects increased utilization of fast muscle fibers [6]. Therefore, the ratio of mobilization of slow muscle fibers for workload was higher during passive cycling exercise than during active cycling exercise.

V. Conclusion

This study demonstrated that aerobic energy metabolism was utilized for muscle contraction during passive cycling exercise. It was possible that changes in blood flow velocity contributed to differences in oxygen uptake between passive and active cycling exercise.

Acknowledgment

This study was supported by JSPS KAKENHI Grant Number 26840159.

References

1. Wray S, Cope M, Delpy DT et al. (1988) Characterization of the near infrared absorption spectra of cytochrome aa3 and haemoglobin for the non-invasive monitoring of cerebral oxygenation. Biochim Biophys Acta 933: 184-192.
2. Owen-Reece H, Smith M, Elwell CE et al. (1999) Near infrared spectroscopy. Br J Anaesth 82: 418-426.
3. Saitoh T, Ooue A, Kondo N et al. (2010) Active muscle oxygenation dynamics measured during high-intensity exercise by using two near-infrared spectroscopy methods. Adv Exp Med Biol 662: 225-230.
4. Jones AM, Poole DC. (2005) Oxygen uptake dynamics: from muscle to mouth–an introduction to the symposium. Med Sci Sports Exerc 37: 1542-1550.
5. Saitoh T, Ferreira LF, Barstow TJ et al. (2009) Effects of prior heavy exercise on heterogeneity of muscle deoxygenation kinetics during subsequent heavy exercise. Am J Physiol Regul Integr Comp Physiol 297: R615-R621.
6. Gerdle B, Henriksson-Larsen K, Lorentzon R et al. (1991) Dependence of the mean power frequency of the electromyogram on muscle force and fibre type. Acta Physiol Scand 142: 457-465.

Author Index

Keyword Index

MIX
Papier aus verantwortungsvollen Quellen
Paper from responsible sources
FSC® C105338

If you have any concerns about our products,
you can contact us on
ProductSafety@springernature.com

In case Publisher is established outside the EU,
the EU authorized representative is:
Springer Nature Customer Service Center GmbH
Europaplatz 3, 69115 Heidelberg, Germany

Printed by Libri Plureos GmbH
in Hamburg, Germany